Dynamic Web Programming
and HTML5

D1260533

Dynamic Web Programming
and HTML5

Paul S. Wang

Vincennes University
Shake Learning Resources Center
Vincennes, In 47591-9986

CRC Press
Taylor & Francis Group
Boca Raton   London   New York

CRC Press is an imprint of the
Taylor & Francis Group, an **informa** business

A CHAPMAN & HALL BOOK

CRC Press
Taylor & Francis Group
6000 Broken Sound Parkway NW, Suite 300
Boca Raton, FL 33487-2742

© 2013 by Taylor & Francis Group, LLC
CRC Press is an imprint of Taylor & Francis Group, an Informa business

No claim to original U.S. Government works

Printed in the United States of America on acid-free paper
Version Date: 20121023

International Standard Book Number: 978-1-4398-7182-9 (Paperback)

Library of Congress Cataloging-in-Publication Data

Wang, Paul S.
  Dynamic Web programming and HTML5 / Paul S. Wang.
  pages cm
  Includes bibliographical references and index.
  ISBN 978-1-4398-7182-9 (pbk.)
  1. Web site development. 2. Internet programming. 3. HTML (Document markup language) I. Title.

QA76.625.W375 2013
006.7'6--dc23

2012031642

Visit the Taylor & Francis Web site at
http://www.taylorandfrancis.com

and the CRC Press Web site at
http://www.crcpress.com

Contents

Preface

Ever since the early 1980s, the Web has grown, expanded, and evolved. HTML5 is a new standard and it promises to revolutionize the Web. Today, the Web is omnipresent and part of our daily lives. In addition to Web access from the home and office, modern users carry it with them on laptops, tablets, and smartphones. Commercial outfits are finding it necessary to offer Web access to their customers, often without charge.

Organizations large and small as well as individuals are increasingly dependent on the Web. The need for well-trained and competent Web developers and maintainers is ever increasing. This book provides a comprehensive and up-to-date guide to Web programming, covering new Web standards as well as emphasizing dynamism and user friendliness.

The text helps you master Web development with a complete set of well-selected topics. Hands-on practice is encouraged; it is the only way to gain experience with the technologies and techniques for building superb sites.

An overview of the Web and Internet provides a whole-forest view. The chapters lead you from page structuring (HTML), page layout/styling (CSS), user input processing (forms and PHP), dynamic user interfaces (JavaScript), database-driven websites (PHP+MySQL), and more, all the way to mobile website development.

There are many examples and complete programs ready to run. A companion website provides appendices, information updates, an example code package, and other resources for instructors as well as students. See page 609 for details.

Standard Web Technologies

The exciting new HTML5 with its associated open Web platform standards is a major focus of this textbook. Topics include the HTML5 markup language, new elements for structuring Web documents and forms, CSS3, the HTML5 Document Object Model, and important JavaScript APIs associated with HTML5.

Dynamic page generation and server-side programming with PHP is another major focus. Topics include page templates, form processing, session control, user login, database access, and server-side HTTP requests.

Also covered are more advanced topics such as XML, AJAX, Web services, PHP+MySQL, the LAMP (Linux Apache MySQL and PHP) server environment, SVG, MathML, and mobile websites.

Many examples are given to illustrate programming techniques and to show how things work together to achieve practical goals.

User Friendly and Comprehensive

There is both breadth and depth in this book's presentation. After the overview in Chapter 1, the user is guided through a well-selected set of topics covering the type of detailed material appropriate for an intensive one-semester course, or a comprehensive two-semester sequence, at the advanced undergraduate or beginning graduate level.

Concepts and Skills

The text balances teaching specific Web programming languages, APIs, and coding techniques with in-depth understanding of the underlying concepts, theory, and principles. Both are important for a well-rounded training that can withstand the test of time and rapidly evolving technologies.

Flexible Usage

This book is for people who wish to learn modern standard Web technologies and their practical applications. The book does not assume prior knowledge of Web development, but has the depth to satisfy even those with good experience.

Compared to other Web programming books, this text is not a thick volume. However, it is comprehensive and in depth. Many examples are given to illustrate concepts and usage. It has more than enough material for a one-semester course. An instructor can cover all the chapters in sequence or choose among them depending on the class being taught.

For an introduction to Web programming course, the materials on LAMP Web hosting, XML, MathML, and SVG can safely be omitted. For an advanced Web programming course, the first few chapters can be assigned for reading at the beginning of the class. This will provide more time for the hard-core programming topics later in the book.

For a two-semester sequence, the text can serve the programming part very well. An instructor may consider adding some artistic design topics also. The author has a co-authored text that does an excellent job of integrating visual communication design with Web programming.

For an introduction to database course, the PHP-MySQL part of this book can be a great boost.

For those who wish to develop and maintain their own websites, this book can be a valuable ally.

Live Examples Online

Throughout the book, concepts and programming techniques are thoroughly explained with examples. Instead of using contrived examples, however, every effort has been made to give examples with practical value and to present them as complete webpages ready to run on your browser. You can experiment with the live examples on the book's companion website (the DWP website) at

`http://dwp.sofpower.com`

The live examples are organized by chapter and cross-referenced to in-text descriptions with a notation such as **Ex:SlideShow** that also appears in the book's index.

Furthermore, all programs covered in the book are collected in an *example code package* ready to download from the DWP website. See page 609 for instructions on obtaining access to the DWP website.

Appendices Online

The table of contents lists all the appendices, but to reduce the volume of the book, we are keeping the appendices online at the DWP website.

The appendices in PDF are easier to use and search. The appendix "Secure Communication with SSH and SFTP" is a webpage so we can supply download and usage hyperlinks. This arrangement also allows us to add and improve the appendices in the future.

Easy Reference

You will find a smooth readable style uncharacteristic of a book of this type. Nevertheless, it is understood that such books are used as much for reference as for concentrated study, especially once the reader gets going with Web development. Therefore, information is organized and presented in a way that also facilitates quick and easy reference. The DWP website provides ample resource listings and information updates. The in-text examples are also cross-referenced with live examples online. The book also has a thorough and comprehensive index and should remain a valuable aid for any Web developer.

Acknowledgments

I would like to thank the editors Randi Cohen, Karen Simon, and others at CRC Press for their help and guidance throughout the writing and production of this book. Their work is much appreciated.

During the planning and writing of this book, several reviews have been conducted. Much appreciated are the input and suggestions from the reviewers:

- Anselm Blumer, Tufts University

- Charles Border, Rochester Institute of Technology

- David Hill, Business Applications Manager, Davey Tree Expert Company

- Rex Kenley, Programmer/System Analyst, Davey Tree Expert Company

Several chapters from the book draft have been used in Web development classes here at Kent State, and I would like to thank the students for reading the materials and providing feedback.

Finally, I want to express my sincere gratitude to my wife Jennifer, whose support and encouragement have been so important to me through the years.

To my darling granddaughter Emma Clarke, the *tigger girl*, I dedicate this book.

Paul S. Wang

王 士 弘

Kent, Ohio

Introduction

The Web is one of the greatest things in modern life. Its global impact has brought us numerous advantages large and small. There is no doubt that the Web will continue to make life easier and better for humankind.

As the Web expands, matures, and moves in different directions, the need for the services of skilled Web developers explodes. People are needed in all types of organizations to create, design, implement, evaluate, rebuild, and manage sites.

This textbook focuses on Web programming and covers standard and widely used Web technologies for modern dynamic websites. Many examples show how individual technology elements work and how they combine for practical purposes. This intermediate to advanced textbook is best suited for people with at least some Web programming experience.

Standard Web Technologies

Enabling technologies for the Web include networking protocols, data encoding formats, clients (browsers), servers, webpage markup and styling languages, and client-side and server-side programming. The Web can deliver text, images, animation, audio, video, and other multimedia content. Standard and proprietary media formats, tools, and players are also part of the Web. The World Wide Web Consortium (W3C) is a nonprofit organization leading the way in developing open Web standards.

This text focuses on a set of core Web technologies recommended by the W3C:

- HTTP—The Hypertext Transfer Protocol employed by the Web. The current version is HTTP 1.1.

- HTML5—The new standard markup language, and its associated technologies, for coding regular and mobile webpages.

- CSS3—The current Cascading Style Sheet standard, offering improved styling, transformation, animation, and styling subject to media conditions (media queries).

- JavaScript—A standard scripting language for browser control and user interaction.

- DOM—Document Object Model, an application programming interface (API) for accessing and manipulating in-memory webpage style and content.

- PHP—A widely used server-side active page programming language and tool that is open and free.

- MySQL—A relational database system that is freely available and used widely for online business and commerce.

- DHTML—Dynamic HTML, a technique for producing responsive and interactive webpages through client-side programming.

- SVG, MathML, and XML—Scalable Vector Graphics (for 2D graphics) and Mathematics Markup Language are part of HTML5. They are important applications of XML, the Extensible Markup Language.

- AJAX—Asynchronous JavaScript and XML providing client-side JavaScript-controlled access to the Web and Web services.

- Web Services—Combining HTTP and XML or JSON (JavaScript Object Notation) to serve data on the Web to other programs.

- HTML5-related APIs—JavaScript-based APIs introduced by HTML5 and other projects for a good number of useful purposes.

- LAMP—Industry standard Web hosts supported by Linux (OS), Apache (Web server), MySQL (database), and PHP (active page).

This book provides well-organized and concise coverage of these technologies and, more important, how they work together to enable serious dynamic website development.

HTML5: A New Web Standard

Among the topics covered in this text, the new and exciting HTML5 is a major focus. The term HTML5 can narrowly mean the new version of the HTML language. But often HTML5 is used in the broad sense of an *open Web platform* encompassing an array of Web standards, centered around the HTML5 markup language, that work together to support the modern Web.

HTML5 aims to reduce the need for browser plug-ins (such as Flash), better handle errors, provide markups to replace scripting, and be device independent. HTML5 promises to revolutionize the Web, making it not only more functional, dynamic, and powerful, but also easier to program and deploy on different platforms such as mobile devices.

The MIT *Technology Review* had a cover article titled "The Web is Reborn" about HTML5. It said this about HTML5: "The technology will enshrine the best of the Web and create new possibilities."

In terms of highlights,

- HTML5 introduces new elements to better organize and structure Web documents.

- New `audio` and `video` elements promise to make sound and video media as easy to deploy as still images.

- A dynamic drawing area, called a `canvas`, makes interactive and dynamic graphics possible.

- HTML5 fill-out forms are more powerful and easier to use/program.

- New features empower Web applications with audio/video playback control, rich-text editing, drag-and-drop programming, client-side file access, browser cache, and local storage for offline use.

- HTML5 APIs provide increased power, dynamism, and efficiency for Web development.

- MathML and SVG enjoy native support in HTML5, making mathematical formulas and 2D graphics markup in a webpage much easier.

Browsers are competing to implement HTML5 features and APIs. You may still find some differences among all browsers. However, HTML5 is supported well enough now by major browsers that we feel comfortable recommending switching to HTML5 for developing or redesigning websites.

Website Development

The central topic of this book is how to develop websites that are highly functional and dynamic. The theme is how HTML5 and other Web programming technologies combine effectively for practical Web projects.

The reader first gets an overview of the Web and its current status, the website development process, and the technologies, techniques, and tasks involved. Then, different topics are presented in a logical sequence. The material balances theory, concepts, tools, and best practices to provide both fundamental understanding and developmental abilities.

PHP is another major focus in this book. It is key for dynamic websites: providing page templates, processing user input, controlling sessions, managing user login, resizing images, detecting user agent types, generating code, and connecting to databases, just to name a few important applications.

E-business and e-commerce are increasingly relying on databases for their Web operations. We provide a good introduction to relational databases, SQL, the MySQL database system, and examples of their application in practice such as online selling and payment processing.

Mobile Websites

The Web is increasingly mobile. According to Morgan Stanley Research, sales of smartphones will exceed those of PCs in 2012. IMS Research expects smartphone annual sales to reach $1 billion in 2016 (half the mobile device market).

Chapter 13 describes the differences between desktop and mobile websites, and illustrates techniques to combine HTML5, CSS, JavaScript, PHP, and HTML5 APIs for developing functional and attractive mobile websites. Ways to make existing sites mobile friendly and approaches for detecting and redirecting mobile visitors are also presented.

Concepts and Programming

As its title suggests, the book is about Web programming. Therefore, helping the reader learn different languages, APIs, coding, and techniques to combine them for practical purposes is our goal. Many realistic examples, complete and ready to run at the DWP website, are used to illustrate and demonstrate Web programming.

However, we do realize that true programming abilities rest more in a good understanding of concepts, theory, and principles than language syntax or specific API information. Thus, you will find a healthy dose of conceptual materials to go with any programming presentation.

Hopefully, the knowledge and skills gained from this book can withstand the test of time and remain useful as the Web evolves and advances.

The DWP Website

This book has a website that offers reference listings, useful resources, online versions of the figures (in full size and color), demo examples cross-referenced to in-text descriptions, and an example package ready to download. In the textbook, demo examples are labeled with "**Ex:ExampleName**" so they easily correlate with the online versions. A list of all demo examples can be found in the index.

The website is at

http://dwp.sofpower.com

In the text, we refer to this website as the *DWP website.*

Making the book less voluminous, we also keep the appendices online at the DWP website.

Intended Use

The text is designed for a one- to two-semester course on Web programming. It is ideal at the undergraduate level for computer science, computer engineering, and computer technology students. It can also be used at the beginning graduate level in other departments. Familiarity with programming and the computer as an operating environment is required.

We recommend a lab-oriented class format where students form Web development teams. Each team can define and propose its own project, which must be ready for presentation at the end of the semester. Sample team organizations, project milestones, requirements, together with report and presentation guides can be found at the DWP website.

The text is also suitable for custom training courses for industry or self-study by information technology (IT) professionals. A shorter course may omit Chapters 10 to 13, as appropriate. An advanced course may proceed at a faster pace, assign one or two chapters for reading, and select more substantial programming projects from the exercises.

Students should have basic computer skills and sufficient programming experience. The materials are comprehensive enough to remain interesting and challenging even for people with Web development experience.

Chapter 1

The Web: An Overview

Ever since its beginning in the early 1980s, the Web has been evolving and maturing. Today, the Web's impact on how industries, governments, institutions, and individuals conduct business and socialize, is ever expanding. It is not surprising that the need for well-trained people to develop and program websites is increasing at a rapid pace.

The Web involves networking, protocols, servers, clients as well as languages, services, and user interfaces. Thus, the Web is a rather complicated arena, and mastering Web development involves in-depth understanding and working experience with all the involved technologies and how to combine them for specific functions and effects.

Even for people who will not become Web professionals, a good understanding of the Web, how it relates to the Internet, and what makes it tick will be valuable in many a workplace.

The overview in this chapter sets up the scene and the context within which to study and apply the specific languages and programming techniques for building well-developed and dynamic websites.

1.1   Web Is Part of the Internet

For the vast majority of computer users, getting online means browsing the Web. That is how important the Web has become in our daily lives. But of course the Web is just one among many services available on the Internet. These include email, file transfer, remote login, audio and video streaming, and many more.

The Internet spans the globe by connecting computer networks together. Such networks include *local area networks* (LANs) in office buildings, college campuses, and homes, as well as *wide area networks* (WANs) that cover whole cities or even countries. The Internet enables each connected computer, called a *host*, to communicate with any other hosts.

In addition to host computers, the network infrastructure itself involves dedicated computers that perform network functions: hubs, switches, bridges, routers, and gateways. For programs and computers from different vendors, under different operating systems, to communicate on a network, a detailed set of rules and conventions must be established for all parties to follow. Such rules are known as *networking protocols*. We use different networking services

1

for different purposes; therefore, each network service defines its own protocol that adds specific features to existing, more general, underlying protocols. Protocols govern such details as

- Address format of hosts and processes

- Data format

- Manner of data transmission

- Sequencing and addressing of messages

- Initiating and terminating connections

- Establishing services

- Accessing services

- Data integrity, privacy, and security

Thus, for a program on one host to communicate with another program on a different host, both programs must follow the same protocol. The *Open System Interconnect* (OSI) *Reference Model* (Figure 1.1) provides a standard layered view of networking protocols and their interdependence. The corresponding layers on different hosts, and inside the network infrastructure, perform complementary tasks to make the connection between the communicating processes (P1 and P2 in Figure 1.1).

FIGURE 1.1: Networking Layers

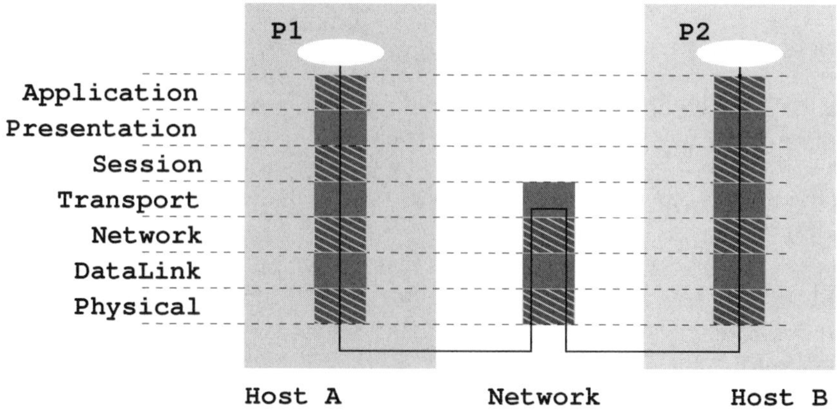

The Internet uses the Internet Protocol Suite. The basic IP (*Internet Protocol*) is a *network layer* protocol. The TCP (*Transport Control Protocol*) and UDP (*User Datagram Protocol*) are at the *transport layer*. The Web is a service that uses an *application layer* protocol known as HTTP (the *Hypertext Transfer Protocol*; Section 1.16).

Networking protocols are no mystery. Think about the protocol for making a telephone call. You (a client process) must pick up the phone, listen for the dial tone, dial a valid telephone number, and wait for the other side (the server process) to pick up the phone. Then you must say "hello," identify yourself, and so on. This is a protocol from which you cannot deviate if you want the call to be made successfully through the telephone network, and it is clear why such a protocol is needed. The same is true of a computer program attempting to talk to another computer program through a computer network. The design of efficient and effective networking protocols for different network services is an important area in computer science.

1.2 IP Addresses and Domain Names

On the Internet, each host has a unique *IP address*, represented by a 32-bit number (four bytes). For example, `tiger`, a host at Kent State, has the IP address `131.123.38.172` (Figure 1.2). This *dot notation* (or *quad notation*)

FIGURE 1.2: IP Address

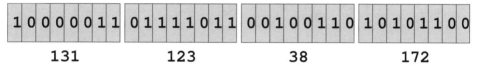

gives the decimal value (0 to 255) of each byte.[1] Similar to a telephone number, the leading digits of an IP address are like country codes and area codes while the trailing digits are like local numbers. A host may be configured with a fixed IP or may obtain an IP address assigned to it on-the-fly as it boots up and attempts to connect to the network initially.

Because of their numerical nature, the dot notation is easy on machines but hard on users. Therefore, each host also has a unique *domain-based name* composed of words, rather like a postal address. For example, the domain name for `tiger` is `tiger.cs.kent.edu` (at Department of Computer Science, Kent State University). And the domain name for the author's IT consulting website is `sofpower.com` where the companion website for this text book (`sofpower.com/dwp`, the *DWP website*) is located.

With the domain names, the entire Internet host namespace is recursively divided into disjoint domains. The address for `tiger` puts it in the `kent` subdomain within `edu`, the *generic top-level domain* (generic TLD) for educational institutions. Other generic TLDs include `org` (nonprofit organizations), `gov` (government offices), `mil` (military installations), `com` (commercial outfits),

---

[1]To accommodate the explosive growth of the Internet, the next-generation IP (IPv6) will support 128-bit addresses.

net (network service providers), info (information sources), name (individuals), uk, (United Kingdom), cn (China), and so forth. In June 2011, the Internet Corporation for Assigned Names and Numbers (ICANN) began to allow the registration of user-owned TLDs by corporations and governments; such private TLDs are expensive to register and maintain.

Within a local domain (e.g., cs.kent.edu), you can refer to computers by their host name alone (e.g., loki, dragon, tiger), but the full address must be used for hosts outside.

Figure 1.3 shows the Internet domain name hierarchy. Each node of the DNS tree has a name (case insensitive) and the name of the root node is a single period (.). A full domain name is formed by connecting all names leading from the host name to the root of the DNS tree. Following are some example full domain names:

```
tiger.cs.kent.edu.
dwp.sofpower.com.
```

FIGURE 1.3: The Domain Name Hierarchy

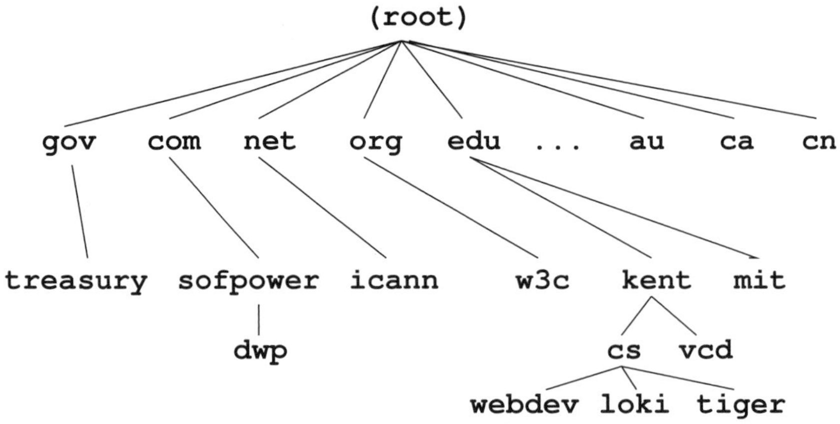

Often, the trailing period (.) can be omitted.

The ICANN accredits *domain name registrars*, which register domain names for clients so they stay unique. All network applications accept a host address given either as a domain name or an IP address. In fact, a domain name is first translated to a numerical IP address before being used.

Data on the Internet are sent and received in *packets*. A packet envelops transmitted data with address information so the data can be routed through intermediate computers on the network. Because there are multiple routes from the source to the destination host, the Internet is very reliable and can operate even if parts of the network are down.

1.2.1 Domain Name System

The Internet *Domain Names System* (DNS) consists of the set of all domain names for hosts connected on the Internet. This set of names changes dynamically with time due to the addition and deletion of hosts, regrouping of local work groups, reconfiguration of subparts of the network, maintenance of systems and networks, and so on. Thus, new domain names, new IP addresses, and new domain-to-IP associations can be introduced in the DNS namespace at any time without central control. The DNS provides a distributed database service (the DNS name service) that supports dynamic update and retrieval of information contained in the namespace (Figure 1.4).

FIGURE 1.4: Domain to IP

domain name ⟶ DNS Server ⟶ IP address

A network client program (e.g., Firefox®) will normally contact a DNS name server to obtain address information for a target domain name before making contact with a server. The DNS name service also supplies a general mechanism for retrieving many kinds of information about hosts and individual users.

1.3 Web Clients and Servers

Most commonly, a network application involves a *server program* and a *client program* (Figure 1.5):

FIGURE 1.5: Client and Server

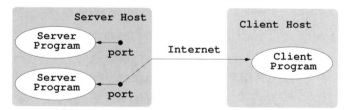

- A *server* program provides a specific service on a host (the server host) that offers such a service. Example services are email (SMTP), secure remote login/file transfer (SSH/SFTP), and the World Wide Web (www). Each *standard Internet service* has its own unique *port number* that is identical across all server hosts. The port number together with the IP address of a host identifies a particular server program (Figure 1.5) anywhere on the network. For example, SMTP uses port number 25, SSH/SFTP 22, and www 80.

- A *client* program on a computer connects with a server on another host to obtain its service. Thus, a client program is the agent through which a particular network service can be obtained. Different agents are usually required for different services. Example clients are Mozilla Thunderbird® and Microsoft® Outlook® for email, OpenSSH for secure login/file transfer.

A *Web host* is a computer that stores files and other resources to deliver to the Web. Web hosts run *Web servers*, programs ready to receive HTTP requests from Web clients. The Web server will process each request and send back an HTTP response to the requesting client. A *Web client* is a user agent program that allows users to access the Web. The most common Web client is the Web browser. Major Web browsers with their December 2010 market share, according to *net applications*, are Internet Explorer® (57.08%), Firefox® (22.81%), Chrome™ (9.98%), Safari® (5.89%), and Opera™ (2.23%).

You would use a Web browser to access the Web where information and services are placed on Web hosts running Web server programs. There are different Web servers, but the most popular is Apache™ from apache.org.

According to netcraft.com, among all Web servers, over 60% are Apache™, and a majority of Apache servers run on Linux systems. A Linux-Apache Web hosting environment usually also supports PHP for *active pages* and MySQL® for database-driven websites. The Linux, Apache, MySQL, and PHP combination (known as LAMP) works well to support *Web hosting*, the business of placing websites on the Web.

Figure 1.6 shows a typical client-server relationship for the Web using the HTTP protocol.

FIGURE 1.6: Web Client and Server

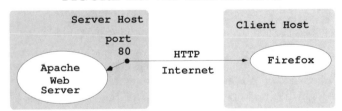

1.4 URLs

The Web uses *Uniform Resource Locators* (URLs) to identify (locate) many kinds of resources (files and services) available on the Internet. A URL may identify a host, a server port, and the target file stored on that host. URLs are used, for example, by browsers to retrieve information, and in HTML documents (Section 1.6) to link to other resources.

A full URL usually has the form

scheme://*server*:*port*/*pathname*?*query_string*

The *scheme* part indicates the information service type and therefore the protocol to use. Common schemes include `http` (Web service), `ftp` (file transfer service), `mailto` (email service), `callto` (Skype[TM] or similar service), `file` (local file system), `https` (secure Web service), and `sftp` (secure file transfer service). Many other schemes can be found at `www.w3.org/addressing/schemes`.

For URLs in general, the *server* identifies a host and a server program. The optional port number is needed only if the server program does not use the default port (for example, 80 for `HTTP` and 443 for `HTTPS`). The next part of the URL, when given, is a file *pathname*. If this pathname has a trailing `/` character, it represents a directory rather than a data file. The pathname can also lead to an executable program that dynamically produces an HTML or other valid file to return. A URL targeting a program can also have a trailing *query string* to provide input values to that program.

URLs are critical in Web operations. You can enter any valid URL into the `Location` box of any Web browser to reach the target resource. When a URL specifies a directory, a Web server usually returns an *index file* (typically, `index.html`) for that directory. Otherwise, it may return a list of the filenames in that directory.

The `http` URL scheme is especially important for the Web. It is used in webpages to link to other webpages, inside or outside the website. The crosslinkages among webpages globally form a worldwide web structure. Because of its importance, many applications, including email readers, PDF readers, text/document editors, presentation tools, and shell windows, recognize the `http` URL and when you click on it, will launch the default browser.

URLs are contained in webpages to create links to image, audio, video, and any other type of files anywhere on the Web. Within an HTML document, you can link to another document served by the same Web server by giving only the *pathname* part of the URL. Such URLs are *relative URLs*. A relative URL with a `/` prefix (for example, `/file_xyz.html`) refers to a file under the *server root*, the top-level directory controlled by the Web server. A relative URL without a leading `/` points to a file relative to the location of the document that contains the URL in question. Thus, a simple `file_abc.html` refers to that file in the same directory as the current document. When building a website, it is advisable to use a URL relative to the current page as much as possible, making it easy to move the entire website folder to another location on the local file system or to a different server host.

Webpages must be placed in the *document space* of a Web server running on some host computer before they become available on the Web. Files inside a Web server document space must also be given the correct *access permissions* before the Web server can deliver them onto the Web. This usually means, in the document space, folders must be *executable* (may be entered) and files must be *readable* (may be retrieved) by the Web server.

1.4.1   URL Encoding

According to the URL specification (RFC1738), only alphanumerics (0–9 and upper- or lower-case a–z), the special characters

```
$   -   _   .   +   !   *   '   (   )   ,
```

and reserved characters

```
;   /   ?   :   @   =   &
```

used for their reserved purposes (supporting the URL syntax) may be included unencoded within a URL. Other characters are *unsafe* and ought to be encoded. To include unsafe ASCII characters, such as spaces and control characters, they must be URL encoded.

To URL encode an unsafe ASCII character, replace it with the three-character sequence %*hh*, where *hh* is its ASCII code in hexadecimal. For example, ~ is %7E and SPACE is %20. Thus, a link to the file "chapter one.html" becomes

```
<a href="chapter%20one.html">First Chapter</a>
```

The Linux command

man ASCII

displays the ASCII table in octal, decimal, and hex. Hex codes for ISO Latin can be found on the Web.

To include non-ASCII characters in a URL, each such character is first UTF-8 encoded (Section 3.8) into two or more bytes. Each unsafe byte is then %*hh* encoded.

1.5   Handling Different Content Types

On the Web, files of different *media types* can be placed and retrieved. The Web server and Web browser use standard *content type* designations to indicate the media type of each file sent for it to be processed correctly on the receiving end.

TABLE 1.1: Content Types and File Suffixes

Content Type	File Suffix	Content Type	File Suffix
text/plain	txt sh c ...	text/html	html htm
application/pdf	pdf	application/msword	doc, docx
image/jpeg	jpeg jpg jpe	audio/basic	au snd
audio/mpeg	mpga mp2 mp3	application/x-gzip	gz tgz
application/zip	zip	audio/ogg	oga, ogg, spx
video/ogg	ogv	video/webm	webm

The Web borrowed the content type designations from the Internet email system and uses the same MIME (Multipurpose Internet Mail Extensions) defined content types. There are hundreds of content types in use today. Many popular types are associated with standard file extensions as shown in Table 1.1.

When a Web server returns a document to a browser, the content type is indicated via the HTTP message header `Content-Type` (Section 1.16). For example, an HTML document is delivered with the content type `text/html`. The content type information allows browsers to decide how to process the incoming content. HTML, text, images, audio, and video[2] can be handled by the browser directly. Other types such as PDF and Flash are usually handled by plug-ins or external helper programs.

1.6   HTML and HTML5

HTML (the Hypertext Markup Language) is used to structure webpage contents for easy handling by Web clients on the receiving end. From its simple start in 1989, HTML has been constantly evolving and maturing. Beginning with HTML 4.0, the language has become standardized under the auspices of the W3C (World Wide Web Consortium), the industry-wide open standards organization for the Web. Subsequently, by making HTML 4.0 compatible with XML (Section 2.4), XHTML became a widely used new standard. Today, the Web is moving toward HTML5, the next-generation HTML standard, which brings many new features and APIs (application programming interfaces). HTML5 makes it easier to provide dynamic user interactions and promises to transform the Web into an even more useful and powerful tool. A document written in HTML contains ordinary text interspersed with *markup tags* and uses the `.html` filename extension. The tags mark portions of the page as heading, section, paragraph, quotation, image, audio, video, link, and so on. Thus, an HTML file consists of two kinds of information: contents and HTML tags. A browser follows the HTML code to organize, process, and layout the

---

[2]HTML5-compliant browsers support standard audio and video playback.

page content for display and user interaction. Because of this, line breaks and extra white space between words in the content are mostly ignored. Various visual editors or *page makers* are available that provide a GUI for creating and designing HTML documents. For substantial website creation projects, it will be helpful to use *integrated development environments* such as Macromedia Dreamweaver.

TABLE 1.2: Some HTML Tags

Meaning	HTML Tag	Meaning	HTML Tag
Entire page	`<html>...</html>`	Paragraph	`<p>...</p>`
Meta data	`<head>...</head>`	Unnumbered list	`<ul>...</ul>`
Page title	`<title>...</title>`	Numbered list	`<ol>...</ol>`
Page content	`<body>...</body>`	List item	`<li>...</li>`
Level n heading	`<hn>...</hn>`	Comment	`<!--...-->`

If you do not have ready access to such tools, a regular text editor can be used to create or edit webpages. An HTML tag takes the form *<tag>*. A *begin tag* such as `<h1>` (level-one section header) is paired with an *end tag*, `</h1>` in this case, to mark content in between. Table 1.2 lists some frequently used tags.

The following is a sample HTML5 page **Ex: Sports**[3]:

```
<!doctype html>
<html xmlns="http://www.w3.org/1999/xhtml"
      lang="en" xml:lang="en">
<meta charset="UTF-8"/>
<head> <title>A Basic Webpage</title> </head>
<body><section>
   <h1>Big on Sports</h1>
   <p>Sports are fun and good for you ...</p>
   <p> What is your favorite sport? ...
   And here is a short list: </p>
   <ol>
     <li> Baseball </li> <li> Basketball </li>
     <li> Tennis </li> <li> Soccer </li>
   </ol>
</section></body></html>
```

Figure 1.7 shows the Big on Sports page displayed by Firefox.

---

[3]Examples available online or in the example code package are labeled like this for easy cross-reference.

FIGURE 1.7: A Sample Webpage

1.7 Webpage Styling

While HTML takes care of page structure, the way information is actually presented (visually or otherwise) to the end user is controlled by the Web browser used and by *styling rules* associated with the webpage.

Styling is coded in *Cascading style sheets* (CSS) rules and attached to different parts of a webpage. Style rules are usually placed in separate files from the webpage. Isolating page styling from page structure makes it easy for Web designers to reuse styling rules in different pages and to enforce consistent visual styling over an entire website.

For example, if we want to make all level-one headers dark blue, we can use this CSS rule:

```
h1 { color: darkblue }
```

CSS has also evolved through the years to provide more features and functions for various styling needs. The current standard is CSS3.

1.7.1 Text Editing

HTML, CSS, and other files for the Web are plaintext files as opposed to *formatted text files* such as MS Word or WordPad documents. To develop HTML, CSS, JavaScript, and PHP code for the Web, it is recommended that you use a plaintext editor. This is especially true for learning through this book.

Text editors such as Vim and Emacs are powerful and best suited for experienced programmers. Both are free and available for all platforms. Others find notepad++, also available on all platforms, easy to use.

Specialized tools exist for editing HTML, CSS, JavaScript, or PHP code. Such tools provide an environment specific to the particular language and often help generate code for you. These tools can involve some significant initial learning. To gain coding skills, we recommend that you use a plaintext editor and write all the code manually.

1.8　Web Hosting

Web hosting is a service for individuals and organizations to place their websites on the Web. Hence, publishing on the Web involves

1. Designing and constructing the pages and writing the programs for a website

2. Placing the completed site with a hosting service

Colleges and universities host personal and educational sites for students and faculty without charge. Web hosting companies provide the service for a fee.

Commercial Web hosting can provide secure data centers (buildings), fast and reliable Internet connections, specially tuned Web hosting computers, server programs and utilities, network and system security, daily backup, and technical support. Each hosting account provides an amount of disk space, a monthly network traffic allowance, email accounts, Web-based site management and maintenance tools, and other access such as FTP and SSH/SFTP.

To host a site under a given domain name, a hosting service associates that domain name to an IP number assigned to the hosted site. The domain-to-IP association is made through DNS servers and Web server configurations managed by the hosting service.

1.9　Web Servers

A Web host provides storage space for websites to store webpages, pictures, audio and video files, and any programs needed. More importantly, it also runs a Web server through which the stored websites can be accessed on the Web.

There are a number of Web servers but Apache, the open source Web server from `apache.org`, is the most dominant. According to Netcraft (April 2011), major Web servers listed with their market share are Apache^{TM} (60.31%), Microsoft® (19.34%), nginx^{TM} (7.65%), and Google® (5.09%).

While Apache runs on multiple operating systems, it most likely will be running under Linux (63.7% market share as Web server host). Apache also works seamlessly with PHP and MySQL, making it the most popular Web server by far.

The installation, configuration, and maintenance of a Web server is an important part of Web programming. We cover the Apache Web server as part of a LAMP (Linux Apache MySQL and PHP) server host in Chapter 10.

1.10　Domain Registration

To obtain a domain name, you need the service of a *domain name registrar*. Most will be happy to register your new domain name for a very modest yearly fee. Once registered, the domain name is property that belongs to the *registrant*. No one else can register for that particular domain name as long as the current registrant keeps the registration in good order.

ICANN accredits commercial registrars for common TLDs, including .com, .net, .org, and .info. Additional TLDs include .biz, .pro, .aero, .name, and .museum. Restricted domains (for example, .edu, .gov, and .us) are handled by special registries (for example, net.educause.edu, nic.gov, and nic.us). Country-code TLDs are normally handled by registries in their respective countries.

1.10.1 Accessing Domain Registration Data

The registration record of a domain name is often publicly available. The standard Internet *whois* service allows easy access to this information. You can do this on the Web at www.internic.net/whois.html, for example. On Linux/Unix systems, easy access to whois is provided by the **whois** command,

whois *domain_name*

which lists the domain registration record kept at a registrar. For example,

whois kent.edu

produces the following information

```
Domain Name: KENT.EDU

Registrant:
   Kent State University
   500 E. Main St.
   Kent, OH 44242
   UNITED STATES

Administrative Contact:
   Philip  L Thomas
   Network & Telecomm
   Kent State University
   STH
   Kent, OH 44242
   UNITED STATES
   (330) 672-0387
   pki-admin@kent.edu

Technical Contact:

   Network Operations Center
```

```
     Kent State University
     120 Library Bldg
     Kent, OH 44242
     UNITED STATES
     (330) 672-3282
     noc@kent.edu

Name Servers:
     NS.NET.KENT.EDU          131.123.1.1
     DHCP.NET.KENT.EDU        131.123.252.2
     ADNS03.NET.KENT.EDU      128.146.94.250

Domain record activated:    19-Feb-1987
Domain record last updated: 06-Jul-2011
Domain expires:             31-Jul-2013
```

Try the **Ex:** WhoIs demo at the DWP website that can retrieve desired domain registration records.

1.11   DNS

DNS provides the ever-changing domain-to-IP mapping information on the Internet. We mentioned that DNS provides a distributed database service that supports dynamic retrieval of information contained in the namespace. Web browsers and other Internet client applications will normally use the DNS to obtain the IP of a target host before making contact with a server over the Internet.

There are three elements to the DNS: the DNS namespace (Section 1.2), the DNS servers, and the DNS resolvers.

1.11.1   DNS Servers

Information in the distributed DNS is divided into *zones*, and each zone is supported by one or more name servers running on different hosts. A zone is associated with a node on the domain tree and covers all or part of the subtree at that node. A name server that has complete information for a particular zone is said to be an *authority* for that zone. Authoritative information is automatically distributed to other name servers that provide redundant service for the same zone. A server relies on lower-level servers for other information within its subdomain and on external servers for other zones in the domain tree. A server associated with the root node of the domain tree is a *root server* and can lead to information anywhere in the DNS. An authoritative server uses local files to store information, to locate key servers within and without its domain, and to cache query results from other servers. A boot file, usually `/etc/named.boot` on Linux, configures a name server and its data files.

The management of each zone is also free to designate the hosts that run the name servers and to make changes in its authoritative database. For example, the host `ns.cs.kent.edu` may run a name server for the domain `cs.kent.edu`.

A name server answers queries from resolvers and provides either definitive answers or referrals to other name servers. The DNS database is set up to handle network address, mail exchange, host configuration, and other types of queries, with some to be implemented in the future.

The ICANN and others maintain *root name servers* associated with the root node of the DNS tree. In fact, the VeriSign host `a.root-servers.net` runs a root name server. Actually, the letter `a` ranges up to `m` for a total of 13 root servers currently.

Domain name registrars, corporations, organizations, Web hosting companies, and other Internet service providers (ISPs) run name servers to associate IPs to domain names in their particular zones. All name servers on the Internet cooperate to perform domain-to-IP mappings on-the-fly.

1.11.2 DNS Resolvers

A DNS resolver is a program that sends queries to name servers and obtains replies from them. A resolver can access at least one name server and use that name server's information to answer a query directly or pursue the query using referrals to other name servers.

Resolvers, in the form of networking library routines, are used to translate domain names into actual IP addresses. These library routines, in turn, ask prescribed name servers to resolve the domain names. The name servers to use for any particular Linux host are normally specified in the file `/etc/resolv.conf` or `/usr/etc/resolv.conf`.

Try the **Ex:** `NSLookup` demo at the DWP website that looks up domain/IP information from the DNS.

The DNS service provides not just the IP address and domain name information for hosts on the Internet. It can provide other useful information as well. Table 1.3 shows common DNS record and request types.

1.12 Dynamic Generation of Webpages

Webpages are usually prepared and set in advance to supply some predetermined content. These fixed pages are *static*. A Web server can also deliver *dynamic pages* that are generated on-the-fly by programming on the server side. Dynamic pages bring many advantages, including

- Managing user login and controlling interactive sessions

TABLE 1.3: DNS Record/Request Types

Type	Description
A	Host's IP address
NS	Name servers of host or domain
CNAME	Host's canonical name, and an alias
PTR	Host's domain name, IP
HINFO	Host information
MX	Mail exchanger of host or domain
AXFR	Request for zone transfer
ANY	Request for all records

- Customizing a document depending on when, from where, by whom, and with what program it is retrieved

- Collecting user input (with HTML forms), processing such input data, and providing responses to the incoming information

- Retrieving and updating information in databases from the Web

- Directing incoming requests to appropriate pages; redirection to mobile sites is an example

- Enforcing certain policies for outgoing documents

Dynamic webpages are not magic. Instead of retrieving a fixed file, a Web server calls another program to compute the document to be returned and perhaps perform other functions. As you may have guessed, not every program can be used by a Web server in this manner.

A Web server invokes a server-side program by calling it and passing arguments to it (via the program's `stdin` and environment variables) and receiving the results (via the program's `stdout`) thus generated. Such a program must conform to the Common Gateway Interface (CGI) specifications governing how the Web server and the invoked program interact (Figure 1.8).

FIGURE 1.8: Common Gateway Interface

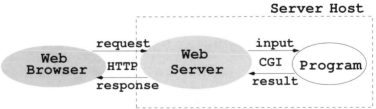

A CGI program can be written in any programming language as long as it follows the CGI specification and can be invoked by the Web server. The Web server and a CGI program may run as independent processes and interact through interprocess communication. Or, the external program can be loaded into the server and run as a plug-n module.

1.12.1 Active Server Pages

The dynamic generation of pages is made simpler and more integrated with webpage design and structure by allowing a webpage to contain *active parts* (Figure 1.9) that are treated by the Web server and transformed into desired content on-the-fly as the page is retrieved and returned to a client browser.

FIGURE 1.9: An Active Page

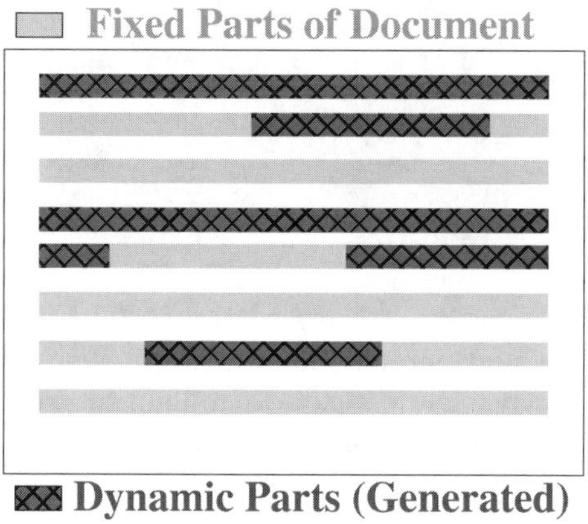

Fixed Parts of Document

Dynamic Parts (Generated)

The active parts in a page are written in some kind of notation to distinguish them from the static parts of a page. The ASP (Active Server Pages, JSP (Java Server Pages), and the popular PHP (Hypertext Preprocessor; Section 3.12) are examples. With PHP, the active parts are enclosed inside the bracket `<?php  ... ?>` and embedded directly in an HTML page or other types of Web document. For example, inside an HTML file you may have code such as

```
<p>Today's date is:  <?php echo(date("l M. d, Y")); ?><p>
```

Here the date is dynamically computed and inserted in the HTML paragraph. We return to PHP in Section 3.12.

When active pages are treated by modules loaded into the Web server, the processing is faster and more efficient compared to external CGI programs. PHP usually runs as an Apache module and can provide excellent server-side programming and support. PHP is an important focus of this textbook.

1.13   Database Access

A database is an efficiently organized collection of data for a specific purpose. Examples of databases abound: employee database, membership database, customer account database, airline and hotel reservations database, user feedback database, inventory database, supplier or subcontractor database, and so on. Relational database systems (RDBMS) support the management and concurrent access of *relational databases.* A relational database is one that uses tables (Figure 1.10) to organize and retrieve data.

FIGURE 1.10: A Database Table

Last	First	Dept	Email
Wang	Paul	CS	pwang@kent.edu

In the age of the Internet, online access to databases is increasingly important for businesses and organizations. Using a Web interface to provide such online access has become the norm. In addition to providing access to databases, many websites also employ databases for their own purposes, such as user accounts, product inventory, blogging and forum support, just to name a few.

There are a number of popular relational database systems. In this text, however, we will focus on the freely available MySQL[4]. The same as other systems, MySQL uses the standard SQL (Structured Query Language) for access and update of information in databases (Figure 1.11). The PHP language provides very good support for database connection to MySQL, for issuing queries and for detecting errors. The PHP-MySQL combination makes Web access to databases easy to achieve.

FIGURE 1.11: Database Function

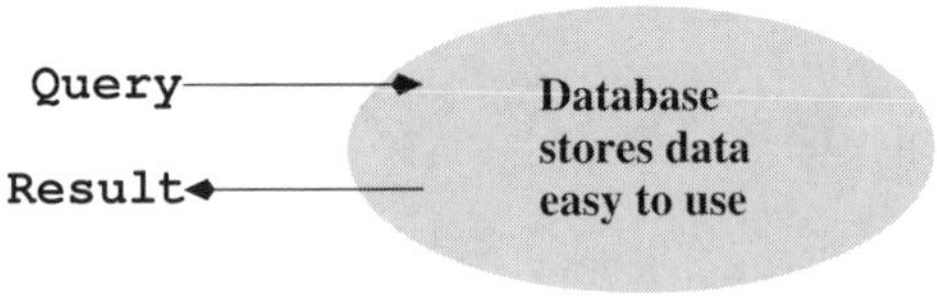

Query ⟶ **Database stores data easy to use**
Result ⟵

---

[4]MySQL is now part of Oracle.

1.14 Client-Side Scripting

Modern browsers make the Web useful for everyone by providing a convenient user interface that usually supports keyboard and mouse interactions as well as video and audio presentations.

The actions of a Web browser can be defined and controlled by programming within a webpage. Such programming can supply customized user experiences and make webpages more responsive and useful for end users. The programs execute within the browser, which runs on the client host, the computer used to access the Web. For all major browsers, *JavaScript* is the standardized scripting language for client-side programming. Because the JavaScript language standard has been developed and maintained by the *ECMA* (European Computer Manufacturer Association), the language is also known as *ECMAScript* (`ecma-international.org`).

With JavaScript, a webpage can define reactions to user interface (UI) events, verify correctness/completeness of user input, send and retrieve information from the page's server without reloading the page, change/update and otherwise manipulate the page display, and much more. Because JavaScript runs on the client, it takes advantage of the processing power of the client host and can potentially lessen the load on the Web server.

1.15 Document Object Model

The *Document Object Model* (DOM) is an application programming interface (API) definition specifying how a document processing program such as a Web browser represents a document as a tree-structured object (the DOM tree) and allows programmatic access and manipulation of the DOM tree.

All major browsers provide DOM support, including HTML, XHTML, HTML5, and XML DOM interfaces. Because JavaScript is the standard, it is not surprising that major browsers all implement the required DOM interfaces in JavaScript.

As we will see in later parts of this text, we can combine JavaScript with DOM and CSS, a technique known as *DHTML*, to make webpages responsive and dynamic, resulting in a much-improved user experience. Additionally, DHTML can be combined with server access from JavaScript to achieve a technique known as AJAX (Asynchronous JavaScript and XML) to make webpages even more useful.

1.16 Hypertext Transfer Protocol

Web browsers and Web servers communicate following a specific protocol. That protocol is HTTP, the *Hypertext Transfer Protocol*. It does not matter which browser is contacting what server; as long as both sides use the HTTP protocol, everything will work.

HTTP is an application layer (Figure 1.1) protocol that sits on top of TCP/IP, which provides reliable two-way connection between the Web client and Web server. Understanding HTTP is important for Web programmers. We give a brief overview of HTTP here. Other aspects of HTTP will be explained as the need arises.

In the early 1990s, HTTP gave the Web its start. HTTP/1.0 was standardized in the first part of 1996. Important improvements and new features have been introduced in HTTP/1.1 and it is now the stable version. While HTTP transmits information in the open, HTTPS (HTTP Secure) is a secure protocol that simply applies HTTP over a secure transport layer protocol *Transport Layer Security* (TLS) that is based on the earlier *Secure Sockets Layer* (SSL).

FIGURE 1.12: An HTTP Transaction

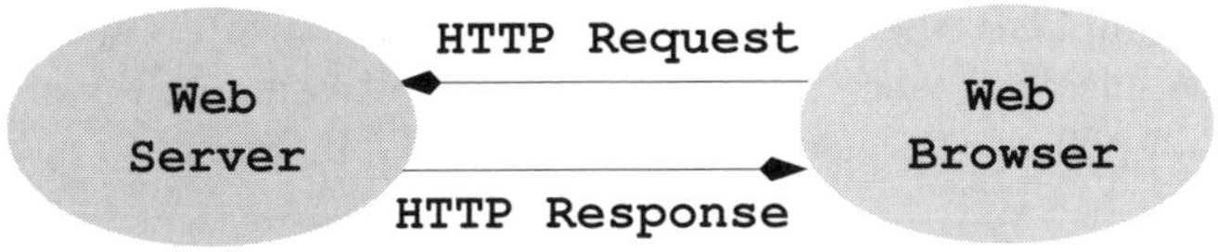

Figure 1.12 illustrates an HTTP transaction. A simple HTTP transaction goes as follows:

1. *Connection*—A browser (client) opens a connection to a server.

2. *Query*—The client requests a resource controlled by the server.

3. *Processing*—The server receives and processes the request.

4. *Response*—The server sends the requested resource back to the client.

5. *Termination*—The transaction is finished, and the connection is closed unless it is kept open for another request immediately from the client on the other end of the connection.

HTTP governs the format of the query and response messages (Figure 1.13).

The header part is textual, and each line in the header should end in RETURN and NEWLINE, but it may end in just NEWLINE.

The initial line identifies the message as a query or a response.

- A query line has three parts separated by spaces: a *query method* name, a local path of the requested resource, and an HTTP version number. For example,

FIGURE 1.13: HTTP Query and Response Formats

---

initial line   (different for query and response)
HeaderKey1: value1   (zero or more header fields)
HeaderKey2: value2

                                            (an empty line with no characters)
Optional message body contains query or response data.
Its data type and size are given in the headers.

```
GET   /path/to/file/index.html   HTTP/1.1
HOST: domain_name
```

or

```
POST   /path/script.php   HTTP/1.1
HOST: domain_name
```

The `GET` method requests the specified resource and does not allow a message body. A `GET` method can invoke a server-side program by specifying the CGI or active-page path, a question mark, and then a *query string*:

```
GET /dwp/join.php?name=value1&email=value2   HTTP/1.1
HOST:  sofpower.com
```

Unlike `GET`, the `POST` method allows a message body and is designed to work with HTML forms for collecting input from Web users. The `POST` message body transmits name-value pairs just like a query string but without length limitations that apply to GET query strings because they are part of the request URL.

• A response (or status) line also has three parts separated by spaces: an HTTP version number, a status code, and a textual description of the status. Typical status lines are

```
HTTP/1.1   200   OK
```

for a successful query or

```
HTTP/1.1   404   Not Found
```

when the requested resource cannot be found.

- The HTTP response sends the requested file together with its content type (Section 1.5) so the client will know how to process it.

When using a browser to access the Web, the HTTP messages between it and the Web servers are kept behind the scenes. But it is possible to expose these messages and gain real experience with HTTP by using the **telnet** (on Windows) or the **nc** (on Linux/Unix/Mac OS X) applications. See the demo **Ex: Http** at the DWP website to reveal a sample HTTP 1.1 response. Also, see the code example package for the BASH script `poorbr.sh` that allows you to manually request any webpage.

1.17   Web Development

The Web is a new mass communication medium. To create a well-designed and effective website is a challenge. It takes expertise in information architecture, visual communication design, mass communication, computer programming, business administration, and consumer psychology, just to name some areas. Creating a great website usually involves teamwork.

Web development tasks include conceptualizing, planning, designing, organizing, implementing, maintaining, and improving a website. The goal is to achieve functionally effective and aesthetically attractive information delivery and exchange.

In this text, the primary emphasis is on the programming aspects of Web development. Some design aspects will be discussed so that a Web programmer will be able to work together with artistic and visual communication designers effectively on a team.

The programming topics in this textbook should be considered in the context of the Web development principles and processes discussed in this section.

1.17.1   Web Development Principles

It is advisable to know, ahead of time, the general principles to follow for creating attractive, functional, and user-friendly websites that will become a great asset for organizations as well as individuals.

- Involve the principals of the website in every step of the planning and development process. Often, after learning what the Web and Internet can do, site owners will have important ideas on what the site can do for their businesses. It goes without saying that most of the site content must be supplied and composed by the principals.

- Make clear on the site entry page, what the website is and what audiences it is intended to serve.

- Know your audiences and anticipate what information they expect to find on your site. Design your site navigation accordingly and provide multiple links from different contexts to important parts of your site.

- Form follows function. Your site architecture, art and visual design, and page layout must enhance its functions and be *user centric*. It is all about making it easy and attractive for your users/customers.

- Avoid extraneous audio, video, blinking text, and other technology show-off features, if they do not serve any good purposes.

- Avoid small fonts and dark backgrounds. Information on the site must be easy to read and understand.

- Make your site fast to load and responsive to user actions.

- Use fluid layout that responds well to varying screen sizes and resolutions.

- The site should be easy to maintain. Information that needs changing/updating must be easily editable by key business personnel within the organization that owns the site.

- Use open standards and test your site under all major browsers.

1.17.2   Web Development Process

Many tasks are involved in creating a website. The overall *website development process* can be summarized as follows:

Requirement Analysis and Development Plan—What are the requirements for the finished site? What exactly will your finished website achieve for the client? What problems does your client want you to solve? Who are the target audiences of the website? Can you realistically help the client? What is the scope and nature of the work? What design and programming tasks are involved? What resources and information will be needed and what problems do you foresee? Who will provide content information for the site and in what formats? What resources are needed or available: textual content, photos, imagery, audio, video, logos, corporate identity standards, copyrights, credits, footers, and insignia?

Answer the preceding questions and make a plan. Create and group content, functional, and look-and-feel requirements, and set clear goals and milestones for building and developing the site.

Site Architecture—Decide on an appropriate architecture for the site. The site architecture is influenced by the nature of the information being served and the means of delivery. Ordinary sites involve static pages with text, images, and online forms. Specialized sites may involve audio, video, streaming media, and dynamically generated information or access to databases.

Website *information architecture* (IA) deals with the structuring, the relationship, the connectivity, the logical organization, and the dynamic interactions among the constituent parts of a website.

Within each webpage, consider the placement, layout, visual effect, font and text style, and so forth. These are also important but may be more "interior decoration" than architecture. However, architecture and interior/exterior decoration are intimately related.

The site architecture phase produces a blueprint for building the website. The blueprint is a specification of the components and their contents, functionalities, relations, connectivity, and interactions. Website implementation will follow the architecture closely.

An important aspect of site architecture is the navigation system for visitors to travel in your site. The goal is to establish site-wide (primary), intrasection (secondary), and intrapage (tertiary) navigation schemes that are easy and clear.

Content-only Site Framework—Follow these steps to prepare a skeletal site as a foundation for making adjustments and for further work to complete the site:

- Content: Create a content list or inventory; prepare content files ready to be included in webpages.

- Site map: Draw a relationship diagram of all pages to be created for the site, give each page appropriate titles, show page groupings and on-site and off-site links, distinguish static from dynamic pages, identify forms and server-side support. Major subsections of the site can have their own submaps.

- Skeletal site: Conceive an entry page, home page, typical subpages and sub-subpages, textual contents (can be in summary form), HTML forms with textual layout and descriptions of server-side support, structure of the file hierarchy for the site, and well-defined HTML coding standards for pages.

- Navigation: Follow the site architecture and site map to link the pages, use textual navigation links with rough placements (top, left, right, or bottom), avoid dead-end pages, and avoid confusing the user.

Visual Communication and Artistic Design

- Design concepts: features, characteristics, and look-and-feel of the site; the design must reflect client identity and site purpose.

- Storyboards: simple layout sketches based on content-only site for typical pages, HTML forms, and HTML form response pages; header, footer, margins, navigation bar, logo, and other graphical elements to support the delivery of content; client feedback and approval of storyboards.

- Page layout (for pages at all levels): content hierarchy and grouping; grids, alignments, constants, and variables on the page; placement and size of charts, graphs, illustrations, and photos; creative use of space and variations of font, grid, and color; style options and variations.

- Home page/entry page: visuals to support the unique function and purpose of entry to the site and home page as required by site architecture.

Site Production

- Page templates: Create templates for typical pages at all levels. Templates are skeleton files used to make finished pages by inserting text, graphics, and other content at marked places in the templates. In other words, a template is a page frame with the desired design, layout, and graphics ready to receive text, links, photos, and other content. A template page may provide HTML, style sheets, JavaScript, `head`, `body`, `meta`, `link`, and `script` tags, as well as marked places for page content. Templates enable everyone on the project team to complete pages for the site. Advanced templates may involve dynamic server-side features.

- Prototype pages: Use the templates to complete typical pages in prototype form, test and examine page prototypes, present prototype pages to the client, and obtain feedback and approval. Make sure that the layout system has been designed with enough versatility and flexibility to accommodate potential changes in content.

- Client-side programming: Write scripts for browsers and possibly other Web clients that will be delivered together with webpages to the client side. These scripts may include style sheets and JavaScripts. Client-side programs can make webpages more interactive and responsive.

- Server-side programming: Write programs for form processing, dynamic page generation, database access, e-business, and e-commerce features. Make sure these follow the site architecture, user orientation, and visual design.

- Finished pages: Following page prototypes, add text, graphics, photos, animations, audio, and video to templates to produce all pages needed; make final adjustments and fine-tuning.

Error Checking and Validation—Apply page checking tools or services to finished pages to remove spelling errors, broken links, and HTML coding problems. Check page loading times.

Testing—Put the site through its paces, try different browsers from different access locations, debug, fine-tune, and check against architecture and requirements.

Deploying—Release the site on the Web, make its URL known, and register the site with search engines.

Documentation—Write a description of the website, its design and functionalities, its file structure, locations for source files of art and programming, and a site maintenance guide.

Maintenance—Devote time to the continued operation and evolution of the website.

1.18   For More Information

- IPv6 is the next-generation Internet protocol. See `www.ipv6.org/` for an overview.

- HTML5 is the new and coming standard for HTML. See the specification at the W3C website.

- The DNS is basic to keeping services on the Internet and Web running. Find out more about DNS at `www.dns.net/dnsrd/docs/`.

- HTTP is basic to the Web. See RFC 1945 for HTTP 1.0 and RFC 2068 for HTTP 1.1.

1.19   Summary

The Web is one of the most important services on the Internet. Browsers, on client hosts, follow the HTTP/HTTPS protocol to access Web server programs running on server hosts. The free and open Apache is one of the most popular Web server programs.

Each website has its own domain name and server document space. Static pages are served as-is. Dynamic pages, created on-the-fly with CGI or active page technologies, can respond to incoming information and access databases.

HTML marks up the structure of webpages, CSS specifies layout and display rules, and URLs provide navigation and linking to pages and services. MIME content types help indicate media types for processing. The DNS maps domain names to IP addresses.

A detailed Web development process makes clear best practices and sets up the stage for Web programming. Server-side programming will involve PHP, MySQL, and Apache management. Client-side programming will involve JavaScript, DOM and other HTML5 defined APIs.

Exercises

1.1. What is a computer network? Name the major components in a computer network.

1.2. What is a networking client? What is a networking server? What is a networking protocol?

1.3. What addressing scheme does the Internet use? What is an IP address? What is the quad notation?

1.4. Consider the IP address

    123.234.345.456

Is there anything wrong with it? Please explain.

1.5. What is the relation between the Web and the Internet? What is the relation between HTTP and TCP/IP?

1.6. What is DNS? Why do we need it?

1.7. What do name servers do? Why do we need them?

1.8. What are the major components of the Web? Why is HTML central to the Web? What is HTML 4.0, XHTML, and HTML5?

1.9. What is an *active page*? What is PHP?

1.10. What is a database system? How is it related to the Web?

1.11. What is the difference between a Web server and a Web browser? Is the Web server a piece of hardware or software? Explain.

1.12. Name the most popular Web server and Web browser. Do you know their market shares?

1.13. How does a webpage get from where it is to the computer screen of a user?

1.14. What is a URL? What is the general form of a URL? Explain the different URL schemes.

1.15. What is URL encoding? Please explain in detail.

1.16. What are content types? How are they useful?

1.17. What is the difference between a static webpage and a generated webpage?

1.18. What scripting language is used by webpages to control the Web browser? What is DOM?

1.19. What is an HTTP transaction? What is an HTTP query? What is an HTTP response?

1.20. Take the domain name `sofpower.com` and find its IP address. Use this IP address instead of the domain name to visit the site. Write the bit pattern for this IP address.

1.21. Search on the Web for ICANN. Visit the site and discover its mission and services.

1.22. Find the domain record for `sofpower.com`. Who is the owner of this domain name? Who are the administrative and technical contacts?

1.23. Find the DNS record for `sofpower.com`.

Chapter 2

Webpage Markup with HTML5

The Hypertext Markup Language (HTML) is a markup language used to organize the contents in a webpage.

HTML has evolved and HTML5 is the new standard. HTML5 not only supplies a language for webpage markup, but also defines a good number of APIs (application programming interfaces) for JavaScript programs to interact, control, and manipulate the webpage in the context provided by the user agent (UA), which is usually a browser. It also specifies how browsers must support HTML5 features. Unless otherwise indicated, HTML descriptions and codes in this text will follow HTML5.

HTML provides various *elements* such as <h2> (second-level header) and <p> (paragraph) with which to structure the page content. HTML enables you to organize text, graphics, pictures, sound, video, and other media content. When a Web browser receives an HTML document, it can then properly format, render, and manage the user interface for the page.

HTML provides elements for headings, sections, paragraphs, lists, tables, headers, footers, links, figures, images, audios, videos, and forms (for user input). Furthermore, it supports image maps (multiple clickable areas in the same image), embedded objects (for content to be processed by browser plugins), internal frames (page-in-page) and canvases (programmable graphics drawing areas), and so on.

The major part of a website is usually a set of HTML documents. Learning and understanding HTML are fundamental to Web programming.

To create HTML files, you may use any standard text editor such as vi, emacs, WordPad (MS/Windows), or SimpleText (Mac/OS). When saving the edited file, make sure to choose UNICODE plaintext encoding in UTF-8. Rich text or other formatted encoding will result in unwanted and erroneous code in the HTML file. Specialized tools for creating and editing HTML pages are also widely available. After creating an HTML document and saving it in a file, with a .html or .htm suffix, you can open that file (by double-clicking the file or using the browser File>Open File menu option) to look at the page.

Learning HTML is not hard. It involves knowing some general rules that apply to all elements and getting to know how to use individual elements as well as how elements combine for practical purposes.

We begin HTML5 in this chapter and continue in Chapter 3 to cover more advanced aspects of HTML5. The two chapters combine to provide a comprehensive and in-depth introduction to HTML5 markup. Other aspects of HTML5 are described in later chapters. Unless specifically noted otherwise, we will use the terms HTML and HTML5 interchangeably in this textbook.

2.1   HTML5 Page Structure

HTML is a *markup language* that provides *elements* to structure page content. Web browsers can then present the page based on the content markup. Any HTML element *xyz* normally consists of an *open tag* <xyz> and a *close tag* </xyz> to enclose desired content between them. HTML tags are always enclosed in angle brackets (< >) to set them apart from contents of the page.

We recommend the following basic structure for HTML5 documents:

```
<!DOCTYPE html>
<html xmlns="http://www.w3.org/1999/xhtml"
      lang="en" xml:lang="en">
<head>
<meta charset="utf-8"/>
<title>Great Company: homepage</title>
<!-- other head elements as appropriate -->
</head>
<body>  <!-- page content begin -->
   . . .
   . . .
<!-- page content end -->  </body>
</html>
```

The DOCTYPE line indicates a page coded in HTML5. Next comes the open tag for the html element where the primary language (Section 3.13) for the page is set to English (en) and the xmlns provides the XML namespace. The closing tag for html comes at the end of the page.

You may copy the first two lines verbatim for your webpages. Change the language setting if the page is in some language other than English.

The html element has two *child elements*, head and body. Inside the head element you would place *meta information* about the page such as page title, description, keywords, and so on (See Section 3.10). You place the page content to be displayed to the end user in the body element, between the <body> and </body> tags. In HTML source code, comments and notes for page developers are placed within <!-- and -->.

We already know that HTML tags come in pairs: a *start tag* and an *end tag*. They work just like open and close parentheses. The end tag is always the start tag with an extra SLASH (/) prefix to the tag name. Between the start and end tags go the contents of the element, which can be text and other allowed HTML elements, known as child elements.

Most elements have begin and end tags, but not all. For *void elements* such as line break (`<br />`) or horizontal rule (`<hr />`), no content or end tag is allowed.

The `head` element contains informational elements for the entire document. For example, the `title` element (always required) specifies a page title that is

1. Displayed in the *title bar* of the browser window

2. Used in making a bookmark for the page

2.2 Creating a Webpage

Let's create a very simple webpage (**Ex:** `FirstPage`) following the template fromSection 2.1. Using your favorite text editor, type the following:

```
<!DOCTYPE html>
<html xmlns="http://www.w3.org/1999/xhtml" lang="en"
    xml:lang="en">
<head>
<meta charset="utf-8"/>
<title>My Sample Webpage</title>
</head>
<body style="background-color: cyan; margin: 50px">
<h2>Hi everybody!</h2>
<p>My Name is (put your name here) and today is
<time>(put in the date yyyy-mm-dd)</time>.</p>
<p>HTML5 is cool.</p>
</body>
</html>
```

Save it into a file named `MyPage.html`. The content of `body` consists of a second-level header and two short paragraphs given by the `p` element. The date is enclosed in the `time` element (Section 2.11.1). With the `style` *attribute* for the `body` element, the page background color is set to `cyan` and a margin of 50 pixels for all sides is specified.

From your favorite browser, select the `Open File` option on the `File` menu and open the file `MyPage.html`. Now you should see the display of your webpage (Figure 2.1). For more complicated webpages, all you need to know is more HTML elements and practice in how to use them.

2.3 HTML5 Elements and Entities

There are more than 100 different elements in HTML5. Elements for meta information are placed in the `head` element. Elements for page content are

FIGURE 2.1: Sample Webpage

placed in the <body> element. HTML5 distinguishes between two broad types
of content elements: *flow elements* that occupy their own vertical space in a
page and *phrasing elements* that act like words and phrases. Flow elements
act like paragraphs, lists, and tables and can contain other flow elements,
phrasing elements, and texts. Phrasing elements can contain other phrasing
elements and texts.

A practical way to classify HTML5 elements is:

Top-level elements: html, head, and body.

Head elements: elements placed inside head, including title (page title),
style (rendering style), link (related documents), meta (data about
the document), base (URL of the document), and script (client-side
scripting). These elements are not part of the page display.

Block-level elements: flow elements behaving like paragraphs, including
article, h1–h6 (headings), header, footer, section, p (paragraph),
figure, canvas (dynamic drawing area), pre (preformatted text), div
(designated block), ul, ol, dl (lists), table (tabulation), form (user-
input forms), and video (video). A block element occupies 100% of the
available width to it and will be *stacked vertically* with preceding and
subsequent block elements. In other words, when displayed, a block-level
(or simply block) element always starts on a new line, and any element
immediately after the block element also begins on a new line.

Inline elements: phrasing elements behaving like words, characters, or
phrases that flow horizontally to fill the available width before starting new lines. Usually, inline elements are placed within block elements.
Inline elements include a (anchor or link), audio (sound), br (line
break), code (computer code), img (picture or graphics), em (emphasis), nav (navigation), samp (sample output), span (designated inline

scope), **strong** (strong emphasis), **sub** (subscript), **sup** (superscript), **time** (time/date), and **var** (variable name).

The **display** mode of a flow element can be set to **block** or **inline**, which will affect its content layout.

When an element is placed inside another, the containing element is the *parent* and the contained element is the *child*.

As we have seen, comments in an HTML page are given as

```
<!-- a sample comment -->
```

where the comment may consist of multiple lines. Text and HTML elements inside a comment tag are ignored by browsers. Be sure not to put two consecutive dashes (--) inside a comment. It is good practice to include comments in HTML pages as notes, reminders, or documentation to make maintenance easier.

In an HTML5 document, certain characters, such as < and &, are used for markup and must be *escaped* to appear literally. HTML provides *entities* (*escape sequences*) to introduce such characters into a webpage. For example, the entity **<** gives < and **&** gives & (Section 3.7). Characters not on the regular keyboard can also be included directly or using HTML5-defined character references (Section 3.9).

2.4 Evolution of HTML

In 1989, at the European Laboratory for Particle Physics (CERN), Tim Berners-Lee defined a very simple version of HTML based on SGML, Standard Generalized Markup Language, as part of his effort to create a network-based system to share documents via text-only browsers.

The first common standard for HTML was HTML 3.2 (1997). HTML 4.01 became a W3C (the World-Wide Web Consortium) recommendation in December 1999. HTML 4.01 begins to clearly separate the document structure and document presentation aspects of HTML, and specifies a clear relationship between HTML and client-side scripting (JavaScript).

Starting with HTML 4, you can specify the presentation style of individual elements and also attach presentation styles to different element types with *style sheets*.

HTML 4.01 is great but not based on XML, a popular and standardized way to define additional elements and use them in documents. Making HTML follow the strict syntax of XML brings important advantages:

- XHTML elements can be used together with other elements defined by XML.

- XHTML pages can be processed easily by any XML tool.

In January 2000, W3C released XHTML 1.0 as an XML reformulation of HTML 4.01. XHTML 1.0 is basically HTML 4.01 written under the strict XML syntax.

HTML5 combines the XHTML, HTML 4, and CSS3 standards, introduces new elements and APIs, as well as incorporates MathML (Mathematical Markup Language) and SVG (Scalable Vector Graphics) into HTML (Figure 2.2). HTML5 can also easily be written in an XML compliant way. The release of the HTML5 standard promises to bring significant advantages to Web developers and benefits to end users.

FIGURE 2.2: HTML5 Integration

2.5 HTML5 Features and Advantages

As you go through this textbook, you will gain an in-depth understanding of HTML5 and what advantages it brings. But, it is perhaps worthwhile to summarize and highlight the ten most important HTML5 features here before we get too deep into details.

1. New elements such as `header`, `footer`, `article`, `section`, `menu`, `nav`, and `aside` to better organize webpages.

2. With the new `audio` and `video` elements, HTML5 makes sound and video media as easy to place in a webpage as a still image, thus eliminating the need for proprietary technologies, such as Flash and Silverlight, or browser plug-ins.

3. With the new `canvas` dynamic drawing area element, interactive graph-

ics and animation can easily be deployed on the client side with JavaScript control.

4. Browser support for drag-and-drop API.

5. Browser support for form input and input checking.

6. Websites can list files to be cached by the browser for offline use, either for browsing offline or for supporting a Web application running offline.

7. With HTML5 *Web Storage*, Web applications can store sizable data (up to 5MB per Web domain) locally on the browser side. Such local storage can be *per session* (lost if browser is closed) or persistent (not lost even if you close the browser or shut down the computer) but private (not transmitted back to the Web).

8. Browser support for editing (by the end user) of webpage content.

9. Enabling easy definition of background tasks and location-dependent information presentation with the *Web workers* and the *geolocation* APIs, respectively.

10. Native support of mathematical formulas and 2D graphics markup with MathML and SVG, respectively.

Browsers are moving rapidly to support all the new HTML5 features. A nice website for testing your Web browser is `www.html5test.com`. Also, the *Technology Review* magazine article "The Web is Reborn" by Bobbie Johnson (`technologyreview.com/web/26565/`) is a good read.

2.6 Webpage Syntax

The *syntax* of HTML5 is no more complicated than HTML 4 or XHTML. In fact, it is simpler and easier. The following suggested syntax rules will help you write *polyglot* documents that are compliant with HTML5 and XHTML5 (the XML version of HTML5) at the same time. All HTML5 code examples in this textbook will be polyglot.

- All tags begin with < and end with >. The *tag name* is given immediately following the leading <. Make sure the tag is spelled correctly. Unrecognized tags are ignored by browsers. Any *attributes* are given following the tag name in the form:

  <tag *attribute*$_1$ ="*value*"   *attribute*$_2$ ="*value*" ... >

  You may use single quotes (`'`), instead of double quotes (`"`), for the value part of any attribute. Be careful; forgetting to close a quote can result in a blank page display.

- Tag names and attributes are lowercase. Attributes are always given in either of the two forms:

  attribute_name="value"
  attribute_name='value'

  where the value is case sensitive and can be empty. For *Boolean attributes*, those that are either on or off, use either of these forms for "on":

  attribute_name ="attribute_name"
  attribute_name=""

  and omit the attribute for "off".

- Unrecognized tags and attributes are ignored by browsers.

- Most elements involve start and end tags. Other elements, such as `<br/>` (line break) and `<img .../>` (image), do not have closing tags and are known as *void elements*. The SLASH (/) at the end is optional for HTML5 but needed for polyglot documents.

- Elements must be *well-formed*. This means no missing opening or closing tags and no improper element nesting. For example,

  ```
  <p>Learning <strong>HTML5</p></strong>
  ```

  overlaps the tags and is not properly nested. Browsers may tolerate such ill-formed code. The correct nesting is

  ```
  <p>Learning <strong>HTML5</strong></p>
  ```

- Attributes can be required or optional and can be given in any order. If an attribute is not given, its initial (default) value, if any, is used.

- Extra white space and line breaks are allowed between the tag name and attributes and around the = sign inside an attribute. Line breaks and white space within attribute values are also allowed but should be avoided because they may be treated inconsistently by browsers.

- The `body` element may contain only flow (block) elements. Freestanding text (not enclosed in block elements) or inline elements are not allowed directly in the `body` element.

Certain tags are only allowed within their permitted context. For example, a `<tr>` (table row) element can only be given inside a `<tbody>` element. Learning HTML5 involves knowing the elements, their attributes, where they can be placed, and the child elements they can contain.

2.7   HTML5 Core Attributes

All HTML5 elements admit the following *core attributes*:

- id—Uniquely identifies the element in a page. All ids in a document must be distinct. Among other uses, a URL ending in *#some_id* can lead directly to an element inside a document.

- style—Gives presentation styles for the individual element. For example, the code used in **Ex: FirstPage** (Section 2.2)

  ```
  <body style="background-color: cyan">
  ```

  gives the color value cyan to the style property background-color for this element. Several style properties separated by semicolons can be given. The style attribute is a direct but inflexible way to specify presentation style. Although this attribute is sometimes necessary, better and much more versatile methods for assigning styles can be found in Chapter 4.

- class—Specifies a *style class* (Section 4.6) or a space separated list of style classes for the element. For example, class="fineprint" or class="footnote fineprint". Thus, you may place HTML elements in different classes and associate presentation styles to all elements belonging to the same class (Chapter 4).

- title—Provides a title for the element. This may be used for tool-tip displays by browsers.

- hidden—Prevents the element from being displayed by a browser when set to true.

Other core attributes include contenteditable, draggable, dropzone, spellcheck, and so on. These will be explained later.

2.8   Webpage Architecture

HTML5 allows flexible markup of pages for your website. A typical webpage is organized into the following parts inside the root element html.

The head element contains child elements: the page title (title), the page character encoding with a meta tag

```
<meta charset="UTF-8"/>
```

both are required by HTML5. Usually there will be additional elements inside head to add styling, scripting, and other meta information (Section 3.10).

The body element provides the page content, often organized into a header part for the top banner and a horizontal navigation bar at the top of the page. After the header, the page may also have a vertical navigation bar on the left side. The flow (block) element `nav` is used for navbars that organize links. The established convention is to have site-wide links on top and page-specific links on the left side of a page.

Then, there are one or more articles (`article`) for the main content, followed by a `footer` at the end of the webpage. Each article may contain one or more sections (`section`) that consist of headings (`h1` through `h6`), paragraphs (`p`), tables, figures, audio, and video. Each paragraph can also involve text, pictures (`img`), audio, and video. The footer often provides information on copyright, author, and links to terms of use, privacy policy, customer service, and so on.

An `aside` flow (block) element can set aside related information, such as links to references, outside resources, and advertisements, that are not the primary content of the page.

Let's first look at elements related to headings and paragraphs.

2.9 Headings and Paragraphs

HTML offers six *heading elements*, `h1` through `h6`, for level-one to level-six section headings. Headings are flow (block) elements that are usually displayed with a bolder font followed by a blank line. Use `h1` for top-level section headings, and `h2` for subsections within top-level sections, and so on. Unless otherwise specified, browsers use increasingly smaller fonts to render deeper level headings.

It is advisable to use `h1` for the most prominent heading, such as the headline of an article. The following (**Ex: GreenEarth**) is an example (Figure 2.3):

```
<article>
<h1>The Green Earth Project</h1>
<section>
<h3>Project Background</h3> <!-- 1st section -->
   <p>Put first paragraph here</p>
   <p>Put second paragraph here</p>
<h4>A Successful Past</h4><!-- subsection -->
   <p>Another paragraph here</p>
</section><section>
<h3>Current Status of Green Earth</h3><!-- 2nd section -->
   <p>Another paragraph here</p>
</section><section>
<h3>Future Goals</h3><!-- 3rd section -->
</section></article>
```

The flow (block) element `p` (a paragraph) may contain texts and phrasing elements. It is typically displayed with a leading and a trailing blank line. The

FIGURE 2.3: Sections and Paragraphs

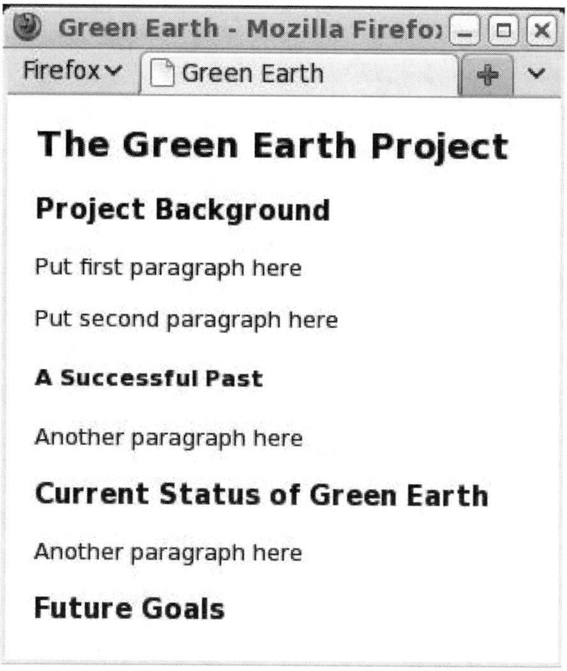

paragraph content will be formatted to fit the available width[1]. Line breaks are inserted automatically (*line wrapping*) where needed to render the paragraph. Extraneous white spaces between words and lines within the source text of the paragraph are normally ignored (*white-space collapsing*). If you need a line break at a specific point in the text or page, you can use the `<br />` tag to call for a line break. For a long-running text without spaces, such as an email or Web address, you can insert the void element `<wbr />` to indicate a *line break opportunity*. The browser will do a line break indicated by `wbr` only if necessary. For example,

```
<p>Please visit www.somelong.<wbr />andcomplicated.com.</p>
```

Inside a paragraph, you can place other phrasing (inline) elements such as `q`, `em`, `mark`, `strong` (see next subsections), `img` (Section 2.17), `video`, and `audio`. Remember always to use the tag `</p>` to end a paragraph.

By default, browsers usually display headings and paragraphs left-aligned and without indenting the lead line. You may use line-break elements (`<br/>`) to call for a line break between phrasing (inline) elements.

The ` ` numerical character reference (Section 3.7) is a nonbreaking space. Use it instead of a regular SPACE character between two words that

---

[1]On the computer screen, width is horizontal and height is vertical.

must be kept together on one line or use several nonbreaking spaces to add more horizontal spacing between two words.

2.9.1 Quotations

The flow (block) element `blockquote` contains one or more flow elements, typically paragraphs, that are quoted from other sources. The optional attribute `cite` can specify a URL leading to the source.

Browsers usually display a quoted block with increased left and right margins (**Ex:** `Quote`). In other words, quoted material is normally indented on the left and right; for example (Figure 2.4),

FIGURE 2.4: Block Quote

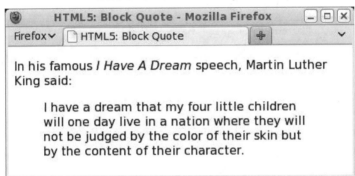

```
<p>In his famous <em>I Have A Dream</em> speech,
Martin Luther King said:</p>
<blockquote cite="http://www.mlkonline.net">
I have a dream that my four little children will one day
live in a nation where they will not be judged by the color
of their skin but by the content of their character.
</blockquote>
```

Inline quotations can be structured with the phrasing element `q`. For example (**Ex:** `InlineQuote`):

```
<p>Confucius: <q>Don't employ a person due to words or
dismiss words due to the person.</q></p>
```

Browsers display inline quotations by supplying language-appropriate quotation marks (Figure 2.5). Nested quotations are handled correctly as well.

The flow (block) element `hr` gives you a horizontal rule to separate sections in a document. A horizontal rule provides a strong, but crude, visual indication of the start of contents of a different nature.

It is normally not necessary to place any rules between paragraphs or

FIGURE 2.5: Inline Quote

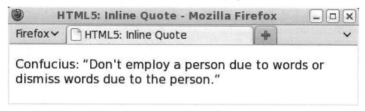

sections of an article. But used appropriately in selected situations, horizontal rules can increase the clarity of a page.

For `hr`, browsers usually render a narrow full-width horizontal line. In HTML, the `<hr />` element admits nothing other than core attributes (Section 2.6). The `style` attribute is used to control the length, width, color, border, and other display styles for the horizontal rule. For example (**Ex: Hrule**),

```
<hr style="height: 4px; width: 50%;
    margin-left: auto; margin-right: auto" />
```

gives a centered rule 4 pixels thick. Add the property `background-color: blue` to get a solid blue rule. Use `margin-left:0` (`margin-right:0`) for left- (right-) adjusted alignment.

The `border` property gives control and flexibility for drawing lines on any of the four sides of `hr` or other elements (Section 4.13.1).

2.10 White Space and Line Wrapping

In HTML, *white space* separates text into words, and user agents such as browsers can flow the words onto a rendered page in a format appropriate to the language of the document or speak the words using text-to-voice software. HTML regards the following as white-space characters:

- SPACE: ASCII 32 (entity `&#x0020;`)

- TAB: ASCII 9 (entity `&#x0009;`)

- FORMFEED: ASCII 12 (entity `&#x000C;`)

- Zero-width space: a non-ASCII character (entity `&#x200B;`)

Words can be separated by one or more white-space characters but will only result in at most one rendered interword space. Hence, browsers perform *white-space collapsing*.

In addition, a RETURN (ASCII 13), NEWLINE (ASCII 10), or a RETURN-NEWLINE pair is considered white space. These are line breaks in the HTML source code, but they have no relation to the displayed line breaks in a web-page. An important function browsers perform is to flow text into lines, fit lines in the available display space, and wrap lines around when necessary. Line wrapping can only happen at word boundaries. Thus, no word will be chopped across lines. When the display window is resized, the text lines are reflowed to fit.

Only white space separates words. Tags do not. Thus, it is safe to use code such as

```
<p>The HTML<strong>5</strong> standard.</p>
```

You can manage displayed line breaks as follows:

- To force a line break, use the `<br/>` element.

- To keep two words on the same line, use the nonbreaking space (` ` or the non-polyglot ` `) instead of a regular SPACE.

- To mark places where a long words can be broken across lines, you may use the non-polyglot *soft hyphen* (`&shy;`), which is rendered as a HYPHEN (-) only at the end of a line. Browsers generally do not break a word that is hyphenated in the source code.

- To indicate where long words can be broken across lines without adding a hyphen, use the `<wbr/>` element.

2.10.1 Preformatted Text

Sometimes text lines are *preformatted* with spacing and line breaks. By en-closing such material in the flow element `pre`, the existing spacing and line breaks will be preserved. For example (**Ex: Pre**),

```
<figure style="width: 12em; background-color: cyan">
<pre>
          North

  West              East


          South
</pre></figure>
```

results in the display shown in Figure 2.6. The figure width is set to twelve em (the width of the M character in the current font). Without the enclosing `pre` tags, the four words would be flowed onto a single line with only a single space between any two words. By default, browsers use a constant-width font, such as `Courier`, to display `pre` contents. But the font used can be controlled by setting the font properties (see next subsection) of `pre`.

FIGURE 2.6: Preformatted Text

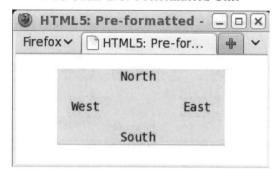

2.11 Phrasing Elements

Frequently useful phrasing (inline) elements include

a: a link (Section 2.15)

br: an explicit line break

cite: a citation

em: emphasis, usually displayed in italics

strong: strong emphasis, usually displayed in boldface

mark: stronger emphasis with highlighting

code: computer code, usually displayed in a monospaced font

del: deleted words displayed with a line through them

sub: subscript (e.g., x₀)

sup: superscript (e.g., x²)

samp: sample computer output

span: a general phrasing element that can contain other phrasing (inline) elements, providing a simple way to attach presentation styles to enclosed elements; for example,

```
<span style="font-weight: bold; color: blue">
Important point</span>
```

var: a variable

kbd: keyboard text

2.11.1   Formatted Time

HTML5 provides a `time` element for specifying, in standard machine readable form, a precise date and/or time in the proleptic Gregorian calendar. Here is the general form:

```
<time datetime="date_time">text</time>
```

For example, you may use

```
<p>Fireworks start at <time datetime="2011-07-04T19:00">
7pm on Independence Day</time></p>
```

The *date_time* string can be a date (`yyyy-mm-dd`) or a time (`hh:mm:ss`), or a date and time connected with the character `T`. So far, the date/time strings refer to local time. To specify a global time, you can add a time zone at the end of the local date_time string. The time zone string is either the single letter `Z` (indicate UTC or coordinated universal time, formerly known as Greenwich mean time, GMT) or an offset `+hh:mm` or `-hh:mm` from UTC. For example,

```
<p>The final NASA space shuttle Atlantis launched on
<time datetime="2011-07-08T11:29-04:00">the morning of
Friday, 08 July 2011</time> in Cape Canaveral, Florida USA.</p>
```

The date_time string indicates July 8, 2011, 11:29 a.m. EDT. It is also possible to omit the `datetime` attribute and place the date_time string as content text of `time`.

To specify the publication date of your webpage or an `article` in your page, do the following:

- Put a `time` element as child of `body` or child of the desired `article`.

- Give `datetime` a date string with optional time string.

- Add the attribute `pubdate="pubdate"`

For example,

```
<body>
<time timedate="2012-07-07" pubdate="pubdate"></time>
...
...
```

2.12   Webpage Presentation Styles

Browsers follow built-in default presentation styles and styles based on user preferences to render webpages. However, it is usually important for the Web developer to control the presentation style to achieve well-designed visual effects.

You control document presentation by attaching *style rules* to elements. There are three ways to attach style rules:

1. Place style rules for individual and groups of elements in separate *style sheets* and then attach them to HTML documents with `<link ... />` in the `head` element (Chapter 4).

2. Include `<style>` elements in the `head` element.

3. Define the `style` attribute for an individual element.

For example,

```
<h1 style="color: darkgreen">The Green Earth Project</h1>
```

renders the heading in dark green. All three ways of attaching style rules can be used in the same page. The `style` attribute takes precedence over styles in the `<style>` element, which takes precedence over those specified in external style sheets.

In Chapters 2 and 3, we will use the simpler and most basic `style` attribute approach. The style knowledge gained here will be directly applicable in Chapter 4 where we discuss style sheets.

A style attribute is given in the general form:

`style="`*property*$_1$`:`*value*$_1$`;` *property*$_2$`:`*value*$_2$`;` `... "`

2.12.1 Foreground and Background Colors

Use the `color` (`background-color`) property to indicate the foreground (background) color for an element. Inside an element, text will be rendered using the foreground color against a background indicated by the background color.

2.12.2 Text Alignment

The `text-align` style property controls how text lines within a block are aligned.

`text-align: left`—lines are left justified

`text-align: right`—lines are right justified

`text-align: center`—lines are centered

`text-align: justify`—lines are justified left and right

For example,

```
<h1 style="text-align: center; color: darkgreen">
The Green Earth Project</h1>
```

centers the headline and renders it in dark green. The code

```
<p style="text-align: center; font-weight: bold">
...
</p>
```

centers all lines in the paragraph and requests boldface font. For many lines, this is seldom useful unless you are rendering a poem. But it is an effective way to center a short line, an image, or some other phrasing element.

To specify the style for a number of consecutive elements that form part of a page, you can wrap those elements in a parent element such as section, aside, or div element and attach the style to the parent element. For example (**Ex: FontSize**), to include some fine print, you can use

```
<footer style="font-size: x-small">
<p> ... </p>
...
<p> ... </p>
</footer>
```

An indentation for the first line in a paragraph can be specified by the style property text-indent. For example,

```
<p style="text-indent: 3em"> ... </p>
```

will indent the first line of the paragraph by a length of three **em**s (one **em** equals the width of M in the current font). To indent entire paragraphs, use the style properties:

```
margin-left: length
margin-right: length
```

For example,

```
<div style="margin-left: 5em; margin-right: 5em">
<p> ... </p>
<p> ... </p>
</div>
```

2.12.3   Style Length Units

In style properties, a length value consists of an optional sign (+ or −), a number, and a unit. Relative length units include

- em—the *font-size* of the current font

- `ex`—the *x-height* of the current font

- `ch`—the size of 0 (zero) of the current font

Absolute length units are `cm` (centimeters), `in` (inches), `mm` (millimeters), `pc` (picas; 1pc = 12pt), `pt` (points; 1pt = 1/72in.), `px` (pixels), 1px = 1/96in.). Absolute lengths are not sensitive to changes in font size or screen resolution, while relative length units are. For a visual display medium, the pixel unit refers to 1/96 inch at a viewing distance of 28 inches (a normal arm's length). This *referential pixel* unit becomes larger at greater viewing distances.

2.12.4   Colors

Color values in style properties can be a color name such as `magenta` or `darkblue`. Currently there are about 150 color names defined in CSS. See `w3.org/TR/css3-color/#rgba-color` for a list. Color values can also be given in a number of standard notations, including RGB (red-green-blue) and HSL (hue-saturation-lightness).

1. `#`*rrggbb*—where the first two, middle two, and last two of the six hexadecimal digits specify red, green, and blue values, respectively (e.g., `#0ace9f`). This is 24-bit color.

2. `#`*rgb*—shorthand for the above notation when the first two, middle two, and last two digits are the same (e.g., `#03c` stands for `#0033cc`).

3. `rgb(`*r* , *g* , *b*`)`—where base-10 integers between 0 and 255 inclusive are used (e.g., `rgb(0,204,108)`). This is the decimal equivalent of notation 1.

4. `rgb(`*r*`%,` *g*`%,` *b*`%)`—where integral percentages are used for the three color components.

5. `hsl(`*h* , *s*`%,` *l*`%)`—where *h* (in 0–360 degrees) indicates the hue on the color wheel (Figure 2.7[2]), *s* a percentage for color saturation, and *l* a percentage for lightness/brightness. Example hues on the color wheel are 0 (or 360) for red, 120 for green, and 240 for blue.

6. `rgba(`*r* , *g* , *b* , *a*`)`—adding an *alpha* opacity value to notation 2 where *a* is a decimal point value with a range of 0 (totally transparent) to 1 (totally opaque). An `rgb(...)` value in notation 2 is the same as `rgba(..., 1)`.

7. `rgba(`*r*`%,` *g*`%,` *b*`%,` *a*`)`—adding an *alpha* opacity value to notation 4 in the same way.

8. `hsla(`*h* , *s*`%,` *l*`%,` *a*`)`—adding an *alpha* opacity value to notation 5.

---

[2]See the DWP website for full-color figures.

FIGURE 2.7: The Color Wheel

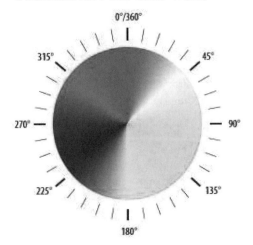

It takes some math to convert between RGB and HSL notations or between decimal and hex RGB notations. Color converters on the Web make it easy to do the job.

2.12.5 Text Fonts

One of the most important design aspects of a website is its readability. The textual content of the site must be easily readable, and the designer's understanding of what factors enhance readability is absolutely essential to Web development. The font type (Figure 2.8), style, and leading (line separation) can affect the readability and the look and feel of the entire site.

You can specify style properties for font family, style, variant, weight, and size for HTML elements. For example (**Ex:** FontFamily),

```
font-family: Times
font-family: Arial, Helvetica, sans-serif
```

You may list more than one name, in order of preference, for the `font-family` property. In this example, if the browser does not have `Arial`, it will check for `Helvetica`, and so on.

FIGURE 2.8: Some Fonts

Times   Arial   Helvetica
Courier   Monospace

It is a good idea to list a generic font family at the end of your preference list. The following generic font families are known:

- `serif`—for example, `Times`

- `sans-serif`—for example, `Arial` or `Helvetica`

- `cursive`—for example, `Zapf-Chancery`

- `fantasy`—for example, `Western`

- `monospace`—for example, `Courier`

Multiword font family names must be enclosed in single or double quotation marks.

Each major browser supports a good set of fonts. But there is no uniformity. It is possible for a website to supply its own fonts by providing the URLs to font files through the `@font-face` (Section 4.22) style rule.

The `font-style` property can be set to `normal` (the default), `italic`, or `oblique` (slanted). The `font-variant` can be set to `normal` (default) or `small-caps` (SMALL CAPITALS).

The `font-weight` property controls how heavy (bold) the font type is. For example,

```
font-weight: normal
font-weight: bold
font-weight: bolder
font-weight: lighter
```

The setting `bolder` (`lighter`) increases (decreases) the boldness relative to the current setting. The absolute weights $100, 200, \ldots, 900$ can also be used. Usually 400 is normal and 800 is bold. The exact meaning of these weights are browser and font dependent.

FIGURE 2.9: Font Sizes

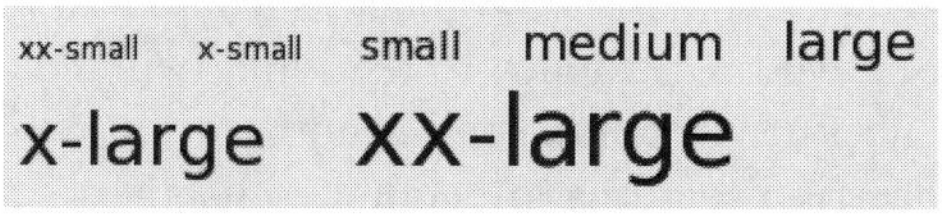

The `font-size` property can be set to a predefined size (Figure 2.9):

```
xx-small    x-small     small
medium      large       x-large     xx-large
```

or a specific size given in

pt (points; 1 pt = 1/72 in.)

pc (picas; 1 pc = 12 pt)

For example, `font-size: 16pt`.

Alternatively, you can set font size to a value relative to the current font size of the parent element.

smaller

larger

xx% (a percentage of the current font size)

It is advisable to set the basic font size in `body` to a predefined size that is correct for different browsers running on different display devices. Inside `body`, headings and fine print can use percentages to get a larger or smaller font size.

The vertical spacing between text lines can be important for readability. Browsers have default settings for vertical spacing depending on the font size. You can control line spacing by setting the `line-height` style property to a *number*, a *percentage*, or a fixed length.

The number and percentage specify a multiple of the current font size. Few places call for a fixed line height independent of the font size. The DWP website uses `line-height: 150%` for 1.5 line spacing.

One point to keep in mind is that users can increase or decrease the text size with browser settings. Thus, it is not possible to assume that your page will be displayed in a predetermined font size. The Web designer must take this into account when laying out a page.

An important website design consideration is what font and sizes to use for headers, running text, links, and fine print. Once the font and sizes have been determined, ensure they are consistently applied throughout the pages in your site. Any deviation must be for a specific design purpose. Otherwise, the unity of the site will suffer.

2.13   Itemized Lists

In articles, sections, paragraphs, and so on, itemized lists can organize and present information for easy reading. Web users like to find information quickly and will usually not have the patience to read long-winded passages. Itemized lists can highlight important points and send visitors in the right directions quickly. Often, navigation sections are coded as lists.

Three flow (block) elements are available for itemized listings:

- Bullet list: The `ul` element provides an *unordered list* where the ordering of the items is unimportant. A `ul` is usually presented as a set of bulleted items.

- Ordered list: The `ol` element offers a *numbered list* where the ordering of the items is important. An `ol` is typically displayed as a sequence of enumerated items.

- Definition list: The `dl` element is handy for a *definition list* where each term (`<dt>`) is given a definition or description (`<dd>`).

List elements may only contain list items. List items in `ol` and `ul` are given as `li` elements, which can contain other block elements such as headings, paragraphs, and lists. List items are usually displayed indented. A list given inside an `li` is nested in another list and is further indented. Here is a simple example (**Ex: List**):

```
<ul>
<li>Tropical Fruits
    <ol><li>Pineapple</li><li>Banana</li><li>papaya</li></ol>
</li>
<li>Cereals
    <ol><li>Barley</li> <li>Rice</li> <li>Wheat</li> </ol>
</li>
<li>Vegetables
    <ol><li>Broccoli</li> <li>Onion</li> <li>Yam</li> </ol>
</li></ul>
```

A version of this list is shown in Figure 2.10.

FIGURE 2.10: Lists

List items in a definition list (`dl`) are terms and descriptions:

```
<dl>
  <dt style="font-style: italic">HTML5</dt>
  <dd>Hypertext Markup Language, a W3C Standard<br /><br /></dd>
```

FIGURE 2.11: A Definition List

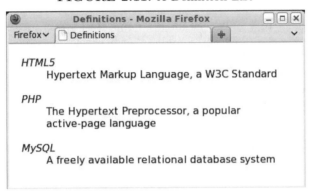

```
<dt style="font-style: italic">PHP</dt>
<dd>The Hypertext Preprocessor, a popular active-page language
    <br /><br /></dd>
<dt style="font-style: italic">MySQL</dt>
<dd>A freely available relational database system</dd>
</dl>
```

The definition term (`dt`) contains inline elements, and the definition description (`dd`) can contain inline and block elements. There is normally no automatic indentation for definition list items (Figure 2.11).

The definition list elements can also be used to present conveniently FAQs and interview dialogs (**Ex: Defs**).

2.14 List Styles

The default *list item marker* for `ul` and `ol` is defined by the browser. Switching item markers for nested lists is also done automatically. But you can take control of the list item marker by the `list-style-type` property. For example,

```
<ul style="list-style-type: circle"> ... </ul>
<ol style="list-style-type: upper-alpha"> ... </ol>
```

specify the marker type for the entire list. Specifying `list-style-type` for `li` allows you to control the item marker for each item separately.

Available list style types are `disc` (solid circle), `circle` (open circle), `square`, `none` (no marker), `decimal` (Arabic numerals), `lower-roman` (lowercase Roman numerals), `upper-roman` (uppercase Roman numerals), `lower-alpha` (lowercase English alphabet), `upper-alpha` (uppercase English alphabet). By giving an inappropriate style type, you can actually display a `ul` with numbered markers or an `ol` with bullets.

Use the property `list-style-image: url(`*imageURL*`)` for any custom list marker image identified by the *imageURL*. It overrides nonimage list markers. The `list-style-position` property takes the value `outside` (the default) or `inside`. The latter makes the list marker part of the first line of each list item.

You can also change the color of the item markers with the `color` style property as the following example demonstrates:

```
<p>The following list has inside positioning</p>
<ul style="list-style-position:inside">
  <li style="list-style-type: square; color: green">
    <span style="color: black">First item in the list with a
        green square marker.</span></li>
  <li style="list-style-type: square;  color: red">
    <span style="color: black">Second item in the list with a
        red square marker.</span></li>
  <li style="list-style-type: square; color: blue">
   <span style="color: black">Third item in the list with a
        blue square marker.</span></li>
</ul>
```

See the DWP website for a live demo (**Ex:** `MarkerStyle`).

The style property `list-style` provides a shorthand notation to specify, in order, the type, position, and image of a list style. For example,

```
list-style: circle inside
```

2.15 Links in Webpages

The ability to support links distinguishes webpages from other documents. Embedded links can take a reader directly to other resources and are most responsible for constituting the Web. In a webpage, you use anchor elements (`<a>` ... `</a>`) to specify URLs to other documents inside and outside your site and create clickable links to them.

The phrasing (inline) element `a` contains an anchor that can be any phrasing element such as text, image, button, or video clip. A link can also be attached to a specific part of an image via an *image map* (Section 2.20).

A link is given in the form

```
<a href="URL">anchor</a>
```

The enclosed phrasing element *anchor* is linked to a resource located with the given *URL*. The URL can link to any document or service allowable. For example,

```
<a href="bio.html">Brief Bio</a>
```

links the text *Brief Bio* to the document `bio.html`. Links to remote documents should be specified with full URLs, and links to local documents should be given by relative URLs (Section 1.4).

Here are some sample links:

```
<a href="http://www.w3.org/">The W3C Consortium</a>
<a href="../pic/dragonfly.jpg" type="image/jpeg"          (1)
   title="dragonfly.jpg">Picture of Dragonfly</a>          (2)
<a href="sound/cthd.mp3" type="audio/mpeg">               (3)
   Tan Dun, Yo Yo Ma - Crouching Tiger,
   Hidden Dragon - Theme</a>
```

The notation `../` leads to the parent directory (folder) of the current document (line 1). The `title` attribute (line 2) supplies text that a browser can use as a ToolTip display. The `type` attribute (line 3) helps inform the browser of the MIME type of the target resource. Here the link is to an `mpeg` audio file in MP3 format. It is usually not necessary to specify the `type` because when the file is retrieved by the browser, it will come with the correct MIME type information supplied by the Web server.

It is important to specify the URL correctly and to use the required URL encoding as explained in Section 1.4.1.

It is also possible to link directly to a specific point within the same document or another document. To do this, the target element in the destination document must have an `id`.

```
<h3 id="products">Our Quality Products</h3>
```

A link in the form

```
<a href="URL#products"> ... </a>
```

leads to the element identified by `products` in the document given by the *URL*. If the *URL* part is omitted, the link leads to an element in the same HTML document.

Webpages often take advantage of this feature to give a set of links at the beginning of a long article to serve as an active table of contents. For example,

```
<article>
<nav><ul>
<li><a href="#product">Products</a></li>
<li><a href="#service">Services</a></li>
<li><a href="#testimonial">Testimonials</a></li>
</ul></nav>

<section>
<h3 id="product">Our Quality Products</h3>
```

...

```
</section><section>
<h3 id="service">Responsive Services</h3>
```

...

```
</article>
```

In HTML5, the `id` attribute uniquely identifies an element in a page. It is an error to have multiple elements with the same `id`.

2.15.1   Internal and External Links

In building a website, links are used for two major purposes: to organize pages within the site (internal links) and to reach resources on the Web (external links).

Following an internal link, a visitor stays within a site and its navigation system. By clicking an external link, the visitor goes to another site. Hence, a well-designed site should make a clear distinction between these two types of links.

It is recommended that each external link

- Is clearly indicated as going off site. Often a webpage will include external links, such as references and advertisements, in an `aside` element displayed on the right-hand side of the page.

- Is displayed in a new browser window or tab so the visitor can come back by closing that new window or tab. A simple way is to use the attribute `target="_blank"` to cause the referenced page to display in a new window/tab:

  ```
  <a href="http://www.w3.org/" target="_blank">
  The W3C Consortium</a>
  ```

2.15.2   Site Organization and Navigation

Now let's consider using links to organize pages within a website.

- Organize the pages for a site into a hierarchy of files and directories (folders) stored on the hard disk of the server host. Avoid nonalphanumeric characters in file and directory names. Otherwise, the file name must be URL encoded before becoming part of a URL.

- Place the site entry page (usually, `index.html`) in the *server root* directory.

- Use subdirectories such as `images/`, `videos/`, `css/` (for style sheets), js/ (for JavaScript code), products/, services/, contractors/, members/, and affiliates/ to organize the site. The index.html page within each subdirectory is usually the lead page for that part of the site.

- Keep the organization simple and avoid using more than three levels of subdirectory nesting.

- Design a navigation system that is clear, easy to use, and effective in getting visitors where they want to go in your site.

- Use relative URLs exclusively for linking within the site and make sure the link is in one of these forms:

  1. Relative to the *host page* itself (href="*file*" or href="*dir/file*")
  2. Relative to the server root (href="/*path-to-file*")

If all links are of the first kind, then the pages of the site can be moved as a group to a different location in the file system or to a different hosting computer without change. If you have both types of relative links, then the pages can be moved to the server root on another host without change.

In creating the content-only site, consider establishing pages with these parts:

1. *Major navigation*—Links to the main page, and first-level pages. If the top banner of a page includes a logo of the business or site, link the logo image to the site entry. Often it is good to avoid a splash entry page and display the main page of the site as the entry page.

2. *Minor navigation*—Links to subpages of this page and links to directly related sibling pages.

3. *In-page navigation* —Links to parts of this page when appropriate.

4. *Draft page content*—Includes text, pictures, and other media types.

2.15.3 Linking to Services

In addition to webpage references, other frequently used types of links in practice are

- Email links—A link in the form

  <a href="mailto:*email-address*?SUBJECT=*line*">

  tells the browser to launch a program to send email to the given address using the indicated subject line. The subject line (from ? on) is optional. For example,

```
<a href= "mailto:pwang@cs.kent.edu?SUBJECT=
   Web%20Design%20and%20Programming">contact Paul</a>
```

Note spaces (%20) and other nonalphanumeric characters should be URL encoded.

Generally, the `mailto`[3] URL may have zero or more & separated *header=value* pairs. Useful headers include `to` (additional recipient address), `cc`, and `body` (message body). For example,

```
<a href="mailto:wdpgroup-request@cs.kent.edu?
   SUBJECT=join&BODY=subscribe">Joint web design
and programming email listserv group</a>
```

provides an easy way to join a listserv.

- Download links—A link in the form

```
<a href="ftp:host:port/path-to-file">
```

tells the browser to launch an FTP program to connect to the given *host* and to download the specified *file* by anonymous FTP. This is useful for downloading large files such as programs and compressed (ZIP or GZIP) files. If *port* is not given, then the standard port 21 for FTP is assumed. For example,

```
Download <a href="ftp://monkey.cs.kent.edu/package.zip">
            <code>package.zip</code></a> (35439 bytes).
```

An FTP URL can also supply username, password, and file location information for file retrieval (see `www.w3.org/Addressing/schemes`).

- Telephone/SMS/Fax links—Links in these forms

```
<a href="tel:phone_number">
<a href="sms:phone_number">
<a href="fax:phone_number">
```

are useful for mobile phone and tablet devices.

- VOIP call links—A link in the form

```
<a href="callto:screen_name or phone_number">
```

asks the browser to launch Skype[TM] or a similar program to make voice-over-IP calls or to conduct voice/video conference.

---

[3]See RFC2368 for more information on the `mailto` URL scheme.

2.15.4 Display Style for Links

Visual browsers pay special attention to the presentation of links. Usually, different display styles (Section 4.7) are used for textual links to indicate whether the link is not visited yet, under the mouse (`hover`), being clicked (`active`), or visited already (`visited`). A link is usually underlined in blue. As you click the link, it turns red. A visited link becomes reddish-purple. The user can control these colors through browser preference settings. An image anchoring a link will by default be displayed with a distinct border. The appearance of links can be controlled by style settings.

Web users are accustomed to seeing links underlined. Therefore, avoid underlining regular text because it can cause confusion. Image links, on the other hand, are almost always presented without the default border (Section 2.17). Web users understand that clicking an image often leads to another page.

A consistent set of link styles and colors is important for site design. Style sheets give you much control over the styling of links (Section 4.2).

2.16 Navbars

Most webpages will have a site navigation system consisting of several navbars (navigation bars), a horizontal navbar as part of the top banner, a vertical navbar on the left side, and a page-end navbar. Simpler pages may opt for either the top navbar or the left-side navbar. External links to other sites are usually set aside on the right-hand side of the page.

The top navbar normally links to major parts of the site. On a page, the left navbar links to subpages of the page and actions a user can take from that page. The page-end navbar serves to mark the end of a page as well as links to fine-print materials such as copyright and privacy information. It can also provide links to go to the next/previous step, to continue an activity, or to move to the top of the page.

The flow element `nav` creates navbars. To create a horizontal navbar, you can simply place the desired links inside a `nav`. Figure 2.12 shows an example whose source code is as follows (**Ex: Navbar**):

```
<header>
<section style="margin-left: 50px">          (1)
<h1>SuperStore.com</h1>
<h3>Shop and Save</h3>
</section>
<nav style="background-color: darkgrey;       (2)
          padding-left: 40px">                (3)
<a style="color:#fff; margin:10px" href="gr/"> (4)
Groceries</a>
<a style="color:#fff; margin:10px" href="hw/">
Hardware</a>
```

FIGURE 2.12: A Sample Navbar

```
<a style="color:#fff; margin:10px" href="au/">
Automotive</a>
<a style="color:#fff; margin:10px" href="of/">
Office Supply</a>
</nav>
</header>
```

The page headlines are indented 50 pixels (line 1). The navbar is in dark-grey (line 2) and the links are in white (line 3). Inside the navbar, the links are nicely separated by a spacing of 10 pixels (line 4). Also, the headlines and the first link are visually aligned vertically by adding the right amount of padding on the left side of the nav element (line 3).

In this example, we have repeated the same style attribute for each link. Later we will discuss how to avoid such repetition and to make links respond to mouse-over action.

See Section 2.21 for an example with a left-side vertical navbar.

2.17   Pictures and Images in Webpages

Clicking on a link such as `<a href="hat.jpg">My hat<a>`, you display a stand-alone image in a separate page. To include an image within a page together with other content, use the void phrasing element img. Thus, the img element lets you embed a picture or image inside a webpage. Here is the form:

```
<img src="hat.jpg" alt="A nice hat"
    style="width:160px; height:200px" />          (A)
```

The src attribute gives a URL to download an image file to be displayed. Thus, the image file replaces the img element. Such an element is known as a *replaced element*.

An image file is usually in JPEG, PNG (Portable Network Graphics), or GIF. These *raster image formats* store a fine grid of *pixels*, or picture elements, to represent the image. They also employ data compression to reduce the size of the file while preserving image quality. SVG, the Scalable Vector Graphics format (Chapter 12), is also supported by major browsers.

The required `alt` attribute provides alternative text to use for nonvisual browsers and when the image file is not available.

An `img` element may also take optional `height` and `width` attributes, given in number of pixels. But we recommend using the style properties (line `A`) instead of these optional `img` attributes. Also, the image width and height settings should normally be placed in a separate style sheet (Section 4.1) rather than inline as on line `A`.

The image width and height information is important to allow browsers to reserve the correct room for the image and continue to render the page without altering the layout after the image is loaded. For a page with multiple images, this can make the page appear on the screen much faster. A size different from the original image can be given, and the image will be scaled up or down to fit the specified area. If only the width or height is specified, the image scaling preserves the image aspect ratio. Because the file size of a larger image is significantly bigger than a smaller image, it is advisable to resize the original image to an appropriate size and not to rely on the browser to scale a large image way down.

To find the width and height of an image, use any image processing tool such as Photoshop^{TM}. On Linux systems, you can use the command

display *file*`.jpg`

to find the image size, resize the image, and perform other useful image processing tasks. **Gimp**, the GNU Image Manipulation Program, available under all major operating systems is a powerful and free tool for image processing.

If you add a `title` attribute to `img`, then the title text will be displayed as a tooltip.

2.17.1   Clickable Image Links

Pictures, logos, and icons are often used as navigation and/or information links. This is simply done by putting an `a` element around an image. Then, the image is clickable and will lead to the linked destination.

To make the user aware that an image is clickable, we can set a tooltip via the `title` attribute for the link or make the image *rollover* (change in response to mouseover; Section 4.11).

Here is an example (**Ex: ImgLink**) whose display is shown in Figure 2.13.

```
<a title="Go to Paul's homepage"
   href="http://www.cs.kent.edu/~pwang">
<img src="http://www.cs.kent.edu/~pwang/paul.jpg"
```

FIGURE 2.13: A Clickable Image

```
        alt="photo of the author Paul S. Wang"/>
</a>
```

For pictures that are large and time consuming to load, a smaller "thumbnail" image can be created as a link to the full image. Use your favorite image processing tool to crop and/or scale down the picture to create a thumbnail. An online photo gallery or portfolio usually displays arrays of thumbnails leading to larger images.

2.17.2   Text around Images

To cause text to flow around an image (**Ex: Float**), you *float* the image to the left or right margin and add appropriate spacing between the image and text. For example,

```
<p>For this green monkey, the new Chevy Volt is just      (1)
<img src="2012volt.jpg" alt="My dream 2012 Chevy Volt"
    height="110" style="float: left;                      (2)
    margin-right: 1em; margin-bottom: 8px;
    margin-top: 8px" />
the car I have been waiting for.  ...  </p>               (3)
<p>I love this car. On a recent trip to ... </p>          (4)
```

The image of a car (line 2) is floated to the left side with text in paragraphs (lines 1 and 3) flowing around the image. The image margins provide room between the image and the surrounding text. The display is shown in Figure 2.14. The style properties `float: right`, `margin-left`, and `margin-top` are also available (Section 4.12).

Floating allows you to place text alongside an image or some other floating element such as a table. Sometimes you need to put just a short caption or legend alongside a floating element, stop wrapping, and start a new paragraph beyond the floating element. To end the wraparound and start a new element below the floating element, use the `clear` style property: `clear:left` (clear

FIGURE 2.14: Text Around an Image

For this green monkey, the new Chevy Volt is just the c waiting for. It has great looks and drive most importantly, it is an extended ran totally avoid gas stations for my daily c have to worry about running out of juice

I love this car. On a recent trip to ...

float on the left), `clear:right` (clear float on the right), or `clear:both` (clear float on both the left and right). Here, *clear* means "move beyond." If we add `style="clear: left"` to the p element on line 4 (**Ex:** FloatClear), the display would become that shown in Figure 2.15.

FIGURE 2.15: Text Clearing Image

For this green monkey, the new Chevy Volt is just the c waiting for. It has great looks and drive most importantly, it is an extended ran totally avoid gas stations for my daily c have to worry about running out of juice

I love this car. On a recent trip to ...

2.17.3 Image Alignment inside a Line

When a text line includes images, the relative vertical position of an image with respect to the text can be controlled by setting the `vertical-align` style property. For example,

```
Here is some text and an
image <img src="URL" style="vertical-align: baseline" />
```

aligns the baseline of the image with the baseline of preceding text. This is usually the default alignment. The `vertical-align` style property may be applied to flow and phrasing elements. Settings that align to the preceding text include (**Ex:** ImgAlign)

- `vertical-align: baseline`—Aligns baselines of image and text.

- `vertical-align: middle`—Aligns middle of image with middle of x character in preceding text.

- `vertical-align: text-top`—Aligns top of image with font top of preceding text.

- `vertical-align: text-bottom`—Aligns bottom of image with font bottom of preceding text.

- `vertical-align: xx%`—Raises the bottom of image xx percent of the text *line height*.

Figure 2.16 shows these vertical alignments using a simple yin-yang image.

FIGURE 2.16: Inline Alignments with Preceding Text

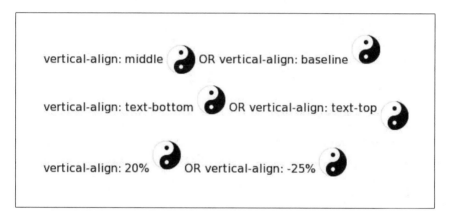

In case a line contains elements of several different heights, you can also align an image with respect to the entire line using

- `vertical-align: top`—Aligns top of image to tallest element on the line, which could be another image or some other tall element in the same line.

- `vertical-align: bottom`—Aligns bottom of image to lowest element on the line, which could be another image or some other element in the same line.

Figure 2.17 shows how the larger yin-yang image is aligned with the top of a smaller one further down the line and how the larger yin-yang image is aligned with the bottom of a previous smaller yin-yang image in the same line. The HTML code is:

FIGURE 2.17: Whole-Line Alignments

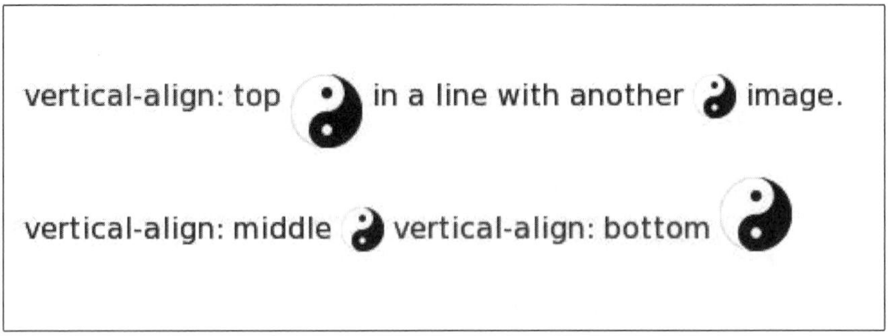

vertical-align: top in a line with another image.

vertical-align: middle vertical-align: bottom

```
<h3 style="color: blue">Alignments with respect
to the whole line:</h3>
<p><span style="color:blue">vertical-align: top</span>
<img alt="yinyang" src="01.png" style="vertical-align: top"/>
in a line with another <img alt="yinyang" src="01.png" width="20"
 style="vertical-align: middle"/> image.</p>

<p>vertical-align: middle <img alt="yinyang" src="01.png"
width="20" style="vertical-align: middle" /><span style=
"color:blue"> vertical-align: bottom</span><img alt="yinyang"
src="01.png" style="vertical-align: bottom"/></p>
```

2.18   Figures

Pictures and diagrams help convey your message on a webpage. The element
`figure` is used to markup figures in your page. Place `img`, `code`, `pre`, `audio`, or
`video` in a figure to separate it from the flow of your textual materials which
can make references to it. To add a caption to your figure, use the `figcaption`
element.

Figure 2.18 shows a displayed figure with caption. The HTML source for
it is as follows (**Ex:** FigCap).

```
<figure style="text-align: center; font-style: italic">        (A)
<img src="dragonfly.jpg" alt="a blue-winged dragonfly" />
<figcaption>
<strong>Fig. 7:</strong> Dragonfly, an insect belonging to
the order Odonata, the suborder Epiprocta or, in the strict
sense, the infraorder Anisoptera. (Wikipedia)
</figcaption>
```

FIGURE 2.18: A Figure with Caption

Fig. 7: *Dragonfly, an insect belonging to the order Odonata, the suborder Epiprocta or, in the strict sense, the infraorder Anisoptera. (Wikipedia)*

```
</figure>
```

In the figure, the picture and the caption text are centered. The caption text is also italicized (line A).

2.19 Image Encoding Formats

Graphical images can be stored in many different digital formats that have different characteristics. In *raster graphics*, images are digitized by recording their colors at a grid of *sampling points* (Figure 2.19). The finer the grid,

FIGURE 2.19: Raster image

the larger the number of sampling points and the better the resolution. Each sampling point results in a *color pixel* (picture element) stored in a *raster image file*. A raster image usually does not need to record each and every pixel individually. Compression methods can significantly reduce the image file size. Display software reproduces the pixels to render a raster image on a computer screen or printer.

In *vector graphics*, images include geometric objects such as points, lines, and curves. Geometric information can be combined with raster information to represent a complete image. Vector image encoding provides accurate information on geometric objects in the image and avoids representing every pixel. Hence, vector images are usually smaller and easier to scale. Display software must understand the geometric information to render vector graphics files.

HTML5 allows Scalable Vector Graphics (SVG) and therefore vector graphics (Section 12.2). Another widely used vector graphics system is Adobe Flash.

HTML5 also provides the `canvas` element for drawing vector as well as raster images under program control (Section 7.17).

2.19.1   GIF, JPEG, and PNG

Currently, the only image compression formats generally accepted on the Web are

- Graphics Interchange Format (GIF)—A raster format suitable for icons, logos, cartoons, and line drawings. GIF images can have up to 256 colors (8-bit).

- Joint Photographic Experts Group (JPEG) format—A raster format usable for color and black-and-white pictures with continuously changing color tones for display. JPEG images can store up to 16.8M colors (24-bit). Images created using a scanner or digital camera are usually stored in TIFF (Tagged Image File Format), JPEG, or GIF.

- Portable Network Graphics (PNG) format—A format designed to replace GIF. PNG really has three main advantages over GIF: alpha channels (variable transparency), gamma correction (cross-platform control of image brightness;), and two-dimensional interlacing (a method of progressive display). Browser support for PNG is increasing steadily, and ideally, PNG will soon replace GIF.

2.19.2   Aliasing and Anti-Aliasing

As you might expect, sampling a continuous image at discrete points may run into problems. Raster images often contain jagged edges (e.g., staircase effect on slanted lines), lost or distorted details (fine features missed or hit by sampling points), or disintegrating texture patterns (Figure 2.20). These

FIGURE 2.20: Aliasing in Raster Images

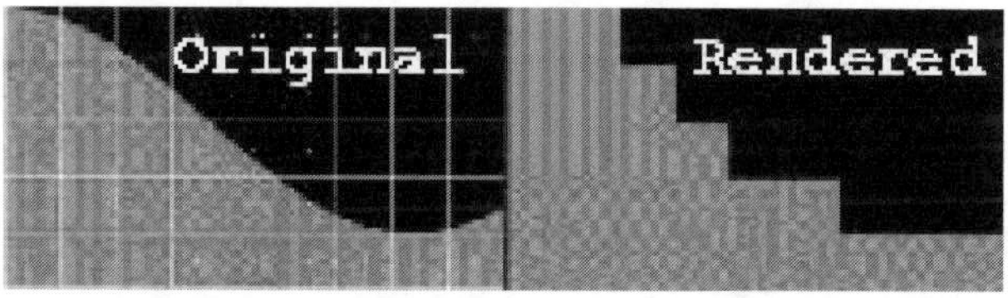

errors are known as *aliasing artifacts*.

Anti-aliasing methods have been developed to reduce the effects of aliasing. One popular anti-aliasing technique involves sampling at a resolution several times (e.g., three times) higher than the target image file resolution. Then compute the value of each image pixel as the weighted average of the values from a number of neighboring sample points (e.g., nine points). There are many fine anti-aliasing algorithms. No matter what the method, anti-aliasing seeks to soften the jagged edges by setting pixel values so that there is a more gradual transition between one color to a very different color.

Anti-aliasing makes lines and edges in a picture look much smoother by using pixels whose color is a blend of the object color and the background color. A graphical object or text font anti-aliased with one background color will not look right on a different background.

2.19.3 Colors in Raster Images

In a raster image, each pixel is a color dot. The size of a raster image file depends on how many colors it uses. You have these choices:

- Monochrome—black and white

- Gray scale—different levels of gray (up to 256 shades)

- Indexed—Each pixel color is indicated by an index into a color palette. The palette may contain a set of up to 256 colors. The smaller the palette, the fewer bits needed for each index.

- High color—thousands of colors, 15 to 16 bits per pixel

- True color—16.8 million colors, 24 bits per pixel

2.19.4 Dithering

Dithering is the attempt by a computer program to approximate a color from a mixture of other colors when the required color is not available. Dithering can occur when reducing the number of colors of an image or displaying a color that a browser on an operating system does not support. In such cases, the requested color is simulated with an approximation composed of two or more other colors that can produce it. The result may or may not be acceptable to the graphic designer. The image may also appear somewhat grainy because the colored area in question consists of different pixel intensities that average to the intended color.

There are several methods or algorithms for color dithering:

- *Pattern dithering* uses a fixed pixel pattern.

- *Diffusion dithering* propagates the error made in replacing a color pixel by a supported color to neighboring pixels.

Pattern dithering is most useful in filling a larger area with a desired color. Diffusion dithering works well for continuous tone pictures.

2.20 Image Maps

An *image map* is an image with *active areas* that, when clicked, lead to designated URLs. For example, you may use a picture of a number of products, a county map of a state, or a group picture of classmates to make an image map.

To create an image map, you associate a map element to an image. A map element contains one or more area elements that define active areas within an image and connect each to a hyperlink. The required name attribute of a map is used to associate it with an image. Even though the map code can be placed anywhere in the body as flow or phrasing content, we recommend that you keep the map next to the img it controls.

An area element can define a rectangle, a circle, or a polygon using coordinates in pixels (or infrequently, in percentages). The upper-left corner of an image is $(0,0)$ following the coordinate system for computer graphics (Figure 2.21).

FIGURE 2.21: Image Coordinates

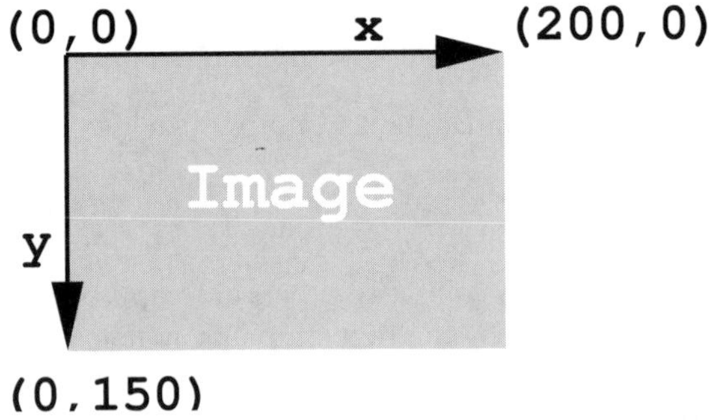

```
<map name="samplemap">
 <area shape="rect" coords="0,0,100,150"      (corners of rectangle)
      href="some-url" alt="item 1" />
 <area shape="poly" coords="0,0,10,32,98,200"  (vertices of polygon)
      href="some-url" alt="item 2" />
 <area shape="circle" coords="0,0,100" (center and radius of circle)
      href="some-url" alt="item 3" />
 <area shape="default"                          (rest of image)
      href="some-url" alt="item otherwise" />
</map>
```

Like images, an `area` requires the `alt` attribute. The special shape `default` stands for the rest of the image not included in the otherwise marked regions. For regions of irregular shapes, the polygon, given as $x_0, y_0, x_1, y_1, \ldots$, is the most useful.

To obtain the coordinates, you can open the target image in any image processing tool (e.g., Photoshop, Gimp) or use an image map editor. On the Web, the Poor Person's Image Mapper is a free service. A `map` is associated with an image with the `usemap` attribute of the `img` element:

```
<img src="img-url"  usemap="#map-name" />      (an image map)
```

Here is a map for planets in our solar system (**Ex:** `Planets`) that links to pages for individual planets:

```
<figure>
<img src="planets.jpg" usemap="#planets"              (I)
     alt="Planets image map" />
<map name="planets" id="planets">                     (II)
  <area shape="circle"                                 (III)
      coords="40,176,7"
      href="mercury.html" alt="Mercury" title="Mercury"/>
  <area shape="circle"
      coords="82,158,10"
      href="venus.html" alt="Venus" title="Venus"/>
  <area shape="circle"
      coords="127,132,11"
      href="earth.html" alt="Earth" title="Earth"/>
  <area shape="circle"
      coords="157,103,10"
      href="mars.html" alt="Mars" title="Mars"/>
  <area shape="circle"
      coords="234,116,27"
      href="jupiter.html" alt="Jupiter" title="Jupiter"/>
  <area shape="poly"
      coords="254,53,327,54,373,102,300,107"
      href="saturn.html" alt="Saturn" title="Saturn"/>     (IV)
```

```
  <area shape="default"                                    (V)
     href="solar.html"
     alt="List of solar system planets" />
</map>
</figure>
```

The usemap attribute (line I) associates the map to the image via the required
name attribute of map[4]. The area for Mercury is a circle (line III) and the area
for Saturn is a polygon (line IV). A default area is also defined (line V). The

FIGURE 2.22: Planets Image Map

map element is not displayed on a webpage and is classified as a *transparent
element*. Figure 2.22 shows the display for this example. The active areas for
Jupiter and Saturn are revealed with dashed lines. The complete example
(**Ex:** Planets) can be found at the DWP website.

---

[4]If present, the id attribute of map must have the same value as name.

2.21   Webpage Layout

A clear and consistent page layout is a must for every website. A well-designed page layout can preserve site visual unity and make information easy to follow/find for end users.

Figure 2.23 shows a sample layout produced with HTML elements and style rules. The top banner often displays a company logo and other graphics, a headline (company or website name), and perhaps also a tagline or two. Beneath the top banner there is a horizontal navbar for top-level pages.

FIGURE 2.23: A Sample Page Layout

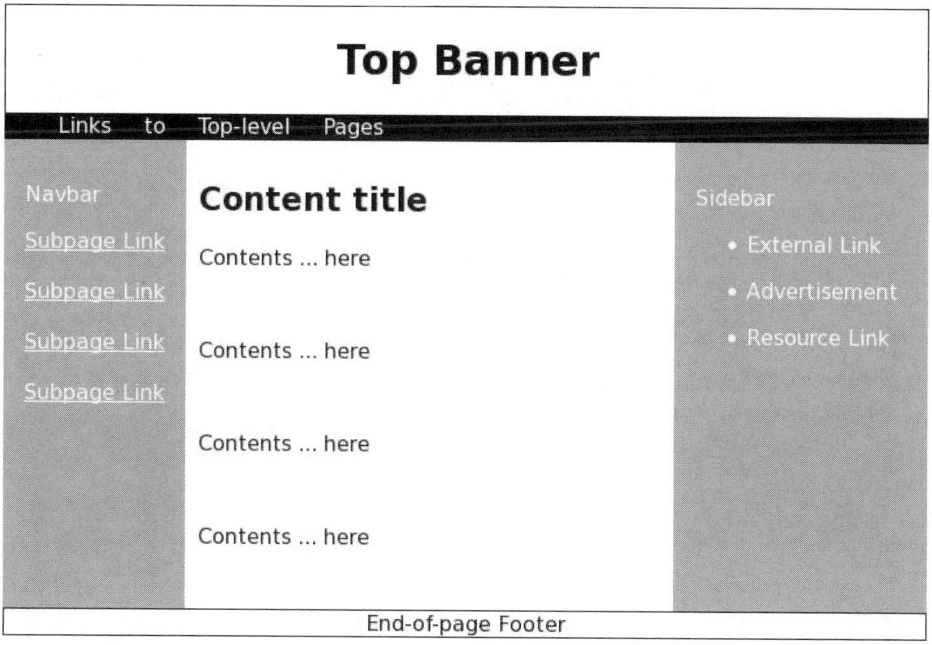

The main content part of the page consists of three columns: a left navbar, a content area, and a right sidebar. The page ends with a footer.

Let's look at the source code for this sample layout. The header

```
<header>
<h1 style="text-align: center">Top Banner</h1>
<nav style="background-color: black; color: white;
     padding-left: 40px">Links to Top-level Pages
</nav></header>
```

is at the top of the page. Next comes a container for the three columns:

```
<div style="background-color: darkgrey">        (1)
<!-- three columns here -->
```

```
<section style="clear: both"></section>          (2)
</div>
<!-- page footer here -->
```

The background color (line 1) and the empty section with clear: both style (line 2) are needed to obtain the full-length darkgrey columns on the left and right.

In between lines 1 and 2 go the three columns. The left column is a nav that is floated left with white text (line 3).

```
<nav style="float: left; padding: 1em; color: white">          (3)
<p>Navbar</p>
<a href="#">Subpage Link</a><br /><br />
<a href="#">Subpage Link</a><br /><br />
<a href="#">Subpage Link</a><br /><br />
<a href="#">Subpage Link</a><br /><br />
</ul>
</nav>
```

The middle column is a section floated left with white background

```
<section style="float: left; padding: 10px;
        width:50%; background-color: white">          (4)
<h2>Content title</h2>
<article class="sectionArticle">
<p>Contents ... here</p><br />
<p>Contents ... here</p><br />
<p>Contents ... here</p><br />
<p>Contents ... here</p><br />
</article>
</section>
```

The right column is an aside floated left with white text (line 5):

```
<aside style="float: left; color: white;          (5)
            padding: 1em">
<p>Sidebar</p>
<ul><li><p>External Link</p></li>
<li><p>Advertisement</p></li>
<li><p>Resource Link</p></li></ul>
</aside>
```

The page footer code has a thin border

```
<footer style="border: thin solid black;          (6)
    text-align: center">End-of-page Footer</footer>
```

The complete example (**Ex: Layout**) can be found at the DWP website.

2.22   Debugging and Validation

Hands-on is the only way to learn HTML. You can use a plaintext editor such as Vim or even Wordpad to do the coding. Or you can use software specially designed for authoring webpages.

A spell checker can help you find typos and spelling errors. Careful proof-reading can catch grammar and other writing errors.

Test your webpage with different browsers under different operating systems. Even when a page looks OK, it may still contain coding errors. That is because Web browsers will ignore elements and attributes not recognized, as well as other problems in your HTML code. Therefore, a page displaying correctly is no evidence that your code is error free. The W3C markup validator (`validator.w3.org`) is useful for debugging your HTML code. And the W3C CSS validator (`jigsaw.w3.org/css-validator`) is handy to check your style code.

Finally, you also need to know that all links in your webpage are correct and not broken. For this, you can use the W3C link checker (`validator.w3.org/checklink`).

2.23   Summary

New webpages should be written in HTML5 which is a W3C-recommended standard that is the successor of both HTML 4 and XHTML. HTML5 makes webpage authoring easier and better structured than before by introducing new elements and APIs.

We recommend that you code HTML5 in a polyglot way with lowercase tag names and attributes, and with attribute values enclosed in quotation marks. All end tags ought to be used. Void elements should end in `/>`.

Each element defines available attributes and allowed child and parent elements. Core attributes applicable to all elements include `id`, `style`, `class`, `title`, and `hidden`.

Phrasing and flow elements are placed in `body` to provide document content. White-space characters separate words. Extra white spaces between words and elements are collapsed. Phrasing content flows to fill lines that can be left adjusted, centered, or right adjusted. Three different lists help organize and present information. Images can be placed inline with text and aligned vertically within the line in multiple ways.

HTML elements and their attributes are primarily used to provide structure to documents. Presentation styles can be attached to elements with the `style` attribute. In Chapter 4, you will see more flexible ways to specify style and how to separate style from structure even more completely.

Use `h1`–`h6` for section headings, `p` for paragraphs, `br` for line breaks, `q` and `blockquote` for inline and block quotations, `hr` for horizontal rules, `time` for formatted time, `ul` for unordered lists, `ol` for ordered lists, `dl` for definition lists (containing dt and dd), img for inline images, map for image maps (containing areas), a for links, and span (div) for attaching style properties. Also, use header, footer, article, section, aside, nav, and figure to organize your webpage.

Style properties are assigned to HTML elements to control page presentation. The style properties mentioned in this chapter include font-family, font-size, font-weight, text-align, color, background-color, vertical-align, list-style-type, text-indent, float, clear, height, width, border, padding, and margin. Chapter 4 covers styles more comprehensively.

Use relative URLs to link pages within your website and full URLs for outside pages. Links may use text or image as an anchor, and areas inside an image can lead to different URLs in an image map. URLs for links may contain a restricted set of ASCII characters. Other characters must be URL encoded before being placed in a URL. In addition to webpages, URLs may link to Internet services such as FTP, Telnet, and email.

Web browsers often tolerate errors in HTML code. Test your webpage on major browsers running under different operating systems. The W3C HTML and link validators can be very useful for debugging HTML code.

Exercises

2.1. What is the relation among HTML 4.01, XHTML, and HTML5? What is a polyglot document?

2.2. What are the major advantages brought by HTML5?

2.3. What does overlapping tags mean? How do you suppose you can guard against such mistakes?

2.4. What is the proper form of a void element? Give three examples of void elements.

2.5. What is the difference between flow elements and phrasing elements?

2.6. How are comments given in HTML files?

2.7. List and explain the core attributes of HTML5.

2.8. What role does white space play in an HTML file? What constitutes white space in an HTML document?

2.9. How is a header, a short line, or a paragraph centered horizontally in a page?

2.10. How are length units specified in a style declaration?

2.11. What are the possible ways to specify colors in a style declaration?

2.12. Name and describe the three different types of list elements in HTML. How do you control the bullet style for lists?

2.13. Explain how a link for sending email is done.

2.14. Explain how a link for remote logon is done.

2.15. What is a full URL? A relative URL? What are the two types of relative URLs?

2.16. What is the form of a URL that links to a local file on the hard disk?

2.17. What image formats are usually used on the Web?

2.18. Can an image anchor a link? What effect will a link have on the image?

2.19. Which attributes are required for `<img>`?

2.20. What is an image map? How is it useful? Where is an image map placed in an HTML file? How is an image associated with its map?

2.21. What is the benefit of organizing a website using document-relative URLs?

2.22. Take the first webpage you created in Section 2.2 and deploy it in on the Web.

2.23. Take the "big on sports" page from Section 1.6 and make it HTML5 using the template given in Section 2.2.

2.24. Take the first webpage you deployed in the previous assignment and make it into a simple personal home page. Introduce yourself to the world, make your résumé available, add a picture of yours, list your interests, talents, and so on.

2.25. Take the first webpage and experiment with font family, font size, and line height settings. View the page with different browsers and different user-selected font size preferences.

2.26. Specify `width` and `length` for an image differently from the image's original size and see the display effects.

2.27. Put `id` on some tags in a page and link to these locations inside the page from another page. Will the links work?

2.28. Take a picture of yours and float it to the right of the a webpage.

2.29. Complete the solar system image map (Figure 2.22) and test it.

Chapter 3

Audio, Video, and More HTML5

Media playing is very important on the Web. Prior to HTML5, there was no standard way to deploy audio or video in a webpage. Users need to install plug-ins to play different media. It was hard to get media to play correctly on different browsers.

HTML5 changes the situation by introducing the elements `audio` and `video`. Now we have an easy way to place standard sound and video in webpages that will work on all platforms. We use working examples to illustrate these new media elements. Explained also is how to use nonstandard media files such as Java applets and Adobe Flash through the `object` or `embed` element.

HTML tables are useful for organizing information in tabular form. The many aspects of table formatting and style control require careful presentation and patient experimentation to learn and master. Many practical examples are provided. A good understanding of tables is also important because the table display model can implement a precise visual layout grid for webpages.

The Web is international. Webpages may be written in UNICODE, a standard 16-bit character code that covers most known languages in use. HTML also provides ways to represent these characters with ASCII character sequences.

Various elements, such as `meta` and `base`, given exclusively inside the `head` element provide critical functions for webpages. Their use is explained in this chapter.

We will learn how to create a separate browsing context with the `iframe` element. It allows us to easily deploy third-party pages or widgets inside our own pages.

Webpages can be composed from different files through *server-side includes* (SSI) and PHP. The chapter concludes with some debugging and validation advice.

Let's begin with audio and video.

3.1   Audio for the Web

3.1.1   Digital Audio

Voice and other audio effects can make a site more functional and user friendly. Technically, audio refers to sound within the human hearing range. An audio signal is naturally a continuous wave of frequency and amplitude. Analog audio must be digitized to play back on a computer.

An analog audio signal is digitized by *sampling* and *quantization*. The continuous sound wave is sampled at regular time intervals, and the values at each sampling point are quantized to discrete levels. The resulting data are stored in binary format as a digital audio file. The higher the sample rate and the greater the bit depth (number of quantization levels), the higher the sound fidelity and the larger the file size.

Let F be the highest frequency of an audio signal. The sampling rate must be at least $2F$ to represent the signal well. This is the so-called *sampling theorem*. Human hearing is limited to a range of 20 to 20K Hz (cycles per second). Thus, the CD-quality sampling rate is often 44.1K Hz. Human speech is limited from 20 Hz to 3K Hz. An 8K-Hz sampling frequency is high enough for telephony-quality audio.

3.1.2   Audio Encoding Formats

Advances in digital audio bring increasingly sophisticated *compression and decompression* (codec) schemes to reduce audio files size while preserving sound quality. For example, the widely used MP3 is the audio compression standard ISO-MPEG Audio Layer-3 (IS 11172-3 and IS 13818-3).

In 1987, the Fraunhofer Institute (Germany) in cooperation with the University of Erlangen devised an audio compression algorithm based on *perceptual audio*, sounds that can be perceived by the human ear. Basically, MP3 compression eliminates sound data beyond human hearing. By exploiting stereo effects (data duplication between the stereo channels) and by limiting the audio bandwidth, audio files can be further compressed. The effort resulted in the MP3 standard. For stereo sound, a CD requires 1.4 Mbps (megabits per second). MP3 achieves CD-quality stereo at 112–128 Kbps, near CD-quality stereo at 96 Kbps, and FM radio-quality stereo at 56–64 Kbps. In all international listening tests, MPEG Layer-3 impressively proved its superior performance, maintaining the original sound quality at a data rate of around 64 Kbps per audio channel.

MP3 is part of the MPEG audio/video compression standards. MPEG is the Moving Pictures Experts Group, under the joint sponsorship of the International Organization for Standardization (ISO) and the International Electro-Technical Commission (IEC). MPEG works on standards for the encoding of moving pictures and audio. See the MPEG home page (`mpeg.org`) for further information.

More recently, Ogg Vorbis (`xiph.org`) offers a fully open, nonproprietary, patent-and-royalty-free, general-purpose audio format that is preferred by many over MP3. Table 3.1 lists common audio formats.

TABLE 3.1: Audio Formats

Filename Suffix, Format	Type	Origin
aif(f), AIFF AIFC	audio/x-aiff	Apple, SGI
mid, MIDI	audio/midi	For musical instruments
mp3	audio/mpeg	MPEG standard
ra or rm, Real Audio	audio/x-realaudio	Real Networks
amr	audio/amr	Mobil phone audio, 3GPP
ogg (sometimes oga) Vorbis	audio/ogg	Open source, Xiph.Org
wav, WAVE	audio/x-wav	Microsoft
wma, Windows Media Audio	audio/x-wma	Microsoft

Actually, Ogg is a *media container* format (Section 3.2.1) rather than a single codec format. Webm is a new media container designed for the Web and is recommended for use with the audio and video elements of HTML. Format converters are freely available (for example **ffmpeg** and **ffmpeg2theora** on Linux) to rewrite audio/video files from one format to another.

3.1.3 The audio Element

HTML5-compliant browsers have built-in support for playing audio rather than relying on third-party plug-ins. You include audio in a webpage with the audio element. In its simplest form, the audio element is given as

```
<audio attributes src="url"></audio>
```

where *url* locates an audio file, such as song.mp3 or song.ogg, or even a live audio stream. Here is an example (**Ex: LiveFM**)

```
<audio src="http://96k.prog.fm:9020/listen.ogg"
       controls="controls" autoplay="autoplay" >
</audio>
```

that plays an Ogg stream from an FM radio station. We have set the controls option to display the audio controls and the autoplay option to cause the audio to play as soon as it is available (Figure 3.1). Using JavaScript, it is also possible to design your own GUI controls for audio/video playback (Section 6.9). Audio formats supported depend on the browser. But .ogg, .mp3, and .wav are likely to be supported. To hedge your bets, you can include alternative formats with the source element:

```
<audio  attributes > <source src=" file.ogg" />
        <source src=" file.mp3" />   ...
</audio>
```

FIGURE 3.1: Audio Controls

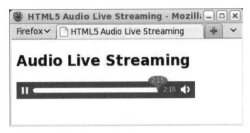

The `type` attribute can specify the media type. A browser will play the first format it supports. For example (**Ex:** PRI),

```
<h1>PRI 06/22/2011</h1>
<audio controls="controls" style="width:324px;">
  <source src="06222011full.ogg" type="audio/ogg" />
  <source src="06222011full.mp3" type="audio/mpeg" />
</audio>
```

makes a Public Radio International sound file available in two formats. Figure 3.2 shows a display where we are 6 minutes and 44 seconds into a 51-minute sound clip.

FIGURE 3.2: Sound File in Play

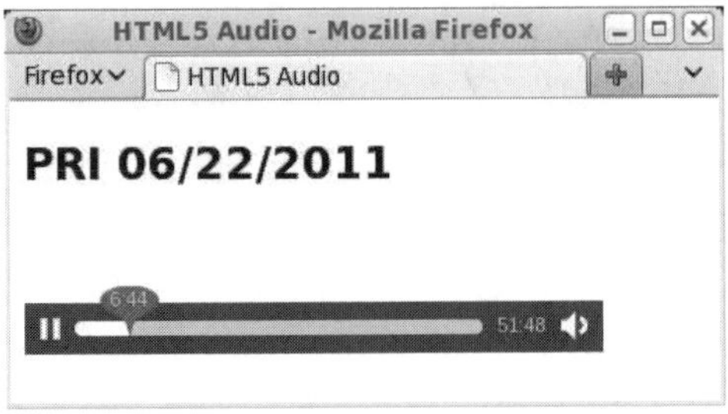

You can use the attribute `loop="loop"` to play a sound file repeatedly. The attribute `preload` can be set to `auto` (browser is free to preload the sound file), `none` (no preload), or `metadata` (preload only the sound metadata).

3.2 Playing Video

A video is a sequence of images displayed in rapid succession that is usually also played in synchrony with a sound stream. For smooth motion, a sufficient *frame rate*, about 30 frames per second (fps), is needed. There are many digital video formats for different purposes, including DVD, Blu-ray^{TM}, HDTV, DVCPRO/HD, DVCAM/HDCAM, and so on.

A video file for the Web usually supplies video tracks, audio tracks, and metadata. Tracks in a video file can also be organized into chapters that can be accessed and played directly. Such files are known as *video containers* and they follow well-designed *container formats*, which govern the internal organization of a video file.

3.2.1   Video Containers

Widely used video container formats include

- MPEG 4—A suite of standard audio and video compression formats from the Moving Pictures Experts Group (content type `video/mp4`; file suffix `mp4`, `mpg4`).

- Mobile phone video—A 3GPP-defined container format used for video for mobile phones (content type `video/3gpp`; file suffix `3pg`).

- Flash Video—A proprietary format used by Adobe Flash Player (content type `video/x-flv`; file suffix `flv`).

- Ogg—A free and completely open standard from the Xiph.Org Foundation. The Ogg video container is called Theora (content type `video/ogg`; file suffix `ogv`). The Ogg audio container is called Vorbis (content type `audio/ogg`; file suffix `ogg`).

- WebM—A new standard developed by `webmproject.org` and announced in mid-2010. The format, free and completely open, is used exclusively with the VP8 video codec and Vorbis audio codec. It is supported by, among others, Adobe, Google, Mozilla, and Opera. Google offers a WebM plug-in component for IE9.

- AVI—Audio Video Interleaved format from Microsoft (content type `video/x-msvideo`; file suffix `avi`).

- RealVideo—An audio and video format by Real Networks. For historical reasons, it uses the seemingly incorrect content type `audio/x-pn-realaudio` and the file suffix `rm`.

3.2.2   Video Codecs

The video and audio tracks in a container are delivered with well-established compression methods. A video player must decompress the tracks before playing the data. Many compression-decompression algorithms exist. Generally speaking, video compression uses various ways to eliminate redundant data within one frame and between frames.

The most important video codecs for the Web include H.264, Theora, and VP8.

- H.264—A widely used standard from MPEG providing different *encoding profiles* to suit devices from smartphones to high-powered desktops. Also known as *MPEG-4 Advanced Video Coding*, the standard enjoys good hardware and software support.

- Theora—A completely free standard from Xiph.org Foundation providing a modern codec that can be embedded in any container. But, it is usually delivered in an Ogg container. Theora is supported on all major Linux distributions, Windows, and Mac OS X.

- VP8—A very modern and efficient standard from On2 (part of Google) giving everyone a royalty-free codec for the Web. VP8 is usually delivered inside the WebM container.

3.2.3   The video **Element**

Use the video element to deploy a video in your webpage. Here is the simplest form:

```
<video {\sl attributes} src=" url">
</video>
```

For example (**Ex:** SimpleVideo),

```
<video controls="controls" autoplay="autoplay"
      src="myvideo.webm"></video>
```

It is a good idea to provide your video in several formats to make it work across browsers and platforms. This is done using the source element inside video. For example (**Ex:** NasaVideo),

```
<video width="480" height="385" controls="controls">
    <source src="nasa.webm" type="video/webm" />
    <source src="nasa.ogv" type="video/ogg" />
    <source src="nasa.mp4" type="video/mp4" />
</video>
```

produces a video display shown in Figure 3.3. To see this video clip (**Ex:** NasaVideo) play, please go to the DWP website.

Attributes for video are width and height (same usage as for img); autoplay, controls, loop, and preload (same as in audio); and audio="mute" (no sound), poster="url" (initial image).

The HTML5 audio and video elements make it easy to include media files in browser supported formats without the need for third-party plug-ins. This is a big improvement over previous HTML versions. If you wish to embed nonstandard media files, you can use the object or the embed element (Section 3.3).

FIGURE 3.3: NASA 25th Anniversary

3.3 Embedding Media

Normally you ought to use the HTML5 `audio` and `video` elements to deploy your media files. However, in special cases you may need to deliver an audio, a video, or some other media file not supported by the HTML5 standard or browsers. In that case, you can use the `object` or equivalently the `embed` element of HTML5.

3.3.1 Embedding Flash

For example, we can deploy a Flash version of the NASA video, `nasa.swf` with the `object` code (**Ex:** `FlashAudio` and **Ex:** `FlashVideo`):

```
<object type="application/x-shockwave-flash"
        data="nasa.swf" width="480" height="385">
<param name="movie" value="nasa.swf" />
<param name="loop" value="false" />
<a href="http://www.adobe.com/go/getflash"> <img          (A)
src="get_flash_player.gif" alt="Get Flash player" /></a>    (B)
</object>
```

The video format `.swf` requires the Adobe Flash Player plug-in. If a browser does not already have the required plug-in installed, the fallback code (lines **A** and **B**) will be used to display a link to the plug-ins download page.

If you use the `embed` element, then the code would be as follows:

```
<embed src="nasa.swf" quality="high"
       width="480" height="385"  loop="false"
       type="application/x-shockwave-flash"
       pluginspage="http://www.adobe.com/go/getflash" />
```

3.3.2　Embedding PDF

On the Web, PDF (the Portable Document Format) is a widely used open standard for document exchange. Clicking a link to a pdf document causes the file to download and display by a browser plug-in or associated application.

But, you can also embed a pdf document as part of a webpage via the `object` or `embed` element. Figure 3.4 shows the display of a page that embeds

FIGURE 3.4: Embedded PDF

a pdf file `html5.pdf` with the following code (**Ex: PDF**):

```
<object style="border: thin solid black"      (1)
        type="application/pdf"                 (2)
        data="html5.pdf"                       (3)
        width="420" height="300">              (4)
<a href="html5.pdf">html5.pdf</a>             (5)
</object>
```

The file url is specified by the `type` attribute (line 3) and the correct MIME type is given on line 2. The `width` and `height` must be given to reserve a display area on the page for the plug-in window (line 4). The fallback code on line 5 is for old browsers with no `object` element support.

3.3.3　Embedding Java Applet

Java applets can supply an interactive graphical tool or a game for a webpage. After an applet is created using Java tools, it can be placed in a webpage with the `object` element.

Figure 3.5 shows a Tic Tac Toe game applet placed in a webpage with the following code (**Ex: TicTacToeApplet**).

```
<object style="float: left; margin: 2em"
        type="application/x-java-applet"          (A)
```

FIGURE 3.5: A Game Applet

```
        width="140" height="140">                  (B)
  <param name="code" value="TicApplet">            (C)
  <param name="archive" value="TicApplet.jar">     (D)
Applet failed to run.  No Java plug-in was found.
</object>
```

The MIME type is given on line A. The screen area dimensions are set to match the game graphics (line B). The `code` parameter gives the Java class name to run (line C), and the `archive` parameter specifies the Java `jar` file that contains all necessary code and images for the game.

3.4   Tables

Displaying information in neatly aligned rows and columns is effective in print media as well as online. The flow (block) element `table` lets us do just that in a webpage. Among other things, a table can be used to

- Present tabular data such as an order receipt

- Organize entries in fill-out forms for user input

A table may involve many aspects: rows, columns, cells, headings, lines separating cells, a caption, spacing within and among cells, and vertical and horizontal alignments of cell contents. The rows and columns can be separated into groupings. A cell can also span several columns or rows. Thus, `table` is a complicated construct that can take some effort to master. But a good understanding and the ability to use `table` effectively is important.

3.4.1 Table Basics

In its most basic form, a `table` has a number of rows, each containing the same number of table entries (or cells). Let's look at the table in Figure 3.6 (**Ex:** `SimpleTable`) with three rows, three columns, and a caption:

```
<table border="1">
<caption>My Contacts</caption>                              (1)
<tbody>
    <tr> <th>Name</th> <th>email</th> <th>phone</th></tr>    (2)
    <tr> <td>Joe Smith</td> <td>jsmith@gmail.com</td>        (3)
        <td>432-555-1000</td></tr>
    <tr> <td>Mary Jane</td> <td>mjane@hotmail.com</td>       (4)
        <td>123-555-3020</td></tr>
</tbody></table>
```

Figure 3.6 shows the displayed table. The `tr` element gives you a table row. We have three rows in this example (lines 2 through 4). You may only place table cells inside a table row.

FIGURE 3.6: A Basic Table

My Contacts

Name	email	phone
Joe Smith	jsmith@gmail.com	432-555-1000
Mary Jane	mjane@hotmail.com	123-555-3020

A table cell is either a `td` or a `th` element. Use `th` for a cell that contains a header for a column or a row. Usually, a header will automatically appear in boldface (line 2). A `td` (`th`) may contain any phrasing or flow elements, including `table`. Hence, you can put table(s) into a table.

The option table caption (`caption`) is automatically displayed centered on top of the table. The `border="1"` (`border=""`) setting causes borders (no borders) around each table cell and the table itself. You can use

```
<table border="1" style="border-collapse: collapse">
```

to collapse two adjacent border lines into one, taking out any spacing, (**Ex:** `Collapse`). The property `border-spacing` specifies the spacing separating borders of adjacent table cells (**Ex:** `Spacing`):

```
<table style="border-spacing: 10px">
<table style="border-spacing: 0px">
```

The `border` and `padding` style properties for `td` can be used to control borders and paddings for individual table cells, for example (**Ex:** `ImgComp`),

```
<td style="border: 0px; padding: 0px">
```

Let's now look at the overall structure of a table. Such a `table` element contains the following child elements in order:

1. An optional `caption`

2. An optional `colgroup` for grouping together columns of the table (Section 3.4.8)

3. An optional `thead` for grouping together rows for the table header

4. One or more `tbody` elements containing rows for the table

5. An optional `tfoot` element for grouping together rows for the table footer

And, in general, each row (`tr`) in a table should contain the same number of cells (`td` or `th`), one for each column of the table.

3.4.2   Content Alignment in a Table Cell

By default, content in a cell is `left` aligned horizontally and `middle` aligned vertically inside the space for the cell. The `text-align` (horizontal alignment) and `vertical-align` style properties can be used to control cell content placement. Possible values for `text-align` are `left`, `right`, `center`, and `justify` (left and right justify lines in a cell).

We have seen all the values for `vertical-align` in Section 2.17.3. For table cell content alignment, possible `vertical-align` values are `top` (top of cell), `middle` (middle of cell), and `bottom` (bottom of cell).

Let's use an HTML table (four rows and four columns) to demonstrate cell content alignments (**Ex:** `Align`). The source code for the table shown in Figure 3.7 is as follows:

FIGURE 3.7: Cell Content Alignments

	text-align:left	text-align:center	text-align:right
vertical-align: top	content	content	content
vertical-align: middle	content	content	content
vertical-align: bottom	content	content	content

```
<table border="1" style="background-color: #def;
                border-collapse: collapse">
<thead><tr><th></th>                                    (A)
```

```
<th><code>text-align:left</code></th>
<th><code>text-align:center</code></th>
<th><code>text-align:right</code></th></tr></thead>
<tbody><tr><th><code>vertical-align:<br/>top</code></th> (B)
    <td style="vertical-align:top">content</td>
    <td style="text-align:center;
            vertical-align:top">content</td>
    <td style="text-align:right;
            vertical-align:top">content</td></tr>
<tr><th><code>vertical-align:<br/>middle</code></th>      (C)
    <td>content</td>
    <td style="text-align:center">content</td>
    <td style="text-align:right">content</td></tr>
<tr><th><code>vertical-align:<br/>bottom</code></th>      (D)
    <td style="vertical-align:bottom">content</td>
    <td style="text-align:center;
            vertical-align:bottom">content</td>
    <td style="text-align:right;
            vertical-align:bottom">content</td>
</tr></tbody></table>
```

The first row supplies headers (line A) indicating the horizontal alignment for cells in each column. The second row (line B) uses `vertical-align:top`, the third row (line C) uses the default `vertical-align:middle`, and the fourth row (line D) uses `vertical-align:bottom`.

3.4.3 Displaying Tables

Tables have an inherent presentational aspect to them. It is important to learn how to use style properties to achieve the desired display in practice.

As a practical example, let's look at a shopping cart (**Ex: Cart**) as a table of purchased items (Figure 3.8):

FIGURE 3.8: Sample Shopping Cart Table

Item	Code	Price	Quantity	Amount
Hand Shovel	G10	5.99	1	5.99
Saw Blade	H21	19.99	1	19.99
			Subtotal:	25.98

```
<table style="background-color: #f0f0f0;                   (1)
            border-radius: 15px;
```

```
                  border-collapse: collapse">
<caption><b>Your Cart</b></caption>
<thead>                                           (2)
<tr style="border-bottom: thin solid black;color:#fc0">  (3)
<th>Item</th> <th>Code</th> <th>Price</th>        (4)
<th>Quantity</th> <th>Amount</th> </tr></thead>   (5)
```

The overall table style (line 1) provides the background color, the rounded corners, and no gap between table cells. The thin black rule and the header text color styles are set on the table row (line 3).

By putting the header row in a `thead` element (lines 2–5) and the rest of the rows in a `tbody` element (lines 6–10), we give the table more structure. It is possible for a table to have several `thead` and `tbody` elements, each containing one or more rows (Section 3.4.8). The header text color is set on line 2, and the horizontal line comes from the bottom borders of the header cells (line 4).

```
<tbody style="text-align: right">                 (6)
<tr>
  <th>Hand Shovel</th>
  <td style="text-align:center">G10</td>          (7)
   <td>5.99</td> <td>1</td> <td>5.99</td></tr>
<tr>
  <th>Saw Blade</th>
  <td style="text-align:center">H21</td>          (8)
   <td>19.99</td> <td>1</td> <td>19.99</td></tr>
<tr>
  <th colspan="4">Subtotal:</th><td>25.98</td></tr>  (9)
</tbody></table>                                  (10)
```

We align text to the right for all cells in the table body (line 6) except for two cells (lines 7 and 8). The `colspan` (column span) attribute (line 9) makes the `Subtotal:` cell span four columns. Similarly, you can use the `rowspan` attribute to make a cell span multiple rows in a table (Section 3.4.6).

In this example, you must have observed an important fact about style rules: a style property in a parent element (for example `tbody`) gets inherited by child (the rows and cells) elements.

3.4.4 Positioning Tables

A table is normally positioned left adjusted by itself on the page. To center a table (or a block-display element with a known width) horizontally on the page, you can use the margin style properties (**Ex:** CenteredTable):

```
<table style="margin-left: auto; margin-right: auto">
```

It asks the browser to automatically compute equal left and right margins to center the element. Add this code to **Ex:** Cart and see how the shopping cart centers.

You can also set the `caption-side` style property to `top` or `bottom` to place the caption above or below the table.

These margin properties can also take on a length (`margin-left: 30px` for example) or a percentage (`margin-right: 10%` for example) to control the horizontal placement of any element.

A table can also be floated left or right with text flowing around it (just like text around images; Section 2.17.2).

3.4.5 Table Layout

Normally, tables use an automatic layout algorithm where the width and height of a table are automatically computed to accommodate the contents in the table cells. You may also explicitly suggest a width using the `width` style property for `table`. The width value can be a length or a percentage (of the available width).

If the suggested width is insufficient, the table will be made wider automatically. If the suggested width is wider than necessary, the excess width is distributed evenly into the columns of the table. The height of a table is determined automatically by the accumulated height of cells and cannot be specified.

By default, all cells in the same table column have the same width. The width of the widest cell becomes the column width. Similarly, all cells in the same row have the same height. The height of the tallest cell becomes the row height. The width and height of each table cell are normally computed automatically to accommodate the cell content.

But you can also suggest a desired dimension for a table cell with the style properties `width` and `height`:

- `style="width: `*wd*`"`—Sets the width of the element to *wd*, which can be a fixed length or a percentage of the available horizontal space.

- `style="height: `*ht*`"`—Sets the height of the element to a fixed length *ht*.

Incidentally, the `width` and `height` style properties apply in general to most HTML elements.

For example (**Ex: TableWidth**), the following is a table taking up 60% of the available width and having a light blue background (line **A**). Cell contents are centered horizontally (lines **B** and **E**). The four columns take up 10, 20, 30, and 40% of the table width, respectively (lines **C** and **D**). Figure 3.9 displays

FIGURE 3.9: Table and Cell Width

10%	20%	30%	40%
10%	20%	30%	40%

this table with two rows.

```
<table border="1" style="border-collapse:collapse;        (A)
      width: 60%; background-color:#def">
<tbody>
<tr style="text-align:center">                            (B)
    <td style="width:10%">10%</td>                         (C)
    <td style="width:20%">20%</td>
    <td style="width:30%">30%</td>
    <td style="width:40%">40%</td></tr>                    (D)
<tr style="text-align:center">                            (E)
<td>10%</td><td>20%</td><td>30%</td><td>40%</td></tr>
</tbody></table>
```

The height of a table cell is the minimum height required to accommodate the cell content. The cell height can be set with its `height` style property (`height: 60px` for example) but will always be large enough for the cell content.

If a table becomes wider than the display window, a horizontal scroll bar will appear, and the user must perform horizontal scrolling to view the entire table. Because horizontal scrolling is one of the most unpleasant tasks for users, it is important to make sure your table fits well in a page and adjusts its width in response to different window sizes.

In summary, you set the the `width` style properties to control table and table cell width. The `height` style property of `td` can only be set to fixed lengths.

The table layout just described is known as automatic table layout and is the default for the `table-layout` style property. If you set `table-layout` to `fixed`, then the cell widths of the first row determines the layout of the entire table. This method can make the layout of large tables easier and faster to compute.

3.4.6 Row and Column Spans

A table cell can span multiple rows and/or columns. You use the `rowspan` (`colspan`) attribute to specify the number of rows (columns) a table cell spans.

To illustrate row and column spans (**Ex: Spans**), let's look at how the table in Figure 3.10 is coded.

```
<table style="width: 120px; text-align: center">    (i)
<tbody>
<tr><td colspan="2"                                 (ii)
    style="background-color:red; height:40px">A</td>
    <td rowspan="2"                                 (iii)
    style="background-color: cyan">B</td></tr>
<tr><td rowspan="2"                                 (iv)
```

FIGURE 3.10: Row and Column Spans

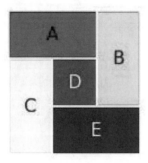

```
   style="background-color: yellow">C</td>
<td style="background-color: green;
    color: white; height: 40px">D</td></tr>          (v)
<tr><td colspan="2"                                    (vi)
    style="background-color: blue;
    color: white; height: 40px">E</td></tr>
</tbody></table>
```

The table has a total width of 120 pixels (line i). On the first row, cell A, which spans two columns (line ii), is followed by cell B, which spans two rows (line iii). Thus, the table has a total of three columns. On the second row, cell C, which spans two rows (line iv), is followed by cell D, which stays in a single row and column (line v). We do not specify the third cell for row 2 because that cell has already been taken. On the third row, cell E spans columns 2 and 3 (line vi), and that completes the table.

This example again shows control of the background color (lines ii and iv) and the foreground color (lines v and vi) for individual cells.

You can practice row and column spanning by making the mirror image of Figure 3.10 ((**Ex: Spans**) at the DWP website.

3.4.7   Rules between Cells

A simple way to get rules (lines) displayed to separate table cells is to use the `border="1"` attribute of the `table` element. Depending on the `border-collapse` style setting (`separate` or `collapse`), you can get rules separating all cells.

Sometimes, we want only vertical rules, only horizontal rules, or some other design for a table. Then, we can use, instead of the `border="1"` `table` attribute, the `border` style properties (Section 4.13.1), on individual cells to piece together the desired rules in a table.

Section 9.12 gives a table-based shopping cart (Figure 9.13) that uses the border-bottom property to obtain a single rule separating the table header from the table body.

3.4.8 Grouping Rows and Columns

Table rows can be grouped by thead (a group of header rows), tbody (a group of table rows), or tfoot (a group of footer rows to be displayed at the end of the table). In addition to added structure, row grouping allows you to specify styles for a group of rows all at once (**Ex:** RowGroup).

Column grouping is also possible with the colgroup and child col elements. Column groupings are given right after any table caption and before everything else in a table. With column grouping, width and alignment settings can be done for all cells in one or more columns at once.

For example (**Ex:** ColGroup),

```
<table border="1" style="width:300px;
      border-collapse:collapse">
<colgroup>
<col span="2" style="width:40%"/>              (I)
<col style="width:20%; background-color: #def"/>   (II)
</colgroup><tbody>
<tr> <td>A</td><td>B</td><td >25.00</td></tr>
<tr> <td>C</td><td>D</td><td >35.00</td></tr>
<tr> <td>E</td><td>F</td><td >45.00</td></tr>
</tbody></table>
```

specifies 40% width (line I) for the first two columns, and 20% width and a light blue background (line II) for the next column (Figure 3.11). Note that

FIGURE 3.11: Column Grouping

A	B	25.00
C	D	35.00
E	F	45.00

col is a void element.

You may also omit the span attribute of colgroup and include one or more child col elements to specify a column group. A col element can supply alignment and width styles for one or a span of columns.

You can easily put one table inside another. All you do is enclose a table in a td element of another table. All the complicated formatting and positioning will be done automatically.

3.5   Using Tables for Page Layout

Webpage designs usually follow some underlying *layout grid* to position various elements of a page for clear visual interpretation.

Figure 3.12 shows a typical layout grid. Clearly we see that the `table` element offers a practical way to create the underlying grid and to make it work for variable window sizes and screen resolutions. The ability to nest tables (**Ex:** `TableNest`) also makes even complicated page layout easy to implement.

FIGURE 3.12: A Sample Layout

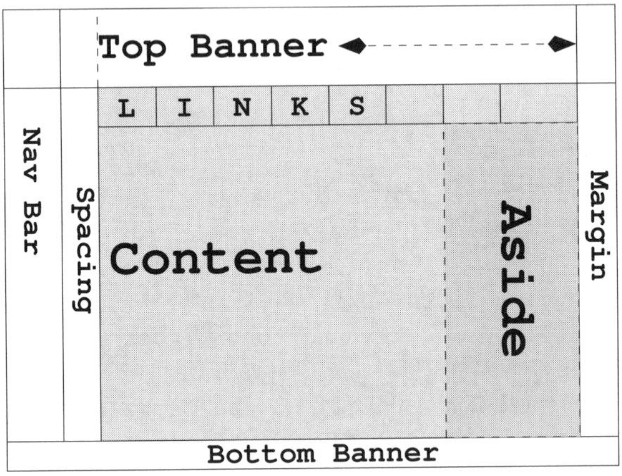

But, because the purpose of HTML5 elements is to provide page structure and not to control visual presentation, we do not recommend using the `table` element directly for page layout purposes. Instead, we can use style rules to create the layout grid and that does not rule out using table styles for the layout (Section 4.10).

3.6   Pagewide Style with `body`

The `body` element supplies the user content of a webpage. To organize page content, you place, inside the `body` container, flow (block) elements such as `header`, `footer`, `section`, `aside`, and so on.

Style properties attached to the `body` element affect the presentation of the entire page. For example,

```
<body style="color:navy; background-color: white"}
```

asks for `navy` characters over `white` background for all child elements in `body`. A child element can override parent style settings by specifying its own style properties, as we have seen in the preceding sections and in Chapter 2.

An image can also serve as the page or element background. Usually, the background image is automatically repeated horizontally and vertically (like floor tiles) to cover the entire page. The background image may be stationary or scroll with the content. Use these style properties to manage the background image:

- `background-image:` *URL*—the *URL* links to an image (GIF, JPG, PNG)

- `background-attachment:` *how*—`scroll` or `fixed`

- `background-repeat:` *how*—`repeat-x` (horizontally), `repeat-y` (vertically), `repeat` (both ways), `no-repeat`

A search on the Web will turn up many sites with background image collections. Use background images only as part of the site's visual design to enhance communication, to increase readability, or to complement functionality. Make sure `background-color` is set to a color similar to the background image in case a browser does not support images.

To set page margins, use the style properties:

```
margin-top: length
margin-bottom: length
margin-left: length
margin-right: length
```

Similar to a printed page, margins give the content breathing room and are important page layout considerations. It is a good idea to use appropriate margins consistently for all pages in a site. For example,

```
<body style="color:navy; background-color: white;
            margin-top: 40px; margin-bottom: 40px;
            margin-left:50px; margin-right: 50px;
            line-height: 150%">
```

sets the left and right margins to 50 pixels and the top and bottom margins to 40 pixels. The margin properties apply to other elements as well.

Other `body` style properties to set in practice include `font-family`, `font-size`, `color`, `line-height`, (affects spacing between lines), and hyperlink styles.

Consistently setting the style of the `body` element (Section 4.2) is important for the design unity of a site.

3.7   HTML5 Entities

HTML5 reserves certain characters such as < for markup purposes. You can not use these characters for themselves without confusing markup processing.

Therefore, HTML5 defines five *entities* (Figure 3.13) to be used for this purpose (**Ex:** Entity). For example, to get a > b & b < c, use the following code:

```
<p>a &gt; b &amp; b &lt; c</p>
```

FIGURE 3.13: HTML5 Entities

Entity	Character	Meaning
<	<	Less-than
>	>	Greater-than
"	"	Double quote
'	'	Single quote
&	&	Ampersand

3.8   Webpage Character Encoding

The ASCII character set, supported by regular English keyboards, contains only 128 characters. HTML uses the much more complete Universal Character Set (UCS), or Unicode[1] character set, defined in ISO10646. UCS contains characters from most known languages. Characters in UCS are put into a linear sequence, and each character has a code position. The ASCII characters are given the code positions 0–127.

An HTML document containing UCS characters can be encoded in different ways when stored as a file or transmitted over the Internet. UTF-8 is a byte-oriented Unicode transformation format that is popular because of its ASCII preserving quality. UTF-8 represents

- ASCII characters (code positions 0–127) with the lower 7 bits of 1 byte (single-byte code)

- Code positions 128–2047 with 2 bytes

- Code positions 2048–65536 with 3 bytes

UTF-8 sets the most significant bit of every byte to 1 for multi-byte codes to distinguish them from single-byte codes. Some bits of the leading byte are used as a byte count. In contrast, UTF-16 encoding uses 2 bytes to represent each Unicode character.

When a Web server returns a document requested by a client (browser), it will specify the correct MIME content type and the encoding used. In addition,

---

[1]www.unicode.org.

an HTML5 document can also specify its own character encoding via this line inside the **head** element.

```
<meta charset="utf-8" />
```

It is advisable to place this character encoding line as the first child of **head**.

3.9 Numeric and Named Character References

On a computer, it is harder to enter non-ASCII characters into a document. With input support for different languages, modern computers do allow you to input UNICODE characters directly into documents. For many users, this requires some learning and practice. However, HTML does provide a character-set-independent way to use UNICODE characters.

A *numeric character reference* specifies the code position with the notation

&#*decimal*;   or   &#x*hex*;

where *decimal* is a decimal integer and *hex* is a hexadecimal integer. A correctly configured browser seamlessly displays character references together with other characters. The Chinese characters displayed in Figure 3.14 are 王 士 弘, the Chinese name of Paul S. Wang (**Ex: Chinese**).

FIGURE 3.14: Numeric Character Reference Display

Paul S. Wang 王 士 弘

We have also mentioned the nonbreaking space character   in Section 2.9.

Numeric references are hard to remember. To make it even easier, HTML5 defines *named character references* giving mnemonic names for special symbols not on the regular keyboard. A named character reference is a short sequence of characters beginning with an ampersand (&) and ending with a semicolon (;). For instance, the nonbreaking space character   is more mnemonic than [2]. As indicated in Section 2.10, the character ­ is also useful.

Figure 3.15 shows the browser presentation of some character references for often-used commercial symbols. See the **Ex: Symbols** example for the source code.

HTML5 character references form a subset of UCS and include Latin-1 characters, mathematical symbols, Greek letters, characters with accents, and other special characters. Figure 3.16 shows how character accents are specified (**Ex: Accents**).

---

[2]Named character references are non-polyglot.

FIGURE 3.15: Commercial Symbols

Reference	Symbol	Description	Reference	Symbol	Description
™	™	Trademark	¢	¢	Cent
€	€	Euro sign	$	$	Dollar sign
¥	¥	Yen sign	©	©	Copyright
£	£	Pound	®	®	Registered TM

FIGURE 3.16: Character Accents

Reference	Character	Reference	Character
é	é	ô	ô
Õ	Õ	à	à
Ä	Ä	Ç	ç

Greek characters are also easy to specify. They are frequently used in mathematical notations. Figure 3.17 shows how to form references for Greek characters and how to add superscripts or subscripts to them (**Ex: Greek**). Good tables for HTML character references are available on the Web, and links can be found at the DWP website.

MathML is a markup language specifically designed to put mathematical formulas on the Web. MathML can now be included in HTML5 directly, and HTML5-compliant browsers provide native support for the display of mathematical formulas (Section 12.8).

FIGURE 3.17: Greek Characters

Code	Display	Code	Display
Α	A	α	α
Β	B	β	β
Σ	Σ	Π	Π
π²	π^2	θ₁	θ_1

3.10   Metadata and Head Elements

So far, we have focused on elements you place inside body to create page content. But HTML also allows you to attach administrative information, or *metadata*, for a page within the head element. The head element is required for each page, and it encloses metadata and administrative elements for a page. Let's now turn our attention to these elements.

The `head` must contain exactly one `title` element and zero or more optional elements:

- `base`—for page location (Section 3.10.4)

- `style`—for in-page style sheet (Chapter 4)

- `link`—for links to related documents such as favicons (Section 3.10.5) or external style sheets (Chapter 4)

- `script`—for in-page or external scripts such as a JavaScript program (Chapter 6)

- `meta`—for various page-related information such as character encoding, keywords, and a summary description

The `meta` element has the general form

```
<meta name="some_name" content="some_text" />
```

and the values for *some_name* and *some_text* can be arbitrary. Well-established conventions for using `meta` are included in this chapter. The `meta` element is also used to place HTTP header information inside a page as illustrated by the page forwarding mechanism in Section 3.10.3.

3.10.1 Search Engine Ready Pages

Search engines use *robots* to continuously visit sites on the Web to collect and organize information into indexed and easily searchable databases. This gives users a quicker and easier way to find sites they want. To attract the right visitors to your site, it is important to provide search engines with the correct information about your site. There are a few simple but important steps you can take to ensure your site is effectively and correctly processed by search engines.

Use these elements inside the `head` element to make your webpage search engine ready:

- `<title>` ... `</title>`—Supplies a precise and descriptive title.

- `<meta name="description" content=" ... " />`—Gives a short and concise description of the content of the page.

- `<meta name="keywords" content="word1, word2, " />`—Lists keywords that a person looking for such information may use in a search.

- `<meta name="robots" content="key1, ..." />`—Tells visiting robots what to do: `index` (index this page), `noindex` (do not index this page), `follow` (follow links in this page), `nofollow` (do not follow links in this page), `all` (same as `index, follow`), `none` (same as `noindex, nofollow`).

For a regular page, you invite the robot to collect information from the page and follow links to recursively index all subpages with

```
<meta name="robots" content="index, follow" />
```

For a page such as terms and conditions of service, copyright notices, advertisements for others, and so on that you do not want indexed as content of your site, use

```
<meta name="robots" content="noindex, nofollow" />
```

to prevent such extraneous information from being indexed as representative of the type of information supplied by your site.

To further help visiting robots, you can place a `robots.txt` file at the *document root* directory of your domain. In the `robots.txt` file, you can indicate that certain parts of your server are off-limits to some or all robots. Commonly, the file is in the form

```
User-agent: *
Disallow: server-root-URL-of-dir1
Disallow: server-root-URL-of-dir2
```

List the URL relative to the server root of any directory that is useless for a robot to index (`/pictures/`, for example). Such a directory may be for images, CGI scripts, administration of your site, internal documentation, password-restricted files, and so forth.

The full details on the Standard for Robot Exclusion can be found at `robotstxt.org/orig.html`.

3.10.2 Prefetching Pages and Resources

The head element

```
<link rel="prefetch" href="url" />
```

tells the browser to load the given `url` ahead of time in anticipation of the user needing it, with high probability, for operations within the current page or for displaying the next page.

This HTML5 feature may not be supported uniformly across all browsers. For example, Google Chrome uses `rel="prerender"` for the same purpose.

3.10.3 Forwarding Pages

Webpages sometimes must move to different locations. But visitors may have bookmarked the old location, or search engines may still have the old location in their search database. When moving webpages, it is prudent to leave a forwarding page at the original URL, at least temporarily.

A forwarding page may display information and redirect the visitor to the new location automatically. Use the `meta` tag

```
<meta http-equiv="Refresh" content="8;  url=newUrl" />
```

The `http-equiv` `meta` tag provides an equivalent HTTP response header (Section 1.16) for the page. The preceding `meta` element gives the response header

```
Refresh     8; url=newUrl
```

The effect is to display the page and load the *newUrl* after 8 seconds.

Here is a sample forwarding page:

```
<head><title>Page Moved</title>
<meta http-equiv="Refresh" content="8; url=target_url" />
</head><body>
<h3>This Page Has Moved</h3>
<p>New Web location is: ... </p>
<p>You will be forwarded to the new location
   automatically.</p>
</body></html>
```

The `Refresh` response header (`meta` tag) can be used to refresh a page periodically to send updated information to the user. This can be useful for displaying changing information such as sports scores and stock quotes. Set the refresh target URL to the page itself so the browser will automatically retrieve the page again after the preset time period. In this way, the updated page, containing the same `meta refresh` tag, will be shown to the user.

3.10.4   Portable Pages

Sometimes a page is downloaded from the Web and saved on your computer. But in all likelihood, the saved page won't display correctly in a browser because it needs images and other files that now can't be found. This happens because the hyperlinks to the needed images and files are relative to the original location of the page.

You can fix this easily by adding the `base` element in the header:

```
<base href="Full_URL_of_original_location" />
```

The `base` provides a browser with an explicit base upon which to interpret all relative URLs in the page. Now your saved page works as if it were in its original location.

If a site expects certain pages to be used after being downloaded, then those pages should have the `base` tag in place already.

3.10.5   Website Icons

A website may define a small icon, such as a tiny logo or graphic, that is displayed by browsers in the URL location box. It is called a *favicon* and is also used in bookmarks and favorite-site listings. Figure 3.18 shows the favicon

FIGURE 3.18: Location Box Showing Favicon

of the NPR site.

Favicons can help brand a site and distinguish its bookmarks. To install a favicon on your site, follow these two steps:

1. Create a `favicon.ico` file for the icon and place it in the document root directory of your Web server. If the document root is not accessible to your site, then place `favicon.ico` at the top directory of your site.

2. Add a header element, for example,

   ```
   <link rel="shortcut icon" href="favicon.ico" />
   ```

   in the `head` element for all `index.html` and other key pages on your site. It tells browsers where the favicon is located if it is not in the document root. It also allows different pages to use different favicons, although this is seldom done. You do not need to insert the `shortcut` link tag in pages if you are able to place the favicon in the document root.

To create the `favicon` file, you may use any raster graphics editor to get a 16-color 16-by-16 graphic. Then save it as an ICO (Microsoft icon format) file. Tools are available to convert `.bmp`, `.gif`, `.png`, and other image formats to `.ico`.

3.11   Subordinate Webpages with `iframe`

The need sometimes arises for a webpage (the parent page) to embed another webpage (a child page, Figure 3.19), often from a third-party site. The embedded page can be advertisement, a Web gadget (shipping calculator, weather, map, for example), or some other self-contained content not directly part of the parent page.

The phrasing element `iframe` is designed for this purpose.

```
<iframe src="url" width="w" height="h"></iframe>
```

displays the webpage at the given *url* inside an inline box of the given width and height.

FIGURE 3.19: Parent and Child Browsing Contexts

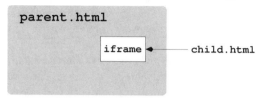

An iframe sets up an independent *browsing context* for the `src` page, similar to navigating to another webpage. This means any HTML, CSS style, or JavaScript code in the iframe will not interfere or conflict with the parent page.

Many Web widgets such as Google Gadgets, are self-contained tools you can add to your Web page. It is a good idea to introduce such an independent tool into your own page by enclosing it in an iframe. However, sometimes `embed` or `object` can also be used. Figure 3.20 (**Ex: StockChart**) shows a stock charting widget (a Google Gadget) that runs in an iframe.

FIGURE 3.20: An `iframe`

Consider the code

```
<section>
<h1>A Stock Chart Widget</h1>
<iframe name="chart" src="gchart.html"      (1)
        width="310" height="290"            (2)
        style="border: 0">                  (3)
</iframe></section>
```

The iframe is given a name `chart` and provides a separate browsing context for the page `gchart.html` (line 1), which is a page for the chart widget. The size of the iframe window is given (line 2), which in this case is enough for the widget. The iframe window has a border by default. We wish not to have that border in this example (line 3).

Iframes are useful to include content from other websites or supplied by users. Examples of user-generated content include forum comments, blog entries, iReport news, and so on. By isolating such content within an iframe, we avoid any problems in the content adversely affecting the parent webpage. You can also define the width and height for the `iframe` to fit your parent page design. Scroll bars are automatically added when necessary.

Navigation links in the parent or child pages can open new pages in either the parent window or an iframe window, as we explain next.

3.11.1   Link Targets

The `target` attribute can be used in any element that provides a link url. If the target is the name of an iframe, then the referenced document is destined to that iframe. In the parent page or a child page, the link

```
<a href="index.html" target="f_one">homepage</a>
```

displays `index.html` in an iframe named `f_one` when clicked.

Without an explicit `target`, the default target for a link is the page or iframe containing the link. The `target` attribute in the `base` element (Section 3.10.4) can set the default target for all untargeted links in a page.

In addition to named frames, there are also *known targets*:

- `_blank`—a new, unnamed top-level window

- `_self`—the same iframe or page containing the link, overriding any `base`-specified target

- `_parent`—the immediate parent of the iframe

- `_top`—the full, original window

The `_blank` target is often useful for displaying links external to a site.

3.11.2   Controlling Iframes

HTML5 allows the parent page to put further restrictions on allowable operations/content from child iframes through the `sandbox` attribute[3]. Specifying `sandbox=""` will

- Always treat the child iframe as from a different Web server

- Disallow link targeting in the child iframe

---

[3]Google Chrome has good support for `sandbox`. Other browsers may still be catching up.

- Disable JavaScript in the child iframe

- Disable HTML forms in the child iframe

- Disable plug-ins in the child iframe

You can re-enable some or all of the disallowed operations by specifying the value of `sandbox` as a list of allowed operations separated by spaces. Allowable operations are

- `allow-same-origin`—If the child iframe comes from the same Web server, treat it as such.

- `allow-top-navigation`—Link targeting in the child frame is allowed.

- `allow-scripts`—JavaScripts in the child iframe may run.

- `allow-forms`—If the child iframe contains HTML forms, they should work normally.

Thus, if we put `sandbox=""` in **Ex: StockChart**, it would stop working because the Google widget depends heavily on JavaScripting. However, if we change that to `sandbox="allow-scripts"`, the widget would work again.

3.12   Website File Organization

Typically, a website consists of a handful of top-level pages, each leading to a set of second-level pages dealing with a specific area. A second-level page may lead to further details or user input forms as third-level pages. All such pages are served by a navigation system and a graphic design that unite the site but make clear visual distinction of the page levels. Such designs usually mean that groups of pages have many parts in common.

For instance, many pages may share a common header, navbar, and perhaps also a footer. Instead of repeating these parts in each page, it is better to separate the common parts into individual files to be shared by all these pages. In this way, any changes in the shared parts will be very easy to make. This is where the *server-side includes* concept comes in (Figure 3.21).

FIGURE 3.21: File Inclusion

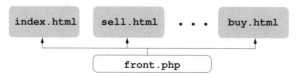

The basic idea of file sharing is simple: You put special instructions inside a webpage for the Web server to include files or other information in the page. The page delivered to the client side results after the inclusions have been made. This can be done with the basic SSI (server-side includes) or with active page technologies. Because we are going to use PHP anyway, that is our choice for including files into a page.

Most, if not all, hosting services also support PHP, an active page language we first mentioned in Section 1.12. Typically, PHP directives can be placed in regular .html files or PHP files .php. PHP directives are given inside php brackets as follows.

```
<?php  . . . ?>   (Full syntax)
<?  . . . ?>      (Shorthand, not recommended)
```

To include another file with PHP, use one of

```
<?php include("file.php"); ?>
<?php include_once("file.php"); ?>
<?php require("file.php"); ?>
<?php require_once("file.php"); ?>
```

The *once* variety avoids including a file a second time. A required file must be found or the page will crash.

For example, here is a page template that uses a front.php and a back.php to construct a page.

```
<?php $page_title="Page Template"; $bg="#def";    (A)
      require("front.php"); ?>
<section id="maincontent">
Place page contents here.
</section>
<?php require("back.php"); ?>
```

Shared files are made more flexible by passing PHP variables to them (line A). Here is a simple front.php.

```
<!DOCTYPE html>
<html xmlns="http://www.w3.org/1999/xhtml"
      lang="en" xml:lang="en">
<head>
<meta charset="utf-8"/>
<title><?php echo $page_title; ?></title> </head>
<body style="background-color: <?php echo $bg; ?>">
```

And you see how the values passed are used to make the front of a page. Effective use of file sharing and inclusion can make websites much easier to construct and maintain.

PHP is a powerful server-side tool, and we will provide more on PHP in Chapter 5 when we discuss form processing.

3.13 Internationalization

As indicated before, it is recommended that all HTML5 pages use UNICODE characters and the UTF-8 encoding declared as the first child of the head element this way:

```
<head>
<meta charset="UTF-8" />
```

The primary language for an HTML5 file is set by the lang attribute of the html element (Section 1.6)

```
<html      ...        lang="language">
```

where *language* is a two-character (case insensitive) language code specified by the International Organization for Standardization (ISO) Code for the Representation of Names of Languages (ISO 639-1). Table 3.2 shows the two-letter codes of some languages. The core attribute lang can be included with any element to indicate a language that may be different from the primary language of the page. It specifies the primary language for the contents of the element and for any of its attributes that contain text.

TABLE 3.2: Some Two-letter Language Codes

Code	Language	Code	Language
ar	Arabic	de	German
es	Spanish	en	English
fr	French	it	Italian
iw	Hebrew	ja	Japanese
ru	Russian	zh	Chinese

The core dir attribute specifies the base direction of *directionally neutral text*—that is, text without an inherent directionality as defined in Unicode or the directionality of tables: dir="ltr" (left-to-right), dir="rtl" (right-to-left), or auto. An rtl table has its first column on the right of the table.

3.14 Summary

Digital audio and video files are encoded and compressed using a variety of different schemes. They are important for reducing file size while preserving the quality of the media. The HTML5 audio and video elements can deploy standard media files for playback by HTML5-compliant browsers without the need for plug-ins.

Standard audio formats include oga (audio/ogg) and mp3 (audio/mpeg). Standard video formats include webm (video/webm), ogv (video/ogg), and mp4 (video/mp4). You can use `source` child elements of `audio` or `video` to provide alternative file formats to make things work in different browsers. Non-standard media files, including PDF, Flash, and Applets, require browser plug-ins and can be delivered with the `object` or `embed` elements.

You can organize tabular data or even lay out webpages with HTML tables, although the latter usage is discouraged. The `table` element lays out its contents in a two-dimensional grid of rows and columns. Table elements include `caption` (table caption), `tbody` (a group of rows), `tr` (table row), `td` (table cell), and `th` (header cell). A cell may span multiple rows/columns. You may control cell spacing, padding, and borders. A table cell may contain another table.

Webpages are usually encoded in UTF-8, a particularly popular way of delivering UNICODE text on the Web. HTML5 defines five entities for special characters (<, >, &, ', ") used in HTML syntax. To make other special symbols easy to enter from the keyboard, HTML also defines a good number of character references for accented characters, commercial and mathematical symbols, non-breaking space, soft hyphen, and so on.

Webpage meta information, loading of related files, and information to help search engines are placed in the `head` element. You can supply favicons and refresh a page automatically or send it to another location using header elements.

The `iframe` element makes it possible to include another webpage as part of your webpage. The `target` attribute on a link can send the linked page to a new window, or some iframe.

Page templates can simplify website construction and maintenance, and PHP makes file sharing and inclusion simple and flexible.

Exercises

3.1. What is sampling and quantization in digital audio? How do these relate to file size?

3.2. What is MP3? What is ogg? What is the difference?

3.3. Consider the `source` element. Does it provide alternatives for a media file or can it also be used to supply multiple files to be played?

3.4. Find good ways to convert sound files to `.ogg` and video files to `.ogv`? on your computer system. (Hint: Look at the Linux command **ffm-peg2theora**.)

3.5. Explain the meaning of the `width` and `height` attributes for media elements.

3.6. Will it work if you put an audio file as the `src` of a `video` element? And vice versa?

3.7. Which part of an HTML5 file specifies the document language? The document character set? The document character encoding? Show the HTML code.

3.8. Consider UTF-8 and UTF-16 character encodings. Why do we say UTF-8 is ASCII preserving?

3.9. UTF-8 seems to use more bits than UTF-16 for characters at higher code positions. What advantage does UTF-8 have over UTF-16?

3.10. Name four ways to enter the Greek character π in a webpage.

3.11. Consider HTML tables. How do you center a table? Float a table to the left or right?

3.12. Are there situations when a correctly specified `table` has rows containing different numbers of `td` elements? Explain.

3.13. What attributes are available for `table`? `tr`? `td`?

3.14. Is it possible to specify the height of a table? The height of a table cell? How?

3.15. List the elements and information you need to place in a page to make it search engine ready.

3.16. Enter the mathematical formula $\sin(\pi^2)$ into a webpage and display the results.

3.17. College sororities and fraternities often have names with Greek letters. Enter your favorite sorority or fraternity name into a page.

3.18. Take a simple table with a caption and center the table. Does the caption automatically go with the table, or do you need to center the caption separately? Experiment with different browsers and discover for yourself.

3.19. Consider specifying table cell `width` with percentages. What happens if the total adds up to more than `100%`? What happens if one or more cells have no `width` specified while others have percentages? Experiment and find out.

3.20. When centering a block element such as `table`, what is the difference between

```
style="margin-left: auto; margin-right: auto"
```

and

```
style="margin-left: 20%; margin-right: 20%"
```

Experiment and explain.

3.21. Practice row and column spanning by creating an HTML table to present the mirror image of Figure 3.10.

3.22. Access the W3C HTML5 validator and use it on some of your pages. Look at the results and fix any problems.

Chapter 4

Styling with CSS

Document structure and presentation style are two important, but largely independent, aspects of a webpage.

It is important for any website to have a well-designed style based on functionality, information architecture, and artistic visual communication design. As we have seen in the two preceding chapters, *cascading style sheet* (CSS) is used to specify the presentation style for a webpage.

HTML5 promotes the separation of document presentation from document structure. The focus of HTML5 is structuring documents so they can be processed automatically by programs. The HTML5 *Document Object Model*, an integral part of the HTML5 standard, provides exact requirements for browsers to formulate the DOM data structure and specifies an *application programming interface* (API) for accessing and editing documents under JavaScript control (Chapter 6).

CSS, or simply style sheets, on the other hand, supply a way to specify presentation styles independently and associate them with markup elements in webpages. In this way, the same styles can be placed on different elements, and different styles can be attached to the same elements, gaining much appreciated flexibility for authoring, processing, and presenting documents.

With CSS, you can specify display styles for visual browsers, print formats for printers, and device-dependent styles for other media such as page readers for the vision impaired. Furthermore, CSS provides *media queries* (Section 13.4), enabling you to attach different styles to the same HTML elements. This is important because of the wide variety of Web-capable devices today, including desktops, laptops, tablets, e-readers, and smartphones. This chapter focuses on display styles for visual browsers.

Hence, a webpage consists of two basic parts: HTML5 code and CSS code. The latter may be placed in one or more separate files, called *style sheets*. By providing a set of *style rules*, a style sheet indicates the presentation style for various elements in a webpage. By associating a style sheet with an HTML5 page, you control the styling of that page. Without such a style guide, a browser can only use its default styles and any user preferences for displaying the HTML elements. CSS is not an advanced technique that one may choose to ignore for simple applications. On the contrary, HTML5 and CSS must be applied together to create any real webpage (Figure 4.1).

Concepts, rules, usage, and examples of style sheets are presented to give you a comprehensive view of CSS and how it is used in practice. The coverage

FIGURE 4.1: HTML + CSS = Webpage

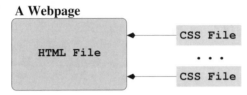

allows you to apply HTML5 and style sheets to implement your desired page layout and visual presentation.

4.1   CSS Basics

CSS is a language, recommended by the W3C and supported by major browsers, for specifying presentation styles for HTML5 and also other documents. CSS consists of the following components:

- *Style declarations* —A style declaration is given in the form

  property : value

  There are many properties (well over 100) for the various presentational aspects of HTML elements, including *font properties, color and background properties, text properties, link properties, box properties, layout properties, transition properties, transform properties, animation properties,* and *classification properties.*

  Listings of CSS properties are available on the Web (see the DWP website). Many properties can be associated with all HTML elements when appropriate. Obviously, putting a font property on an `img` element does nothing. Some, such as `text-align`, apply only to block elements. Others, such as `vertical-align`, apply only to inline elements. A few styles apply only to specific elements. For example, `list-style-type` is only for list items. CSS documents the applicability of each style property. Browsers provide default presentation styles to all HTML elements. By associating your own style declarations, you can control the presentation style of an entire page and how any element in it is displayed.

  A big part of learning CSS has to do with getting to know the available style properties, their meanings, initial values, and possible settings.

- *Selectors* —CSS defines selectors to give you multiple ways to indicate which style properties are assigned to which HTML elements. Assigning style properties by selectors is in addition to defining the `style` attributes for HTML5 elements.

- *Inheritance and cascading rules*—CSS defines how values for properties assigned to an HTML element are inherited by its child elements. For example, because of inheritance, font and background settings for `body` can affect the entire HTML document. In addition, because of no inheritance, margin, padding, and border declarations do not affect child elements. CSS documents the inheritance status of each style property. When conflicting style declarations occur on a single HTML element, cascading rules (Section 4.26) govern which declaration will apply.

With CSS, you define style rules and attach them to HTML elements, thus controlling their presentation style. Hence, each webpage may have an HTML file and a set of style rules. A style rule consists of a selector and one or more style declarations separated by semicolons. Therefore, the general syntax for a style rule is

selector
{ *property*$_1$: *value*$_1$;
    property$_2$: *value*$_2$;
       . . .
    property$_n$: *value*$_n$
}

A simple selector can be just the name or names of the HTML elements that take the style. Figure 4.2 shows the anatomy of a style rule.

FIGURE 4.2: Structure of a Style Rule

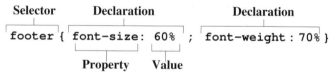

Getting to know the different style properties, their possible values, and the effect they have on page presentation is part of learning CSS. But we already know quite a few things about CSS: length units (Section 2.12.3), color values (Section 2.12.4), font properties (Section 2.12.5), width and height (Section 3.4.5), horizontal alignment (Section 2.12), vertical alignment (Section 2.17.3) of inline elements, and more. We will reinforce and build on this knowledge in this chapter. You will get a comprehensive view of CSS and learn how to apply them in practice.

4.1.1   Style Sheets

A *style sheet* is a file (usually with the `.css` suffix) that contains one or more style rules. In a style sheet, comments may be given between `/*` and `*/`.

You have multiple ways to associate style rules with HTML elements, making it easy and flexible to specify styles. For example, the rule

```
h1 { font-size: large }
```

specifies the font size for all first-level headers to be `large`, a CSS predefined size. And the two rules

```
h2 { font-size: medium }
h3 { font-size: small }
```

give the font size for second- and third-level headers. These rules only affect the font size; the headers will still be bold because that is their default `font-weight`.

To make these three headers dark blue, the rules can become

```
h1 { font-size: large; color: #009 }
h2 { font-size: medium; color: #009 }
h3 { font-size: small; color: #009 }
```

A well-designed site should use a consistent set of font family, size, and line height settings for all its pages. CSS rules can help the implementation of this immensely.

4.1.2   Attaching a Style Sheet

To attach the three rules to a webpage, you can place them in a file, such as `myfile.css`, and put the following `link` element inside the `head` element of the page:

```
<link rel="stylesheet" type="text/css" href="myfile.css" />
```

This external-file approach allows you to easily attach the same style sheet to multiple pages of your website. You can even use style sheets at other sites by giving a full URL for the `href`. Figure 4.3 shows the relation between a

FIGURE 4.3: Attaching a Atyle Sheet

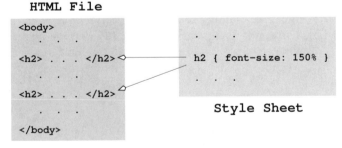

webpage and its style sheet.

A `.html` file may have one or more `link rel="stylesheet"` elements. Additionally, you can include rules directly in a webpage via the `style` element

```
<style type="text/css">
body { font-size: small }
h1 { font-size: large; color: #009 }
h2 { font-size: medium; color: #009 }
h3 { color: #009 }
</style>
```

which is also placed inside the **head** element. This is the **style**-element approach. The external-file approach has advantages. It makes changing the style that affects multiple pages easy. And a browser needs to download the style sheet only once for all the pages that use it, making your site faster to load.

4.1.3 A Brief History of CSS

Realizing the need to separate document structure from document presentation, Robert Raisch produced the *WWW HTML Style Guide Recommendations* in June 1993. The work laid the foundation for the development of style sheets. Later, in 1994, Håkon Lie published *Cascading HTML Style Sheets— A Proposal*, which suggested the sequencing of style sheets applicable to a webpage.

The W3C worked to make CSS an industry standard. In 1996, the W3C made its first official recommendation, *Cascading Style Sheets, Level 1* (CSS1), which allows you to separate style from content and gives much more control over styling than before. The W3C continued to evolve CSS as browser support slowly evolved. In May 1998, the next version of style sheets became *Cascading Style Sheets, Level 2* (CSS2).

CSS2 is a superset of CSS1 and adds support for media-specific style sheets so that authors may tailor the presentation of their documents to visual browsers, aural devices, and so on. This specification also supports content positioning, downloadable fonts, table layout, internationalization, and more.

As CSS became popular and sophisticated, the W3C decided, in 2001, to divide CSS into *specification modules* to make it more manageable. This means different aspects of style such as color, font, positioning, selector, table, box model vertical, box model horizontal, and so on will be specified by separate specifications maintained by different working groups. The modularization work led to CSS3, the current standard (**w3.org/TR/CSS**), which offers cool features such as opacity, rounded border, border image, multi-column layout, drop shadow, transitions, gradients, and animation. Browser support for some advanced features of CSS3 is still not uniform.

4.2 Whole-Page Styling

Basic Web design calls for a unified look and feel for all pages of a site. To help achieve this design unity, we can apply the same page margins, font, foreground and background colors, and heading and link styles to all pages in your site. If some page deviates from the overall style, it should be by design and with good reasons.

Because most style properties set for the body element apply to the entire page through inheritance, CSS makes whole-page styling easy to enforce.

We know that the font type, style, and leading (line separation) can affect the readability as well as the look and feel of the entire site. It is advisable to use standard CSS font sizes (Section 2.12.5) rather than pixel or point settings. In this way, your font will work well under different screen sizes and resolutions. Also, we can use more line separation to increase readability. For example, a website may use the following:

```
body
{  font: small Verdana, Geneva, Arial, helvetica, sans-serif;
   color: black;                 /* foreground */
   background-color: white;      /* (1) */
   margin: 0px  0px  30px  0px;  /* top right bottom left*/
   border: 0px; padding: 0px;    /* (2) */
}
```

The running text is set to small (one notch below medium) for the entire page. The font style property allows you to specify all font-related properties in one place in the general form:

font: *style variant weight size / line-height family*

Only the *size* and *family* are required. (See Section 2.12.5 for font properties and values.)

Normally, line-height is 120% of font-size. To improve readability of textual materials on screen, we recommend

```
h1, h2, h3, p, li { line-height: 150% }
```

to set line spacing to $1.5 \times$ font-size for the indicated elements.

The background-color of an element can be set to a specified color (line 1) or transparent, which is the initial value. It is also possible to specify an image for the background. We will see how to use background images later in Sections 4.13.2 and 4.19. Page margins are set to 0 except for the bottom, which is 30 pixels. We also make sure no border, margin, or padding is used for body (line 2).

```
body { border: 0; padding: 0;   margin: 0; }
```

We have discussed how to specify colors (Section 2.12.4) and fonts (Sections 2.12.5). A website may use the following set of section header styles:

```
h2                              /* in-page heading   */
{  font-weight: bold;
```

```
    text-transform: capitalize;
    color: #666;
    font-size: medium;              /* predefined size   */
}

h2.red {  color: #933; }           /* Page top heading  */

strong.heading                     /* subhead            */
{  font-weight: bold; display: block; }
```

The body, heading, line height, and link styles (Section 4.7) form a good basic set for the overall styling of a page.

4.3   Centering and Indenting

To center text or other inline elements such as images, use the

```
text-align: center
```

declaration. The `text-align` property applies to block-level elements and controls the horizontal alignment of their child inline elements. The possible values are `left`, `right`, `center`, and `justify`, as discussed in Section 2.12.2.

For example, you can use an HTML element

```
<h2 class="center">Topic of The Day</h2>
```

for any text you wish centered and provide a style rule such as

```
h2.center { text-align: center; color: #006600 }
```

to specify the display style. In this way, not every `h2` element but only those with the `class="center"` attribute will be centered. To associate the same rule with all six different headings, you can code it as follows:

```
h1.center, h2.center, h3.center, h4.center, h5.center, h6.center
{ text-align: center; color: #006600 }
```

As mentioned in Section 2.7, the core attribute `class` is used by style rule selectors to associate styles.

By omitting the element name in front of the class name, a selector addresses any HTML element with the given class attribute. For example, the rule

```
.center { text-align: center; color: #006600 }
```

applies to all these elements

```
<h1 class="center">Topic of The Day</h1>
<h3 class="center">Lunch Menu</h3>
<p class="center">Some text</p>
```

The general form of a class selector is

element . class

If *element* is omitted, then the selector matches any element in that class.

To center a block element with a fixed width or a table, use the style rules

```
margin-left: auto; margin-right: auto
```

With this style, a browser automatically computes equally sized left and right margins to center a block element or table.

The style `display: table` can be used with auto left and right margins to center any element on a page.

If a table has a caption, you may need to center it with automargins as well. Older browsers such as IE 6 do not support automargins, and you may have to resort to the deprecated `center` element. **Ex:** `CenterStyle` demonstrates the centering of different elements.

To indent the first line of a paragraph, use

```
p {  text-indent: 3em }
```

To indent entire paragraphs, you can increase the left and/or right margins as desired. For example,

```
p.abstract {  margin-left:  5em; margin-right: 5em }
```

centers a paragraph with the left and right margins each increased by 5em (Section 2.12.3).

4.4   Centered Page Layout

Modern computer screens are wide with high resolution. Often we do not want a webpage to become so wide and horizontal. The technique is to define a narrower webpage with a border or a background color and center it horizontally on the screen (browser window). This design can also make your webpage easier to read and display on different devices, including hand-held ones.

Figure 4.4 shows a simple centered page layout (**Ex:** `CenterPage`) resulting from the HTML structure

```
<body><div class="pagebox">
<!-- page contents go here -->
</div></body>
```

and the style code

```
body { background-color: #def } /* browser window background */
div.pagebox
{ background-color: white;      /* page background */
```

FIGURE 4.4: A Centered Webpage

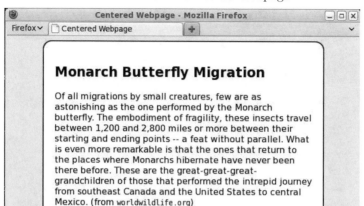

```
margin-left: auto;          /* page centered in window */
margin-right:auto;
width: 75%;                 /* narrowed page width */
padding: 20px;              /* margin around content */
border: 2px solid navy;     /* page border width and color */
border-radius: 16px         /* rounded corners */
```

The CSS3 border-radius property defines the curvature of the rounded corner. See Section 4.13.1 for more on rounded corners.

4.5 Multicolumn Layout

Similar to a newspaper or magazine, a webpage can have a layout with two or more columns. CSS makes this easy to do. Figure 4.5 shows a two-column page (**Ex:** TwoCol) produced with the style

```
body
{  margin: 50px;
   column-count: 2;      /* two columns          */
   column-gap: 2em;      /* gap between columns */
}
```

Instead of a fixed column count, you can specify a column-width and allow the browser to fit as many columns as possible in the available window or parent element width. Adding rules between columns for more visual separation is also easy with the column-rule style property. For example,

```
column-rule: thin solid black    /* 1px thick */
```

FIGURE 4.5: Two-Column Layout

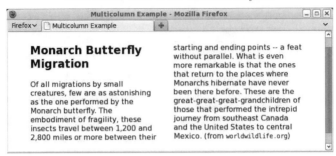

Within a multicolumn layout, child elements are flowed from one column to the next automatically. But, you can set the style `column-span: all` for a child element for it to span all columns. This is useful for displaying a headline. For example,

```
h1 { column-span: all }   /* only other value 1, the default */
```

Other than `body`, you may apply multicolumn layout for flow (block) elements such as `article`, `section`, `div`, and so on.

Multicolumn layout is introduced by CSS3 and support by browsers may not be uniform. You may need to add a browser prefix to the property name:

- `-moz-` for Firefox

- `-webkit-` for Chrome/Safari

- `-ms-` for MSIE

- `-o-` for Opera

To cover more bases, you may want to use, for example,

```
body
{  margin: 50px;
   column-count: 2;           /* two columns         */
   column-gap: 2em;           /* gap between columns */
   -moz-column-count: 2;
   -moz-column-gap: 2em;
   -webkit-column-count: 2;
   -webkit-column-gap: 2em;
}
```

See **Ex:** MultiCol for another example.

4.6 CSS Selectors

Separating style rules from the HTML page they control brings important advantages. However, the price to pay is having to tell which style rule applies to what HTML element. This is exactly what a *selector* does. It associates a rule to a set of HTML elements. Different forms of selectors give you the ability to easily associate rules to elements of your choice:

- *Type selector*—A type selector is the simplest selector. It specifies an HTML element tag name and associates the rule with every instance of that element in the HTML document. For example, the rule

  ```
  h3 { line-height: 140% }
  ```

  is used for the DWP website.

- *Universal selector*—The symbol * used as a selector selects every HTML element, thus making it simple to apply certain styles to all elements, all element in a class (*.class*) or all child/descendant elements.

- *Class selector*—The *.className* selector selects elements in the named class. An element is in class *xyz* if its `class` attribute (Section 2.7) contains the word *xyz*. For example,

  ```
  .cap { text-transform: uppercase }
  ```

  makes elements in the `cap` class ALL CAPS. And

  ```
  .emphasis { font-style: italic; font-weight: bold }
  ```

  makes the attribute `class="emphasis"` meaningful for many elements.

- *Attribute selector*—The selector [*attr*] selects the elements with the attribute *attr* set (to any value). The selector [*attr*="*str*"] selects elements with *attr* set to *str*.

  Instead of =, you may use *= (contains *str*), ^= (begins with *str*), $= (ends with *str*), and ~= (contains the word *str*). For example, the class selector *.someName* is the same as the attribute selector [class~="*someName*"].

- *Id selector*—The #*idName* selector associates the rule with the HTML element with the unique `id` attribute *idName*. Hence, the rule applies to at most one HTML element instance. For example,

  ```
  #mileageChart{ font-family:Courier, monospace; color:red }
  ```

  applies to `<table id="mileageChart">` ... `</table>` only.

- *Concatenated (conjunction) selector*—When a selector is the concatenation of two or more selectors, it selects elements satisfying each and every selector included. For example,

`span.highlight`	(`span` elements in class `highlight`)
`nav.main.mobile`	(`nav` elements in class `main` and class `mobile`)
`table#mileageChart`	(`table` element with id `mileageChart`)

- *Combinator selector*—A sequence of selectors separated by spaces, in the form *s1 s2 s3* ..., selects the last selector contained (a descendant) in the previous selector and so on. For example,

  ```
  nav#left a img { width: 36px }
  ```

  sets the width for all link anchoring `img` in the `nav` with `id="left"`.

  If you use > instead of SPACE to separate selectors, then it means the next selector is a child of the previous selector. If + is used, it makes the next selector an immediate sibling of (next to) the previous selector[1].

- *Pseudo-class selectors*—The *pseudo-class* is a way to permit selection based on conditions at run-time or on the hierarchical structure in the document. The eight most widely used pseudo-classes are the selector suffixes `:link` (a valid link), `:visited` (a visited link), `:hover` (mouse over element), `:active` (element being clicked), `:focus` (UI element gained focus), `:enabled` (UI element usable), `:disabled` (UI element unusable), `:checked` (element selected). The first four are often applied to the a element to control link styles (Section 4.7). Also `:target` selects target of an in-page link action by the user. For example,

  ```
  section:target { border: thin solid black }
  ```

  would add the border to a `section` with `id="foo"` immediately after the user clicks on a link `<a href="#foo">`.

  Additional pseudo-class selectors include `:first-child`, `:last-child`, `:nth-child(`*n*`)` (the nth child of parent; see Exercise 4.12), `:empty` (if no content or children).

- *Pseudo-element selectors*—These are fictitious elements for the purpose of assigning styles or content to well-defined parts of the document. CSS provides pseudo-elements, `::first-line` and `::first-letter`, allowing you to specify styles for the first line and first letter of an element. For example, `p::first-line` means the first line of a paragraph, and `p:first-letter` means the first character of a paragraph.

---

[1] Browser support for > and + in selectors is not uniform.

In addition, ::before and ::after can insert styled content immediately before and after a target element. See Section 4.24 for more on generated content.

For backward compatibility, browsers may also support a single-colon version (:before, for example) of a pseudo-element selector.

Selectors sharing the same properties can be grouped together in one rule to avoid repeating the same rule for different selectors. To group selectors, list them separated by commas. For example,

```
h1, h2, h3, h4, h5, h6 { color: blue }
```

is shorthand for six separate rules, one for each heading. Be sure to use the commas. Otherwise, the selector turns into a contextual selector. Table 4.1 lists

TABLE 4.1: CSS Selector Examples

Selector	Selector Type
`body { background-color: white }`	Element
`*.fine or .fine { font-size: x-small }`	Universal + Class
`h2.red { color: #933 }`	Class
`table.navpanel img { display: block }`	Contextual
`a.box:hover`	Pseudo-class in Class
`{ border: #c91 1px solid;`	
`  text-decoration: none; }`	
`p, ul, nl { line-height: 150%; }`	Element Shorthand

selector examples for easy reference. Complete specifications of CSS selectors can be found at the W3C website.

4.7 Link Styles

Link styles form an integral part of the overall look and feel of a website. You do not style links through the a tag directly. Instead, four pseudo-class selectors allow you to specify visual styles to indicate whether a link is not visited yet, visited already, ready to be clicked (the mouse is over the link), or during a click. For example,

```
a:link { color: #00c;  }      /* shaded blue for unvisited links */

a:visited { color: #300; }    /* dark red for visited links */

a:active                      /* when link is clicked */
{   background-image: none;
    color: #00c;              /* keeps the same color */
```

```
    font-weight: bold;          /* but turns font bold */
}

a:hover                        /* when mouse is over link */
{   background-color: #def;    /* turns background gray-blue */
    background-image: none;
}
```

In case you are wondering, these four selectors do inherit any style properties from the a element and can selectively override them.

Sometimes it is useful to have different classes of links (e.g., external and internal links). In that case, you can use selectors in the form

```
a.external:link
a.external:hover
```

Links are usually underlined; to avoid the underline, you can add the rule

```
a { text-decoration: none }
```

Be advised that Web surfers are used to seeing the underline for links. Unless you have another way to distinguish links, do not remove the underline.

4.8 Conflicting Rules

With the variety of selectors, multiple style rules could apply to a given HTML element. For example, the two rules

```
p { font-size: normal }
li p { font-size: small }
```

both apply to a `<p>` element within an `<li>` element. Intuitively, we know which one applies. Because CSS intrinsically relies on such conflicts, there are very detailed and explicit ways to decide which among all applicable rules applies in any given situation (see Section 4.26). Generally, a rule with a more specific selector wins over a rule with a less specific selector.

According to the CSS3 standard, a selector's specificity is calculated as follows:

- Let a be the number of ID selectors in the selector.

- Let b be the number of class selectors, attributes selectors, and pseudo-classes in the selector.

- Let c be the number of type selectors and pseudo-elements in the selector.

Then the number $a\,b\,c$ (in a number system with a large base) is the specificity of the selector. The higher the value, the higher the specificity.

If an element has a `style` attribute in HTML (an in-element style attribute), it is the most specific.

This arrangement allows you to define global styles for elements and modify those styles with contextual rules. Special cases can be handled with class selectors. One-of-a-kind situations can be handled with `Id` selectors or in-element `style` attributes.

4.9 Style Properties

With a good understanding of selectors, we are now ready to tackle the rich set of available style properties and their possible values. You will see that there are many properties to give you fine control over the presentation style, including properties for font, color, text, and the box around an element. Many properties apply generally. Others are for specific display modes: inline, block, table, and so on. Unrecognized properties are ignored.

When displaying a page, the browser presents each HTML element according to its style properties. If a property is not specified for an element, then that property is often *inherited* from its *parent element*. This is why properties set for the `body` element affect all elements in it. Because of inheritance, it is important to make sure that all elements are properly nested and end with closing tags.

When a property has a percentage value, the value is computed relative to some well-defined property. For example, `margin-left: 10%` sets the left margin to `10%` of the available display width. Each property that admits a percentage value defines the exact meaning of the percentage.

With a good understanding of CSS syntax, we are now ready to put that knowledge to use and see how CSS is applied in practice.

4.10 Webpage Layout with CSS

A critical task in developing a new website is creating a good visual design and page layout.

A layout grid is a set of invisible vertical and horizontal lines to guide content placement. It is the primary way designers organize elements in a two-dimensional space. A grid aligns page elements vertically and horizontally, marks margins, and sets start and end points for element placement. A well-designed grid makes a page visually clear and pleasing; it results in increased usability and effective content delivery. A consistent page layout also helps to create unity throughout the site.

A fixed-width (or ice) layout can be easier to implement but a fluid layout that adjusts to varying page width and screen resolution is more desirable.

Given a layout grid, CSS is used to implement the design. There are two common approaches to implement a layout grid: using floating columns and using CSS table display. The float approach does a good job of forming multiple columns but not so much for horizontal or vertical alignments. The table approach is very exact and can be less complicated to implement as well.

4.10.1 Float Layout

As an example that can be used as a template (**Ex:** FloatLayout) for a simple webpage, let's look at the fluid layout shown in Figure 4.6.

FIGURE 4.6: A Float Layout

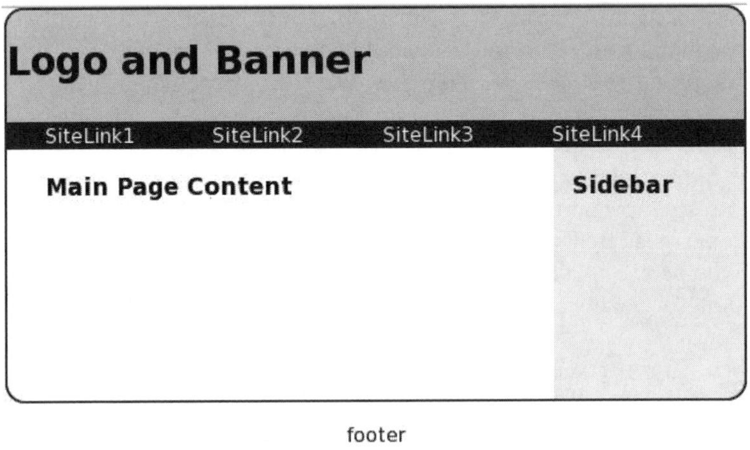

footer

The page structure consists of a centered main div and a footer. The main div, with round-corner borders, has two child elements:

1. A header for top banner and navbar

2. A section containing an article (for the main content), an aside (for the sidebar), and an empty div (line b)

```
<body id="top">
<div id="centerpage">
   <header class="banner">... </header>
   <section id="main">                      (a)
      <article>... </article>
      <aside>... </aside>
      <div style="clear: both"></div>      (b)
   </section>
</div>
<footer>
```

```
<p style="text-align: center">footer</p>
</footer>
```

The main `div` is styled this way:

```
div#centerpage
{ margin-left:auto; margin-right:auto; /* centering */
  width: 80%;  /* fluid page width */
  border: 2px solid darkblue;  /* border */
  border-radius: 16px;  /* rounded corners */
  overflow: hidden;
}
```

An overflow property can take care of the case when the available width or height becomes insufficient for the content of an element. See Section 4.15 for more details.

Let's now look at the HTML and CSS for each part inside the main `div`. The `header` corresponds to the top banner and navbar.

```
<header class="banner">
<section class="logo">Logo and Banner</section>
<nav> <a href="#">SiteLink1</a>              (c)
      <a href="#">SiteLink2</a>
      <a href="#">SiteLink3</a>
      <a href="#">SiteLink4</a></nav></header>
```

The logo `section` is a container for the top banner graphics and, for the purpose of this example, styled as follows:

```
header.banner { background-color: #bcd; }

header.banner > section.logo
{ font-size: xx-large; font-weight: bold;
  height: 60px; padding-top: 30px;
}
```

The horizontal navbar (line c) has the following style:

```
header.banner nav
{ background-color:darkblue; /* color of navbar   */
  padding-left: 2em;          /* lead spacing       */
  white-space: nowrap         /* links on one line */
}

header.banner > nav a:link
{ text-decoration: none;      /* no underline       */
  color: white;               /* links in white     */
  margin-right: 60px;         /* spacing the links  */
```

```
}

header.banner > nav a:hover    /* (d)                    */
{ text-decoration: underline;/* mouseover effect  */
}
```

We used `white-space: nowrap` to keep the links on a single line. Otherwise, they may be broken into another line when the user resizes the browser window. The `a:hover` rule (line d) adds an underline on mouseover. Users appreciate such responsiveness in webpages.

Moving on to the main `section` (line a) styles:

```
section#main
{  overflow: hidden;                /*                  (e) */
   background-color: #def;          /* for sidebar (f) */
}

section#main > article
{  width: 69%; float: left;         /* column one (g)  */
   background-color: white;
   padding-left: 2em;               /* left margin     */
}

section#main > aside
{  float: left;                     /* column two      */
   margin-left: 1em;top-margin: 2em;   /* margins         */
}
```

The two children, `article` and `aside`, are floated to the left, forming two columns. This is a well-known technique for creating multicolumn webpages. Our fluid design (lines e and g) here responds well to page resizing. Here is an important detail: the height of the two floated columns (`article` and `aside`) are made equal only by adding the all-clearing `div` (line b). Otherwise, the background color (line f) will not show through for the `aside` column.

Such all-clearing `div` can be avoided if we use a table-based layout as done in the next section (Section 4.10.2).

4.10.2 Table Layout

Multiple columns with floats can sometimes be tricky for layout control. Let's reimplement the layout shown in Figure 4.6 with CSS table displays (**Ex: TableLayout**). We do this without using an HTML `table` to maintain the page structure. Instead, we use the CSS3 `display` styles

```
display: table
display: table-row
display: table-cell
```

to achieve the same purpose. In addition to the above, CSS3 also provides `table-column`, `table-caption`, `table-header-group`, `table-footer-group`, and `inline-table` styles. All these cause the so styled element to behave like the corresponding HTML table element. The `inline-table` makes it an inline block (Section 4.12).

The HTML structure remains largely the same. The banner `header` is not changed but uses CSS rules that place the outline blue borders on the `header` element instead of the parent centerpage `div`.

```
div#centerpage
{   margin-left: auto; margin-right: auto;
    width: 80%; border-radius: 16px;
    overflow-x: hidden;
}
```

```
header.banner
{   background-color: #bcd;
    border-top: 2px solid darkblue;      /* borders here */
    border-left: 2px solid darkblue;
    border-right: 2px solid darkblue;
    border-top-left-radius: 16px;
    border-top-right-radius: 16px;
}
```

The two columns are now styled using table cell alignments instead of left floating. The HTML adjustments are

- Removing the all-clearing `div`

- Putting the main `section`, the `article`, and the `aside` each inside its own `div`

```
<div id="contentbox">
<div class="layout_table"><section id="main">
<div class="col1"><article>...   </article></div>
<div class="col2"><aside>...   </aside></div>
</section></div></div>
```

We now put the left, right, and bottom borders on the contentbox (line I). The following CSS code displays two columns using table `display` settings (lines II–IV).

```
div#contentbox                         /* I */
{   border-left: 2px solid darkblue;
    border-right: 2px solid darkblue;
    border-bottom: 2px solid darkblue;
    border-bottom-left-radius: 16px;
```

```
      border-bottom-right-radius: 16px;
}

div.layout_table                     /* II */
{   display: table; width: 100%;
    border-spacing:0px;
}

section#main { display: table-row; }
div.cols { display: table; width: 100%; }
section#main { display:table-row }

section#main > div.col1
{   display: table-cell;
    width: 75%;                      /* III */
    background-color: white;
    border-bottom-left-radius: 16px;
}

section#main > div.col2
{   display: table-cell; width: auto;  /* IV  */
    border-bottom-right-radius: 16px;
    background-color: #def;
}
```

The styles for `article` and `aside` are simplified to

```
div.col1 > article { padding: 2em; }
div.col2 > aside { padding-left: 1em; }
```

When the user narrows the width of the browser window greatly, the float-ing layout may shift the `aside` column beneath the `article` column. The table-based approach does not have that problem. The complete HTML and CSS code files for both layouts ((**Ex**: `FloatLayout` and **Ex**: `TableLayout`) can be found at the DWP website.

4.11   A Vertical Navbar

A clear and logical navigation system contributes much to the quality and user friendliness of a website. The two main questions from a user's viewpoint are: Where am I in the site? How do I get where I need to go? A good navigation system must clearly identify each page and provide consistent visual clues and feedback for links.

A navigation bar (navbar) or panel collects important links in a central place. Visitors expect to find navigation constructs on the top and left sides of pages. Frequently, auxiliary navigation links are also placed on the right and bottom sides. Links can appear "charged" or "armed" when the mouse cursor is positioned over them. This is the so-called *rollover* effect. A consistent design of rollovers can help the user feel in control when visiting a site.

In previous section (Section 4.10) we saw how a horizontal top-side navbar is constructed with HTML and CSS. Let's now look at a vertical left-side navigation panel (**Ex: NavPanel**).

Figure 4.7 shows the navigation panel that we will build. The panel is the first column of a two-column fluid page design. The table styling approach as discussed in Section 4.10.2 is used to form the two columns. The HTML code is

```
<div id="centerpage">  <!-- centered table with border -->
  <section id="main">  <!-- table row -->
    <nav class="leftnavbar">  <!-- first table cell -->
      <span class="self">Main Page</span> <!-- page id -->
      <a href="#">Products</a>
      <a href="#">Services</a>
      <a href="#">News</a>
      <a href="#">Contact Us</a>
    </nav>
    <article id="content">...   <!-- 2nd table cell -->
    </article>
  </section>
</div>
```

FIGURE 4.7: CSS-Defined Navigation Panel

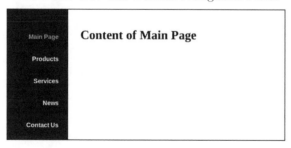

To make this work, the CSS code below provides the desired styles for the page. Note the navigation panel is the first cell (line 3) in a table layout (lines 1, 2, 7). The links inside will be displayed in block mode (line 5) with CSS-defined rollover effects (line 6). The link to the current page is inactive and is displayed in a different color to distinguish it from clickable links (line 4). This navigation feature helps identify the current page clearly. Links will be displayed without the usual underline, which makes the navbar clean looking, and a background color visually defines the left column.

```
div#centerpage                          /* 1 */
{  width: 80%; margin-left: auto;
   margin-right: auto; display: table;
   border: 2px solid darkblue;
}
section#main                            /* 2 */
{display:table-row; width:100%; }

nav.leftnavbar                          /* 3 */
{  display: table-cell;
   background-color: darkblue; color: white;
   font-family: Arial, Helvetica, sans-serif;
   font-weight:bold; width:20%; text-align:right;
   padding-right: 1em; padding-left: 0.5em;
   white-space: nowrap;
}

nav.leftnavbar > span.self              /* 4 */
{  display: block; color: orange;
   border: thin solid transparent;
   margin-top: 1.5em; padding-right: 0.5em;
}

nav.leftnavbar > a:link                 /* 5 */
{  display: block; text-decoration: none;
   color: white; border: thin solid transparent;
   margin-top: 2em; padding-right: 0.5em;
}

nav.leftnavbar > a:hover                /* 6 */
{border: thin solid white }

section#main > article#content          /* 7 */
{  display: table-cell; width: 70%;
   background-color: white; padding: 2em;
}
```

When a link on the navbar happens to link to the page being viewed (the current page), this link must be modified to become inactive and look distinct. This brings two important advantages:

1. The user will not get confused by clicking a link that leads to the same page.

2. The distinct-looking link name helps identify the current page.

The span with class="self" (line 4) fills this need by specifying a different style for such a current-page link. But this also means the navbar code must be modified slightly when it is placed on a different page whose link appears on the navbar. Such modifications can be automated when PHP is used to generate the HTML code (Section 5.16.2).

The rollover effect is achieved by displaying a white border box around the link under the mouse. In order for this to work without any layout movement of the displayed links, potentially affecting other elements on the page, we arranged a transparent border box around each link in the first place. This way, the rollover effect is achieved by changing the color of the border box instead of adding a border box (lines 5 and 6).

See the DWP website for the complete ready-to-run example (**Ex: NavPanel**).

4.12   Page Formatting Model

An understanding of the CSS page formatting model will help you use the style properties effectively. In this model, the formatted elements form a hierarchy of *rectangular boxes*:

- Inline box—A box that forms part of a line such as a group of words, an italic phrase, an `img`, or any phrasing element.

- Line box—A horizontal sequence of inline boxes placed on one full-width line.

- Block-level box—A box containing one or more line boxes and/or block-level boxes grouped into an independent vertical unit. The page layout consists of a vertical stack of block-level boxes.

For example, `<p>`, `<h2>`, `<ul>`, `<li>`, `<table>`, and `<div>` each generates a block box. And `<em>`, `<strong>`, `<a>`, `<span>`, `<img>`, and plain running text form inline boxes.

The *initial containing box* (root box) corresponds to the root of the document tree (`body` or `iframe` for this discussion). In this box, a vertical stack of block boxes of the same width as the root box is formatted.

A box is displayed at a position determined by the formatting of its containing element. The box for each element has a core area to display the content of the element and *padding, border*, and *margin* areas surrounding the content. The `width` and `height` style values refer to the content area of the element. However, the `background-color` of the element does show through in the padding areas. The margin area is transparent, and the `background-color` of the parent element shows through. The used width for the content area of an element is specified by the `width` property value (when possible) or computed using the following formula:

```
width-available-from-parent-box =
    left-margin + left-border + left-padding
    + used-width + right-padding + right-border + right-margin
```

Figure 4.8 shows the block formatting box.

FIGURE 4.8: Block-Level Box Model

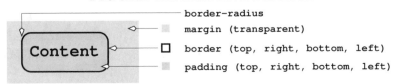

The CSS properties `min-width` and `max-width` (`min-height` and `max-height`) set the minimum and maximum for width (height) of an element to help the browser adjust the element when the browser window is resized.

An inline box is the same, but its height is always set only by the `line-height` property.

Except for table-related elements, the `min-width` (`max-width`) property specifies a minimum (maximum) width value for an element. This value provides a lower (upper) bound in determining the actual content display width of the element. Similarly, there are the `min-height` and `max-height` properties.

The `display` property governs what formatting mode to use for computing the layout of an element. Unless you explicitly change the `display` of an element, the default setting will govern its layout. Among the many display modes, the following are the most important:

- `display: block`—If a parent block element contains only block elements, then the child elements are formatted in *block mode*. Each block box is, by default, as wide as the available width (the content width of its parent box). Thus, the available width becomes the width of the child box computed by the preceding formula. Each block box is just high enough for its contents. The vertical separation between block boxes is controlled by the top and bottom margin properties of adjacent boxes. In normal flow, vertically adjoining margins are collapsed and become one margin whose height is the maximum of the two adjoining margins.

- `display: inline`—If a parent block element contains all inline/phrasing elements and text, then the child elements are formatted in *inline mode*. An inline box flows horizontally to fill the available line width and breaks automatically to form several lines when necessary. Horizontal margin, border, and padding will be respected.

Within a line, inline boxes (Figure 4.9) may be aligned vertically ac-

FIGURE 4.9: Inline Layout Model

cording to their `vertical-align` properties. The rectangular area that contains the inline boxes on a single line is called a *line box*. The line box is just tall enough to contain the inline boxes plus one half *leading* (leading = `line-height` − `font-size`) on top and bottom. Thus, the height of a text inline box is exactly `line-height`, and the height of an `img` inline box is the height of the image plus the leading. Top and bottom margins, borders, and paddings do not contribute to the line box height. A browser may render these outside the line box or clip them.

- `display: inline-block`—An inline block shrinks to fit the size of its content and the resulting block is flowed inline. *Replaced elements* (Section 4.13.4) such as image, audio, and video are formatted in as inline blocks.

- `display: list-item` —Lists (`ul` and `ol`) are formatted with a marker in front of one or more block boxes.

- `display: none`—Contents are not displayed.

To see how all this works, let's look at an example where lines involve text and images:

```
<h3><span style="color: blue">Alignments with      (A)
respect    to the whole line:</h3>
<p><span style=                                 (B)
"color:blue">vertical-align: top</span>
<img alt="preface" src="u_preface.gif"
    style="vertical-align: top" /> in a line with
another <img alt="loon.gif" src="loon.gif"
style="vertical-align: baseline" /> image.</p>
```

Figure 4.10 shows the box hierarchy for this example. Two block boxes of different heights are formatted inside the containing block box. The first block box corresponds to `h3` (line A). The second block box corresponds to `p` (line B). It encloses a line box that contains five inline boxes vertically aligned as indicated. Their bounds are outlined by dotted lines.

If a flow/block element contains both flow and phrasing elements, block boxes will be generated to contain the inline boxes, resulting in block mode formatting. Thus, the code

```
<section>Mixed inline and <p>block elements</p></section>
```

FIGURE 4.10: Visual Formatting Model

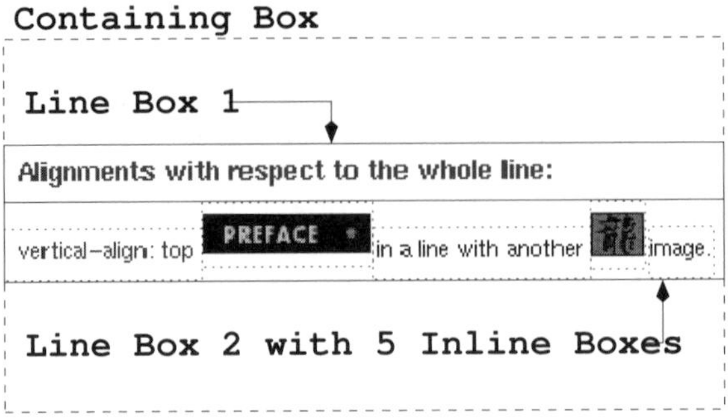

is treated as if it were

```
<section><div>Mixed inline and </div>
<p>block elements</p></section>
```

Two implications of the inline mode are:

1. Unless there is sufficient line box height, any padding or border around inline elements may overlap adjacent lines. Even inline box content may bleed to adjacent lines.

2. A line box adds extra top and bottom spacing equal to one half leading to the vertical extent of its inline boxes. This may result in extra white space above and below inline images, for example.

This discussion pertains to the normal flow of CSS visual formatting. Floating and positioning can alter the normal flow as we will see in Section 4.23.

Also, other values for `display` are available. We have already seen (Section 4.10.2) the use of `display` to render an element in `table`, `table-row`, or `table-cell` mode.

4.13   Margin, Border, and Padding

The size of the margin is set with the properties

`margin-top:` *length*	`margin-right:` *length*
`margin-bottom:` *length*	`margin-left:` *length*

If *length* is `auto`, then it is determined by the available space and by the value of the opposite margin. For example,

```
<div style="width:700px; background-color:#def">
<img src="img/paul.jpg" alt="paul" style="display: block;
          margin-left: auto; margin-right: 5px"/>
```

pushes the image to the right side of the containing `div`.

Padding and border are set similarly, and shorthand notations are available. For example,

```
padding: 2px
```

sets the padding on all four sides to 2 pixels. And

```
margin: 50px 10% 50px 10%          /* top right bottom left */
```

sets the four margins. With such shorthand, there can be from one to four values:

1. One value: for all sides

2. Two values: the first value for top and bottom; the second value for right and left

3. Three values: for top, left-right, and bottom

4. Four values: for top, right, bottom, and left, respectively

These may seem confusing, but there is a uniform rule: The values go in order for top, right, bottom, left. If fewer than four values are given, the missing values are taken from the opposite/available side. If a margin is negative, it causes the box to invade the margin of an adjoining box.

The padding area uses the same background as the element itself. The border color and style are set with specific border properties. The margin is always transparent, so the parent element will shine through. The vertical margin between two displayed elements (e.g., a paragraph and a table) is the maximum of the bottom margin of the element above and the top margin of the element below. Sometimes this behavior is termed *margin collapsing*.

The horizontal width of a box for an element is the element width (i.e., formatted text or image) plus the widths of the padding, border, and margin areas. The vertical height of a box is determined by the font size (or height of image) and the line height setting.

4.13.1   Border Styles

For a border, you can set its style in addition to width and color. Values for the `border-style` property can be `dotted`, `dashed`, `solid`, `double` (two lines), `groove` (3-D groove), `ridge` (3-D ridge), `inset` (3-D inset), or `outset` (3-D outset). The three-dimensional varieties use the border color to create a three-dimensional look. The `border-style` property can take one to four values for the four sides: top, right, bottom, left. A missing value is taken from the opposite/available side.

Values for the `border-width` can be `thin`, `medium`, `thick`, or a length. The value for `border-color` can be any RGB color (no rgba) or `transparent`. Both properties can take one to four values, as usual.

You can also set `border-top-width`, `border-right-width`, `border-bottom-width`, and `border-left-width` independently.

The properties `border-top`, `border-right`, `border-bottom`, and `border-left` each can take a length, a border style, and a color. You may specify one, two, or all three for the particular side of the border. We used

```
border-left: thin solid #000;
```

in our navbar example. The property `border` takes the same three values and applies them to all four sides.

FIGURE 4.11: Four Borders

Figure 4.11 shows a `div` block box with all four borders showing resulting from the code (**Ex:** `Border`).

```
div#border
{   width:40px; height:40px;
    background-color: white;
    border-bottom: 15px solid red;
    border-right: 15px solid blue;
    border-left: 15px solid darkblue;
    border-top: 15px solid darkred;
    display: inline-block;
    vertical-align: middle; margin-left: 2em;
}
```

Just outside of the border, an outline can also be displayed. Browsers often use a faint outline to indicate the extent of an element for easy clicking. You can control the outline display with the properties `outline-width`, `outline-color`, and `outline-style`. Settings for these are the same as border styles. Normally, the outline is displayed at the border but you may use `outline-offset` to move it further outside the border.

We have seen a vertical navbar with rollover effects coded with border styles in Section 4.11.

4.13.2   Elastic Banner

Let's apply the box model to design an elastic banner (Figure 4.12) that responds well with window resizing. The example combines several styling features including `inline-block` to make a fluid banner work. The techniques

FIGURE 4.12: An Elastic Banner

can be applied in general to make banners or other content scale well with changing window sizes.

Let's look at the HTML and CSS code for the elastic banner (**Ex:** `Elastic`) and explain how it works.

```
<section class="banner">                              (1)
<div id="red"></div><div id="text">
<div class="letter">D</div>
<div class="letter">Y</div><div class="letter">N</div>
<div class="letter">A</div><div class="letter">M</div>
<div class="letter">I</div><div class="lastletter">
C</div></div><div id="blue"></div></section>         (2)
```

Here are the features of the elastic banner and how they are achieved with the code:

- Banner structure—A class `banner` `section` containing a `red`, `text`, and `blue` `div`. The characters are in individual `letter` `div`s (lines1 and 2).

- Banner position—A 50-px high, full-width strip placed at page top with no margin, border, or padding (lines 3).

- Banner box layout—A `div` (line 4) containing three inline blocks (red, text and blue `div`s) with no wrapping at white spaces and invisible overflow. The linebreaks in the HTML above is for display on this page. The actual code contains no unnecessary white space.

```
section.banner
{ border:0; margin:0; padding:0;            /* 3 */
  white-space: nowrap; overflow: hidden;    /* 4 */
}
```

- Inline blocks—All blocks inside the banner section are inline with the same (50-px) height (line 5). The red (blue) block has a 19.1% width and a white right (left) border (lines 6 and 7).

- Letter blocks—Child blocks of the text `div` which sets the green background, width, and font attributes. The widths of the red, blue, and text blocks add up to 100% and follow the golden ratio. The line height is equal to the block height to center the letters vertically (line 8).

- Letter spacing—Each letter block has the same 8% width so the letter spacing also expands and shrinks in response to window resizing (line 9).

```
section.banner div                        /* 5 */
{ display: inline-block; height: 50px;
  vertical-align: top;  }

div#red                                   /* 6 */
{  width: 19.1%; background-color: red;
   border-right: solid white 2px; }

div#blue                                  /* 7 */
{  width: 19.1%; background-color: blue;
   border-left: solid white 2px; }

div#text                                  /* 8 */
{  width: 61.8%; background-color: green;
   color: white; font-family: Courier;
   line-height: 50px; font-size: 34px;
   font-weight: bold; text-align: right;
   overflow: hidden; }

div#text > div.letter {  width: 7%; }     /* 9 */
div#text > div.lastletter
{ width: 7%;  margin-right: 8px; }
```

You can experiment with the full example (**Ex: Elastic**) online at the DWP website.

4.13.3   Border Radius

We have already seen usage of margin, border, and padding in many places. We have also seen examples of curved borders obtained through the border-radius property (Section 2.21).

The curve for a rounded corner is defined by an ellipse. To specify a rounded corner use one of the following rules:

```
border-top-left-radius:  x y
border-top-right-radius:  x y
border-bottom-left-radius:  x y
border-bottom-right-radius:  x y
```

giving the lengths for the x radius and y radius. The y is optional if it is the same as x (circular).

To easily specify all four corners, you may use

```
border-radius:  x1 x2 x3 x4  /  y1 y2 y3 y4
```

for the x and y radii of the top-left, top-right, bottom-right, and bottom-left corners. The 1–4 x values define the x radii for all four corners, and the 1–4 y values, preceded by a /, define the y radii for all four corners. If the y values are omitted, they become the same as the corresponding x values. For the set of x or y values, the following applies: If bottom-left is omitted, it is the same as top-right; if bottom-right is omitted, it is the same as top-left; and if only one value is supplied, it is used to set all four radii equally. Note the maximum

FIGURE 4.13: Circle Using Border Radius

value for an x (y) radius is half the total width (height) of the element block box, including any padding and border. For example **Ex:** `Circle`, the circle in Figure 4.13 is produced with the following HTML code:

```
<body style="background-color: #def">
<h1 style="display: inline">CSS Circle</h1>
<div id="circle"></div></div>
```

and using the CSS

```
div#circle
{    width:40px; height:40px;
     background-color: red;
     border: 30px solid blue;
     border-radius: 50px;        /* half of 40+30+30 */
     outline: thin dotted black;
     display: inline-block;
     vertical-align: middle; margin-left: 2em;
}
```

4.13.4   Replaced Elements

A replaced element is an element that is outside the scope of the CSS formatter. Such elements include `img`, `audio`, `video`, `embed`, and `object`. We know that the `img` element is normally replaced by an image specified by its `src` attribute. A replaced element has an intrinsic width, an intrinsic height, and an intrinsic ratio. Replaced elements are normally laid out using the `inline-block` mode.

4.14   Preformatted Content

Sometimes we want to avoid automatic flowing and formatting to preserve the original format or to avoid line breaks in undesirable places.

The style rule `white-space: pre` preserves white space in preformatted text. The style declaration is useful for preformatted inline text such as

```
<p>Here is <span style=
   "white-space: pre">a     hint    :-)</span></p>
```

For preformatted text with multiple lines, the `<pre>` element is the right choice.

Use the rule `white-space: nowrap` to disallow line breaks unless explicitly called for by `<br />`. Also, `white-space: pre-line` collapses sequences of white spaces into a single white space, wraps text when necessary, and on line breaks. The style `white-space: pre-wrap` preserves white space and wraps text when necessary, and on line breaks. The style `word-wrap: break-word` can be set to allow breaking of a long word across lines.

4.15   Controlling Content Overflow

When the width or height of a block containing box is insufficient for the content inside, overflow occurs. Overflow often happens when the browser window is not wide or tall enough and the content becomes visible outside the containing box. This destroys your page layout and must be avoided.

The properties `overflow-x` and `overflow-y` specify how to handle width and height overflow, respectively. Possible values are

- `visible`—makes the overflowing content visible outside the box

- `hidden`—makes the overflowing content invisible outside the box

- `scroll`—adds a scroll bar to begin with

- `auto`—adds a scroll bar only when overflow takes place

- `inherit`—inherits the overflow setting from the parent element

The position for a scroll bar is between the border and the padding of an element.

The shorthand `overflow` property lets you specify `overflow-x` and `overflow-y` at once.

The `text-overflow` property gives you the ability to control how overflow for running text is handled (**Ex: TextOverflow**). The choices are `clip` (hide the overflowed text), `ellipsis` (display '...' instead), or `string` *some-string* (display the given string instead of an ellipsis).

The CSS3 `resize` property controls how users can resize block-level and replaced elements, table cells, and inline blocks. The property can be set to `none` (default), `both`, `horizontal`, `vertical`, or `inherit`.

4.16   Styled Buttons

Let's apply CSS to build buttons that can be the display representation for a link (a), a submit button (Section 5.6), or almost any other element.

FIGURE 4.14: Sample CSS Buttons

The three buttons in Figure 4.14 are defined using border radii, color gradients, drop shadows, as well as :hover and :active actions (**Ex: Buttons**). The HTML code is as follows:

```
<section>
 <a href="#" class="btn black">Corners</a>
 <a href="#" class="btn ends black">Ends</a>
 <a href="#" class="btn circle green"><div id="label">Go</div></a>
</section>
```

Let's begin by looking at the first button, the one with rounded corners. The btn class provides the following styles:

```
.btn
{ display: inline-block;        /* displayed inline (1) */
  vertical-align: baseline;
  text-align: center;          /* label styles    (2) */
  text-decoration: none;
  font: 14px Arial, Helvetica, sans-serif;
  padding: .5em 2em .55em;      /*                 (3) */
  border-radius: .5em;          /* rounded corners (4) */
  outline: 0;                   /* no outline      (5) */
  box-shadow:
     2px 4px 6px rgba(0,0,0,.4); /* drop shadow     (6) */
}
```

```
.btn:active { position: relative; top: 1px; } /* clicked  */
```

A button is displayed as an inline block (line 1) with a centered label (lines 2–3) and rounded corners (line 4). We do not want any outline box (Section 4.13.1) that may result from clicking (line 5). A drop shadow (line 6; Section 4.17) and a gradient background (line 7; Section 4.20) color (line 7) help create a three-dimensional look:

```
.black
{ color: #ddd;
  background:-moz-linear-gradient(top,#aaa,#000) /* (7) */
}
.black:hover
{ background:-moz-linear-gradient(top,#888,#000) /* (8) */
}
.black:active
{ background:
  -moz-linear-gradient(bottom, #888, #000) }     /* (9) */
```

On mouseover, we darken the background color (line 8) by a shade and when the mouse is clicked, we reverse the gradient direction (line 9).

As for the second button, the one with semi-circular ends, we simply need to increase the border radius to half the button height or more.

```
.ends { border-radius: 2em; }
```

The green circular Go button takes a bit more to do. By shaping the border of a square block box and canceling any padding (line 10–11), we make a circle. A centered div positions the button label (line 12):

```
.circle                        /* (10) */
{ width: 6em;   height: 6em;
  padding: 0px;
  border-radius: 4em;          /* (11) */
  vertical-align: middle;
}

div#label                      /* (12) */
{ display: block;
  font-size: 2em;
  margin-top: 0.85em; margin-left: .20em;
}
```

Now we use a radial gradient background color for the circular button (line 13), a color shifting for mouseover (line 14), and a radial gradient origin shifting effect (line 15) for clicking:

```
.green                         /* (13) */
{ color: #fff;
  background: -moz-radial-gradient
    (40% 35%, farthest-side, #a2ffa2, #0d0);
}

.green:hover
{ background: -moz-radial-gradient(40% 35%,  /* (14) */
```

```
                farthest-side, #8f8, #0b0);    }
```

```
.green:active
{ background: -moz-radial-gradient(30% 35%,  /* (15) */
               farthest-side, #8f8, #0b0);    }
```

See Section 4.17 and Section 4.20 for more on drop shadows and color gradients, as well as making these effects work for all major browsers. The complete example (**Ex:** Buttons), ready to run, can be found at the DWP website.

4.17 Drop Shadows

CSS makes it easy to add drop shadows to texts and block boxes with the text-shadow and box-shadow properties. We have seen some use of the latter to create nice-looking and responsive buttons in the previous section (Section 4.16).

4.17.1 Box Shadows

The general syntax for box-shadow is

box-shadow: *dx dy bl s color*

The shadow, in the specified *color* (default black), is displayed with an x-offset *dx* and a y-offset *dy* (shifts from the element position). Using the offsets, you can control placement of the shadow to indicate the desired direction of light. The blur radius *bl*, which gives the width of the blurred shadow edge, and the shadow spread *s*, which scales up/down the size of the shadow, are both optional. The default blur is no blur, and the default spread is no spread. An outside shadow can provide a floating or raised look. An inside shadow, obtained by adding the keyword inset to the end, can also provide a protruding or sunken look.

The buttons shown in Figure 4.15 are styled with

```
a.out { box-shadow: -3px -3px 0 3px #888 inset; }
```

```
a.in { box-shadow: 3px 3px 0 3px #888 inset; }
```

FIGURE 4.15: Inset Shadow Buttons

The shadow color in this example is grey (#888), but many prefer `rgba` values for shadow color because it can better blend in with any background against which the shadow is displayed.

The complete example (**Ex: `Inset`**) can be found at the DWP website.

Multiple shadows can be obtained by simply giving `box-shadow` a list of shadow values separated by commas.

4.17.2   Text Shadows

To get text shadows, use

`text-shadow:` *dx dy bl color*

and the values are the same as for `box-shadow`.

Let's apply text shadows to achieve some interesting effects on text display (**Ex: `TextShadow`**). We can create an outline (a dark border) around each letter by placing a 1-px black shadow with a $\pm$1-px offset in both the x-and y-directions (Figure 4.16, left). The CSS code for four desired text shadows

FIGURE 4.16: Outline and 3D Effects

follows.

```
span.outline      /* Text Outline */
{  color: #aaa;
   text-shadow: -1px 0 black, 0 1px black,
                1px 0 black, 0 -1px black
}
```

The 3D effect (Figure 4.16, right) is achieved by a series of increasingly darker and further offset text shadows.

```
span.threeD       /* 3D Effect */
{  color: #aaa;
   text-shadow: .5px -.5px #999, 1px -1px #888,
   1.5px -1.5px #888, 2px -2px #777,
   2.5px -2.5px #666, 3px -3px #555;
}
```

Figure 4.17 and Figure 4.18 show different text embossing/stamping effects. Against a light gray background (line B), we sandwich a slightly darker gray text (line A) with a 1-px white border to the left and a 1-px dark gray border to the right (line C) to produce the left side of Figure 4.17.

```
span.light
{  color: #ccc; font-size: 25px;              /* A */
   background-color: #d1d1d1;                 /* B */
}


span.emboss1
{text-shadow: -1px -1px white, 1px 1px #333 } /* C */


span.emboss0
{text-shadow: 1px 1px white, -1px -1px #444 } /* D */
```

A darker gray to the left and white to the right (line D) produces a stamped look as shown on the right side of Figure 4.17. As for the dark embossing, we

FIGURE 4.17: Embossing Effects

FIGURE 4.18: Impress Effects

used the following style:

```
span.dark
{  background-color: #474747;
   font-size: xx-large; letter-spacing: -2px;
   font-family: Tahoma, Helvetica, Arial, Sans-Serif;
}


span.press0
{  color: #666;
   text-shadow: 0px 2px 3px #2a2a2a;
}


span.press0:hover {  color: #777; }
```

for text on the left side of Figure 4.18 and the style

```
span.press1
{  color: #222;
   text-shadow: 0px 2px 3px #555;
}
```

for text on the right-hand side of the figure.

4.18   Borders by Custom Images

Web developers often want to create custom layout graphics, fancy boxes, or picture frames for their Web project. This could be done by piecing together images placed in a nine-cell table grid (Figure 4.19) using the `table` element or table styles. Individual images are supplied for each part of the borders and corners. An additional background image may be supplied to the middle cell (number 5), which is for any content to be framed inside.

FIGURE 4.19: Nine-Cell Grid

CSS3 makes this much easier by allowing you to define the border of any box with a reference image. The sides of the reference image are used to draw the border for the box, and the center of the reference image can serve as the background image for the box.

The `border-image` style property is used to specify the reference image and the nine parts:

```
border-image: url(imgfile) offsets border-span center-span
```

In a style sheet, a `url` is either absolute or relative to the location of the style sheet. The *offsets* indicates the t, r, b, l values relative to the respective borders (Figure 4.20). The offset can be given in percent (`10%` for example) or in pixels (`15` for example). Specify less than four, the other values are inferred as usual.

The image parts from the reference image become tiles used to fill the borders of the element box. The two span parameters indicate how to cover the box borders and center, respectively. Each span parameter can be `stretch` (scale to fit), `repeat` (tile to cover), or `round` (scale the tiles so an integral number is used to cover). Let's look at an example.

Figure 4.21 shows a stop button styled with a border image. The HTML code for this example (**Ex:** BorderImg) is

```
<div>This button <div class="redbutton">STOP</div>
uses the border image
<img src="redbutton.jpg" alt="red button image" /></div>
```

And the CSS code is

FIGURE 4.20: Image Borders

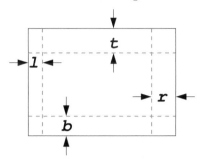

FIGURE 4.21: Border Image Demo

```
div.redbutton
{   display: inline-block;   vertical-align: middle;
    color: white;   font-weight: bold;
    font-size: larger; width: 3.5em;
    text-align: center;
    border-width: 8px 6px;
    border-image: url(redbutton.jpg) 20 18 stretch;
    -moz-border-image: url(redbutton.jpg) 20 18 stretch;
    -webkit-border-image:url(redbutton.jpg) 20 18 stretch;
}

div img { vertical-align: middle;   }
```

Browsers may still be catching up with the CSS3 specifications on image borders.

4.19   Background Colors and Images

The background of any HTML element, transparent by default, can be set to provide a desired backdrop for displaying the element content. The purpose is to provide a visual frame, color contrast for readability, or flexible fill of negative space in a fluid page layout.

The `background-clip` property clips the background and can be set to `border-box` (default), `padding-box`, or `content-box` (**Ex:** Bgclip).

Use the `background-color` property to set a solid background color, which can be completely opaque or have degrees of transparency via an `rgba` color (Section 2.12.4).

To set the background to that defined by an image, use the `background-image` style property . For example, to make `myblue.gif` the background image of a `section` child of an `article`, you may use

```
article > section td.right
{   background-image: url(myblue.gif)   }
```

If necessary, the given image is automatically repeated horizontally and vertically to fill the entire background. To gain control over the repetition, you can set `background-repeat` to `repeat` (horizontally and vertically, the default), `repeat-x` (only horizontally), `repeat-y` (only vertically), or `no-repeat`.

Pictures and sketches as backgrounds can also help create fluidity in a page layout. The element automatically crops the background image depending on the space available to it.

Setting background image for the `body` or `html` element is another way to customize the look of a site. You can use a well-designed tile pattern (Figure 4.22) to make a margin, a watermark, or a special paper effect for your

FIGURE 4.22: Background Patterns and Textures

pages. Patterns 1 and 2, when repeated, make vertical and horizontal stripes, respectively. The other patterns work in both directions. You can find many background patterns and textures online. Be careful using such background tiles for entire pages. They often tend to distract attention and annoy visitors.

The `background-position` property lets you specify the starting position of a background image: `top left`, `top center`, `top right`, `center left`, `center center`, `center right`, `bottom left`, `bottom center`, or `bottom right`. You can put the starting position anywhere on the page with

```
background-position: x y
```

where x and y can be a length, (e.g., `50px`) or a percentage (`0% 0%` is top left and `100% 100%` is bottom right). The background position references the padding box, unless you set `background-origin` to `border-box` or `content-box`.

Combining starting position with repetition gives you many ways to create backgrounds. For example,

```
body
{   background-image: url(tile4.gif);
    background-repeat: repeat-x;
    background-position: bottom left;
}
```

makes a tiled bottom margin at the end of your page.

A background image also has an *attachment* property, `background-attachment`, with either a `scroll` (default) or `fixed` value. With a `fixed` attachment, a background image remains in place as the page content scrolls. Thus, you can place a background watermark or logo, for example, at a fixed position on the screen. Thus, the code

```
body
{   background-image: url(tile4.gif);
    background-repeat: repeat-y;
    background-position: top right;
    background-attachment: fixed;
}
```

gives you a tiled right margin that stays in place while you scroll the page.

Instead of repeating a small image pattern to create the whole background, you can also consider using a big enough image to blanket the background. In such a case, the property

`background-size: ` $w\ h$ ` | ` $w\%\ h\%$ ` | cover | contain`

is useful. If only one of width and height is given, the other is computed to preserve the image aspect ratio. The `cover` (`contain`) option scales down (up) the image to fit in the content area. The DWP website top banner uses this technique.

4.20   Color Gradient

We already know that a background can be provided by an image file. But it can also be provided by a color gradient. Basically, a color gradient image consists of a *start color*, a smooth transition, and a *stop color* along a *gradient line* that passes through the center of the element. Among many other uses, a gradient background can make buttons look three-dimensional, as we have seen in Section 4.16. And it can nicely bridge the background gaps upon window resizing.

4.20.1   Linear Gradients

A linear gradient varies color along a given direction, and a radial gradient varies color from a given center point circularly or elliptically outward in all directions.

FIGURE 4.23: Gradient Concepts

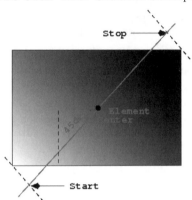

A linear gradient background is specified as follows:

```
background-image: linear-gradient(dir, start, stop)
```

where the arguments are

- *dir*—gives the gradient line direction: `top` (from top edge to bottom edge), `bottom` (bottom to top), `left` (left to right), `right` (right to left), `top left` (top left to bottom right diagonally), and so on. Alternatively, it can take on a degree (`45deg`, for example; Figure 4.23) measured from 12 o'clock (`0deg`) clockwise. Thus, `0deg` is up and `90deg` is right and so on.

- *start*—the start color and an optional starting position (length or percent measured from the starting point defined in Figure 4.23) to begin gradient fading.

- *stop*—the stop color and an optional stopping position where to arrive at the stop color.

The four sample gradient backgrounds shown in Figure 4.24 are produced with the following style code:

```
background-image: /* left to right, red to white */
linear-gradient(left, #f00, #fff 90%)

/* bottom-left to top-right, green to white */
background-image:
linear-gradient(bottom left, #0f0, #fff 90%);

background-image: /* 45-degree up, blue to white */
```

```
linear-gradient(45deg, #00f, #fff 90%)

/* top to bottom, opaque yellow to transparent */
background-image:
linear-gradient(top, rgba(255,255,0,1),
                     rgba(255,255,0,0) 90%)
```

FIGURE 4.24: Color Gradients

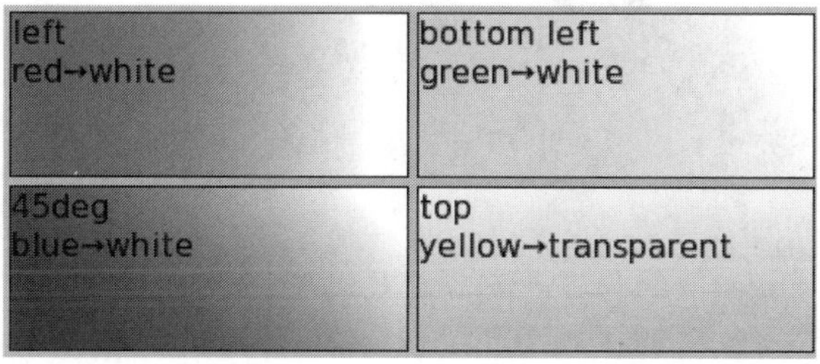

In all four cases, we used a 90% stop position to make the effects easier to see. The full-color example (**Ex: Gradient**) can be found at the DWP website.

It is possible to give intermediate stop colors at specified positions simply by adding them as additional color-position arguments after the start color.

4.20.2 Radial Gradients

A radial gradient stars with a color at a center point and smoothly spreads out toward a stop color in ever larger concentric ellipses[2]. The center point need not be inside the display box of the element itself. The exact *final gradient shape* where the stop color is achieved can be specified explicitly or inferred from the geometry of the element box and the center point position.

The radial gradient syntax is

radial-gradient(*pos, shape, start, stop, ...*)

where *pos* is a **background-position** value as defined before. A center point given by *pos* may lie outside the display box of the element itself. The start and stop colors are given the same way as in linear gradient. The final gradient *shape* can be given explicitly with a pair of space-separated lengths or percentages. The shape can also be given with a pair of keywords:

circle | ellipse *size*

Possible values for *size* are closest-side (or contain), closest-corner, farthest-side, and farthest-corner (or cover).

---

[2]Circles are special-case ellipses.

The final gradient shape will reach the specified *size* while keeping its circular or elliptical shape. The four radial gradients in Figure 4.25, each in a 200×80 box, are produced with the following style code (**Ex:** `Radial`):

FIGURE 4.25: Radial Gradients

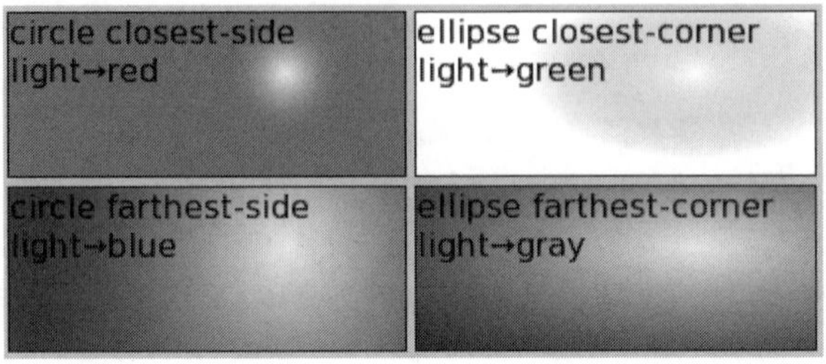

```
background-image: radial-gradient(
   140px 30px, circle closest-side, #fee, #f00)

background-image: radial-gradient(
   140px 30px, ellipse closest-corner, #efe, #0f0, #fff)

background-image: radial-gradient(
   140px 30px, circle farthest-side, #eef, #00f)

background-image: radial-gradient(
   140px 30px, ellipse farthest-corner, #eee, #333)
```

Now let's apply radial gradient to make a rainbow (Figure 4.26). A 400×150 `div` with a black border is used as the frame for our rainbow (**Ex:** `Rainbow`).

```
div.rainbow
{ width: 400px; height: 150px;
  border: thin solid black;
  background-image: radial-gradient(
    200px 320px,            /* center position  */
    circle farthest-side,  /* circular shape    */
    #4d76ce 63%,,    /* sky    */
    #987ee3 65.3%,   /* purple */
    #a772f8 65.8%,   /* indigo */
    #4d76ce 68.2%,,  /* sky    */
    #6f6 72%,        /* green  */
    #f3f320 76%,     /* yellow */
    #f8cb72 80%,     /* orange */
    #ffa0a0 84%,     /* red    */
    #4d76ce 90%);    /* sky    */
```

}

FIGURE 4.26: A CSS Rainbow

We use a circular gradient shape with a center point 170-px below the midpoint of the bottom edge of the display box. The first color is sky blue and it stops at 63%, where a purplish color begins, stopping at 65.3%, and so on. We return to the sky color at 90%. A total of nine color stops are used. See the exercises for ideas on using transparency and adding a linear gradient for the sky to make this rainbow even more realistic.

4.21   Browser Support for CSS

Major browsers support most CSS1 and CSS2 features. When it comes to certain CSS3 features, such as multicolumn formatting (Section 4.5) or gradient background image, support may still be less than uniform.

However, in many cases, all you need to do for different browsers is to add a browser-specific prefix to the feature in question and maybe also use a slightly different syntax.

For example, for `linear-gradient`, you can use the following prefixes:

```
/* Safari 5.1+, Mobile Safari, Chrome 10+ */
background-image: -webkit-linear-gradient(...);

/* Firefox 3.6+ */
background-image: -moz-linear-gradient(...);

/* IE 10+ */
background-image: -ms-linear-gradient(...);

/* Opera 11.10+ */
background-image: -o-linear-gradient(...);
```

Section 4.5 provided an example to combine all these different notations to make your style work across all major browsers. Updated cross-browser coding information can be found on the **resources** page at the DWP website.

The property **background** is a shorthand to specify many background properties at once, in any order.

background: *color image position repeat size attachment*

Any missing values will be set to the default values.

4.22   At-Rules

At-rules are instructions to the CSS parser. They include

- `@import`— At the beginning of a style sheet file (before all style rules), you can include other style files with

`@import url("`*target-sheet-url*`"); @import "`*local-style-sheet*`";`

It is also possible to add media type info at the end of an `@import`. For example,

`@import" "view.css" screen, projection;`

- `@media`—You can use

`@media` *type* ... `{` *rule set* `}`

to target rule sets for particular media types including `all`, `aural` (speech and sound synthesizers), `braille` (braille tactile feedback devices), `embossed` (paged braille printers), `handheld` (small or handheld devices), `print` (printers), `projection` (projectors), `screen` (computer screens), `tty` (teletypes and terminals), and `tv` (television-type device). The W3C proposed CSS3 *media queries* (Mid 2012), allowing you to add *media feature conditions* to media types, thus making it easier to apply different CSS rules to the same media type with different features. For example, you will be able to differentiate a large desktop screen from a small touch screen on a mobile device (Section 13.4).

- `@font-face`—You can supply your own WOFF fonts for your webpage with this at-rule. For example,

```
@font-face
{  font-family: 'Myfontname';
   font-style: italic;
   font-weight: bold;
   src: url('http://.../myfont.ttf');
}
```

WOFF (*Web Open Font Format*) was developed by Mozilla in concert with other font organizations. It uses a compressed version of the same table-based sfnt structure used by TrueType, OpenType, and Open Font Format, but adds metadata and private-use data structures. You can pick up free fonts from `fontsquirrel.com`, among other places, for your website and use `@font-face` to deploy them.

- `@page`—For printed pages, you can specify margins for all pages, all left-hand pages, all right-hand pages, or for the first page. For example,

```
@page { margin: 2.4cm; }
@page :first { margin-top: 7.8cm; }
@page :left { margin-left: 4.8cm; }
@page :right { margin-right: 4.8cm; }
```

4.23 Explicit Element Positioning

CSS uses three *positioning schemes* to place block and inline boxes for page layout:

1. *Normal flow*—This is the normal way inline and block boxes are flowed in the page layout.

2. *Floating*—A floating element is first laid out according to the normal flow and then shifted to the left or right as far as possible within its containing block. Content may flow along the side of a floating element.

3. *Explicit positioning*—Under explicit positioning, an element is removed from the normal flow entirely (as if it were not there) and assigned a position with respect to a containing block.

Use the `position` property to explicitly position an element anywhere on the page layout:

- `position: static`—The element follows the normal flow. This is the default positioning scheme of elements.

- `position: relative`—The element is laid out normally and then moved up, down, left, and/or right from its normal position. Elements around it stay in their normal position, ignoring the displacement of the element. Hence, you must be careful not to run into surrounding elements, unless that is the intended effect. Relative positioning can supply fine position adjustments. For example (**Ex: Relative**),

```
li.spacing
{ position:relative; left: 1em } /* move right 1em */
li.morespacing { position:relative; left: 3em }
```

FIGURE 4.27: Relative Positioning

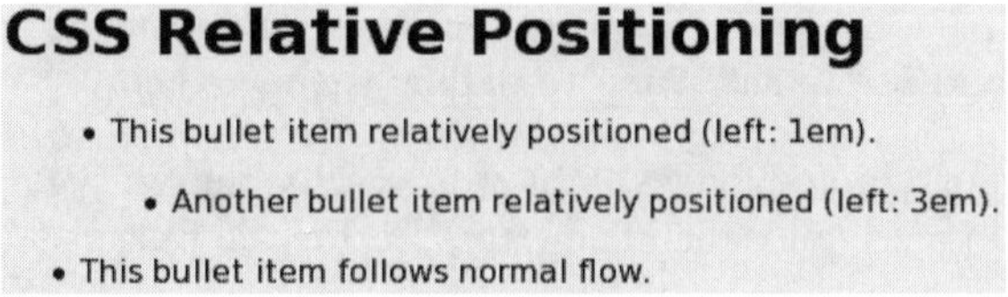

CSS Relative Positioning

- This bullet item relatively positioned (left: 1em).
  - Another bullet item relatively positioned (left: 3em).
- This bullet item follows normal flow.

moves the bullet items to the right in the list (Figure 4.27):

```
<ul><li> <p class="spacing">This text relatively
    positioned (left: 1em).</p></li>
<li><p class="morespacing">Another line relatively
    positioned (left: 3em).</p> </li>
<li> <p>This item follows normal flow.</p></li></ul>
```

Relative positioning is also useful in refining vertical alignments of adjoining elements.

- `position: absolute`—This property applies only to block elements. The element is taken out of normal flow entirely and treated as an independent block box. Its position is given by the `left`, `right`, `top`, and `bottom` properties relative to the first explicitly positioned containing element (with a position other than static) or the html element. The `bottom` (`right`) setting gives the distance between the bottom edge (right edge) of the block box to the bottom (right edge) of the containing box. In this case, ensure there is a well-defined bottom or right edge (Figure 4.28).

FIGURE 4.28: Absolute Positioning

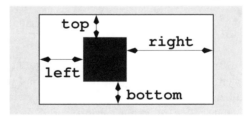

For example, Figure 4.28 shows a centered outer div with `position:relative` containing an inner div with `position:absolute`. The HTML code for this example (**Ex:** Absolute) is

```
<div class="center"><div id="ab">  </div></div>
```

And the CSS is

```
div.center
{   width:200px; height:100px;
    border: thin solid black;
    margin: auto; background-color: white;
    position: relative;
}

div.center div#ab
{   width: 50px; height: 50px;
    background-color: darkblue;
    position: absolute; left: 50px; top: 25px;
}
```

- position: fixed—This property applies only to block elements. An element with fixed positioning is placed relative to a display medium rather than a point in the page. The medium is usually the screen viewport or a printed page. The familiar browser window is the on-screen viewport that can usually be resized and scrolled. A fixed element will be at the same position in the browser window and will not scroll with the page. A block fixed with respect to a printed page will show up at the same position on every page.

For example (**Ex:** FixedNav), we can have a left navbar that does not scroll with the page (Figure 4.29) using

```
@media screen
{ div.navbar
  { background-color: #a70; width: 170px;
    border: 1px #630 solid; padding: 3px;
    text-align: center; position: fixed;
    left: 20px; top: 120px; }
}
```

FIGURE 4.29: Fixed Positioning

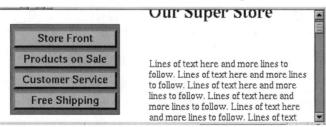

To ensure the navbar will not be repeated on each printed page, consider also including the following statement:

```
@media print  /* for printed pages */
{  div.navbar { position: absolute; }   }
```

It is a good idea to take printing into account when using fixed positioning primarily for the screen.

If no explicit positions are given (via `left`, `top`, etc.), a fixed-position element will first be placed according to normal flow and then fixed to the display medium.

4.23.1   Fixed Positioning, Stack Levels, and Visibility

With fixed positioning, you can easily position page elements such as top banners and navigation bars at fixed positions so they will always be visible and usable no matter how the page is scrolled (**Ex: Fixed**).

In practice, fixed positioning is often used to create pull-down menus for navigation or other purposes. The menus are usually kept hidden (`display: none`) at the end of the webpage and positioned just below its anchor on the menu bar on mouseover. The menu disappears when you leave it or click on its menu bar anchor. The actions require JavaScript control and will be discussed in Section 6.10.

With positioning, elements may overlap. To manage overlapping elements, each element also has a `z-index` property that designates its *stack level* on the z-axis. The x- and y-directions are in the viewport plane, and the z-direction is the depth perpendicular to the viewport plane. An element with a larger stack level is in front of one with a smaller stack level. The element in front will obscure any elements behind it. Because the initial value for `background-color` is `transparent`, the element behind may show through the background areas of the element in front. Setting `background-image` or `background-color` to anything else, nothing will show through.

The `opacity` property controls the transparency of the whole element, including its foreground and background. An opacity value of 0 gives total transparency and a value of 1 gives complete opacity. Figure 4.30 shows an example (**Ex: Caption**) in which the overlapping effect is used to place a caption (line A) on a photograph (line B):

```
<div id="box">
<div id="text1" class="caption">Monarch Butterfly</div>   (A)
<p><img id="image" class="photo" src="monarch.jpg"        (B)
        alt="A butterfly image" /></p></div>
```

The style file `caption.css` places the photograph and the text caption at the desired positions (lines C and D) inside a relatively positioned container box. The caption is in front (`z-index: 1`) of the photograph (`z-index: 0`

FIGURE 4.30: CSS-Based Caption

by default). The caption has translucent white background to let the picture show through (line **F**).

```
div#box { position: relative;
          width: 220px;   height: 280px;   }
img.photo
{   position: absolute;                          /* (C) */
    width: 220px; height: 280px;
}

div.caption
{   position: absolute; left:8px; top:16px;       /* (D) */
    white-space: nowrap;   z-index: 1;            /* (E) */
    padding: 4px;
    text-align: right; font-weight: bold;
    background: rgba(255, 255, 255, 0.5);         /* (F) */
}
```

The CSS property `visibility` can be set to `visible` (the default) or `hidden`. A hidden element is not rendered even though the space it occupies is in the page layout. Thus, you may regard hidden elements as fully transparent.

Unless specified, the `z-index` of an element is the same as that of its containing element. The root containing element (`body`) has `z-index` 0.

By setting the stack level for the top banner and the left navbar to 1 (lines **E** and **F**), they come in front of the other elements, all with stack level 0.

Any element with an explicitly set `z-index` also starts its own *local stacking context* with it being the local root element and having a local stack level 0.

Under JavaScript control, positioning, stack levels, visibility, and other style properties can combine to produce interesting dynamic effects (Chapter 6).

4.24   CSS-Generated Content

The pseudo-elements `::before` and `::after` can insert CSS-supplied content, with desired styles, before and after any given element.

For example (**Ex: Generated**), we can place a blue "An" before, and a red "`for You`" after the h1 head (Figure 4.31):

```
<h1 id="target">Introduction</h1>
```

with the style rules

```
h1#target::before { content: 'An '; color: blue }

h1#target::after { content: " for You"; color: red }
```

FIGURE 4.31: CSS Generated Content

An Introduction for You

The `content` property can supply a text string, in single or double quotes. Except for whitespace characters, all characters are taken literally. Thus, HTML elements, entities, or character references will not work in the content strings. You may use UNICODE characters directly in the text string or include them by their hex code preceded by a BACKSLASH (\).

A practical use of CSS-generated content is in displaying numbers for chapters, sections, figures, tables, and so on. You can use counters to keep track of numbering and insert the values from counters into the displayed content.

As an example (**Ex: SecNum**), we can use `::before` to display section titles with automatically generated section numbers (Figure 4.32). The HTML code has three numbered `sections` inside `article`:

```
<article>
<h1>A Project Report</h1>
  <section class="numbered"><h3>Introduction</h3></section>
  <section class="numbered"><h3>Background</h3></section>
  <section class="numbered"><h3>Project Status</h3></section>
</article>
```

We use three style rules to generate the section numbers:

```
article { counter-reset: sn }                 /* A */
```

FIGURE 4.32: Generated Section Numbers

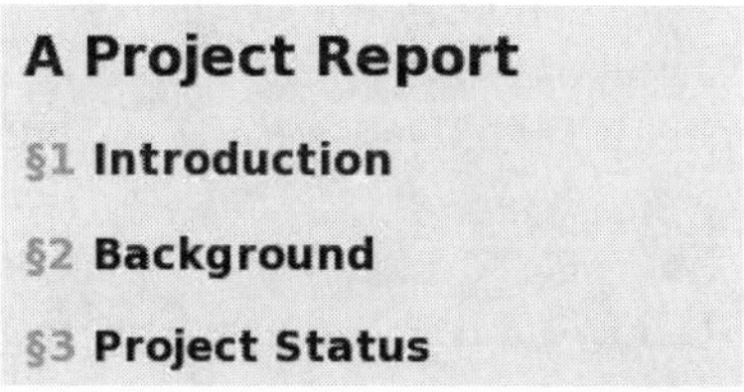

A Project Report

§1 **Introduction**

§2 **Background**

§3 **Project Status**

```
section.numbered { counter-increment: sn }    /* B */

section.numbered h3:first-child::before        /* C */
{   content:"§" counter(sn) " ";               /* D */
    color: darkorange; }                       /* E */
```

A counter sn is set to zero when we display article (line A). The counter is incremented for each numbered section (line B). And before any h3 that is the first child of a numbered section (line C), we place the generated section number (line D) in dark orange (line E). See **Ex:** SecNum at the DWP website.

The ability to use counters and insert generated content with styling brings additional powers to CSS.

The generated content feature also allows us to insert the values of element attributes into the display. For example (**Ex: GenAttr**), the HTML

```
<p>Please visit the
<a href="http://dwp.sofpower.com">DWP website</a>
for ready-to-run examples.</p>
```

with print media style

```
@media print
{   a[href]::after
    {   content: " (" attr(href) ")"; }
}
```

will have the actual URL included after the link anchor as shown in Figure 4.33.

FIGURE 4.33: Print View of Link

Please visit the DWP site (http://dwp.sofpower.com) for re

The notation content: url(...) can insert the content at the given URL. The CSS3 *Generated and Replaced Content Module* draft specifies more useful content generation features.

4.25 Translucent Callout via CSS

As another example, let's combine some of the CSS techniques to create a callout for webpages. Among other uses, CSS-based callout can be very useful when a team of developers is building a website or authoring a document.

FIGURE 4.34: A Translucent Callout

Figure 4.34 shows a "changed" callout placed on a sample page (**Ex: Callout**). To place such a note, we simply add a **span** around a word in the running text, like this:

```
<p>Lorem ipsum dolor sit amet,
<span class="changed">consectetur</span> adipisic
...
```

The following style rule implements the desired translucent note:

```
span.changed::before
{ content: "Changed";                /* Generated content   */
  position: relative; top: -1.8em; /* relative positioning */
  z-index: 30; opacity: 0.7;        /* translucent on top  */
  width: 6em; padding: 0.5em 0em 0.5em 0.5em;
  font-size: larger;  font-weight: bold;
  background-color: gold;
  border-right: 10px solid gold;   /* A  */
  border-bottom: 10px solid white; /* B  */
}
span.changed { position:static; margin-left:-6em;} /* closes gap */
```

Note, we used the right and bottom borders (lines A and B) to form the pointing corner of the callout.

You can try this example (**Ex: Callout**) on the DWP website.

4.26 Understanding the Cascade

A webpage may have several style sheets coming from different origins: the page author, the user (via preferences), and the browser. The user may specify rules such as underline hyperlinks via browser options, and a browser always has its default styles for HTML elements.

A style sheet may also @import other style sheets. The @media may specify different styles for elements, depending on the presentation medium. All of this creates a situation in which multiple style declarations applying to the same element may conflict with one another. In this case, the *cascading order* determines which one applies.

The CSS-defined cascade is quite complicated and detailed. But remembering the following rules will be adequate for most occasions:

- User style rules override the default rules of the browser.

- A declaration from the page author overrides another by the user unless one declaration is designated important (by putting !important at the end of a declaration), in which case the important declaration takes precedence.

- Conflicting rules in author style sheets are selected based on selector specificity (Section 4.8).

- Everything being equal, a rule later in a style sheet overrides one that is earlier. Imported rules are earlier than all rules in the importing sheet.

4.26.1 CSS Validation

Style sheets can get complicated and may contain hard-to-detect errors. This is because unrecognized rules or rules with typos are ignored by browsers.

The Firefox *element inspection* feature, activated by right-clicking on a webpage, allows you to navigate the HTML hierarchy and examine the style details of any desired element. Style settings are also correlated to actual source code lines in style files. You can also check your styles using the CSS validation services at W3C (`jigsaw.w3.org/css-validator/`).

4.27 Additional Style Features

Browser support is moving ahead with due speed to include many of the new features introduced by CSS3, including transitions, animations, transformations, and additional features.

Many of these new features make it much easier to achieve dynamic user-interface effects with much reduced JavaScript coding. We also discuss some of these features in later chapters.

4.28   For More Information

A practical collection of information on CSS has been presented here. Application of CSS in conjunction with JavaScript can be found in later chapters.

The w3schools.com/css3 page provides a listing of all CSS3 properties and simple examples. The CSS working group site (w3.org/Style/CSS) and the CSS specification (w3.org/TR/CSS) at the W3C website provide official information and the standard documentation.

Browsers are still making progress to fully support CSS3. See the CSS2 specification and CSS3 recommendations for other properties and complete information on style sheets.

The DWP website has many useful resources for CSS.

4.29   Summary

A webpage consists of HTML code for document structure and CSS code for document presentation. The separation of structure from presentation brings important advantages. Style sheets can be attached to multiple webpages, and a page can easily switch style by using different style sheets. CSS is used to implement the overall page layout for your pages and used to specify how elements within a page are presented.

A style sheet consists of style rules; each has a selector and a set of declarations. A style declaration specifies a value for a property. There are also special rules such as @import and @media.

CSS is flexible. You can specify style properties for all elements (universal selector *), any particular element (element selector), any element in a class (*.class selector), any particular element in a class (*tag.class* selector), an element as child or descendant of other selected elements (contextual selector), or a single element with a specific id (id selector). The pseudo-class selectors (:link, :visited, :hover, :active, :focus, :target) are handy for creating rollover effects for links and UI elements. And pseudo-element selectors (::first-line and ::first-character) help style text materials. CSS provided content (via the content property) can be inserted for display with ::before, and ::after.

The CSS page formatting model uses block and inline boxes to flow contents onto the displayed page. The top and bottom margins that give the vertical separation for inline boxes are controlled by the line-height only, whereas the vertical separation of block boxes is affected by margin, padding, and border settings. Each of the four corners of a block box can be rounded by specifying the radii of an ellipse for that corner. Inline blocks together with percentage width specifications can help create elastic banners that expand and shrink with changing browser window size.

Both linear and radial gradients can be used to define the background image of an element. Text and box shadows can easily be added. Combining these features plus transparency you can create interesting effects including lighting, 3D buttons, rainbow colors, and so on.

CSS positioning (normal, relative, absolute, fixed) can be used to flexibly place page elements and to fix them with respect to different display media. Cascading rules govern precisely which style rule applies to an element in case multiple conflicting declarations exist for the element.

Innovative use of CSS can bring three-dimensional buttons, rollover effects on image links, stay-in-place navigation bars (fixed positioning), and framelike page layouts.

Exercises

4.1. Is it possible to give an HTML element multiple style classes? If so how is it done?

4.2. How is the space between one text line and the next controlled by CSS? What style property can you use to change that spacing?

4.3. Can the spacing between characters be changed? How is it done?

4.4. How can we prevent line breaking at white spaces? Give two methods.

4.5. What can be done to make an element translucent? Please give two different methods and explain their difference.

4.6. What is the difference between margin, border, and padding? Please explain.

4.7. Is it possible to specify multiple background images for a given element? If so find out how.

4.8. What is a pseudo-class selector, a pseudo-element selector? What is the difference?

4.9. Follow the float page layout (Figure 4.6) and make the main column into two columns resulting in a three-column layout.

4.10. Redo the previous exercise but use CSS table styling instead.

4.11. What does the `:target` pseudo-class do? Create an example to show how it works.

4.12. Generally, the *arg* in the pseudoclass selector `:nth-child(arg)` can take on values `even`, `odd`, and *an+b*. Find out the meaning of the argument expression and explain the meaning of these arguments: 5, $-n+4$, even, odd, $4n + 1$, $n + 3$.

4.13. Find out how well the `border-image` features of CSS3 are supported by major browsers.

4.14. Create an elastic horizontal navbar following the techniques in Section 4.13.2.

4.15. Figure 4.13 showed a circle created using corner radius. Can you create an ellipse using the same ideas?

4.16. Using CSS, create a ball (3D) using radial gradients in a circle.

4.17. Take the rainbow example in Section 4.20.2 and make it even more realistic looking by adjusting the colors.

4.18. Take the rainbow example in Section 4.20.2 and add a linear gradient sky background for it. Hint: Use transparency.

4.19. Use HTML and CSS to display three towers and disks of the *Tower of Hanoi puzzle*, well-known in computer science. Your page tower.html should look similar to Figure 4.35. Your solution needs to be a pure

FIGURE 4.35: Tower of Hanoi Screenshot

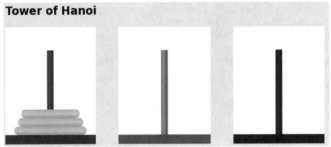

application of HTML and CSS. Each tower and each disk must be represented by a single separate HTML element. And the pole and base of each tower must belong to the same single element. Disks can potentially be placed on any tower as direct child elements of the tower. Also, no CSS floating or positioning (`position: ...`) or Z indexing is allowed. Later in Chapter 7 we will implement the Tower of Hanoi puzzle with drag-and-drop operations. Test and make sure your page works on multiple browsers, including Firefox and Google Chrome.

Chapter 5

Forms, PHP, and Form Processing

When you surf the Web, from a desktop, laptop, notepad, or smartphone, information flows from servers to your browser. But the Web is much more than a one-way street. You can also enter information into the browser to be sent to the server or processed directly by JavaScript programs on the client side.

HTML provides a complete set of form and input elements to obtain user input and upload data, including files, to the server. A `form` element can specify a server-side program to receive and process the uploaded data and provide a response for the user. The Web would be a much poorer place without the ability to collect and process user input. In fact, e-business and e-commerce are dependent on the interactive exchange of information between users and servers.

User input is sent to the server-side program via an HTTP POST or GET query. The receiving Web server invokes a designated program to process the data and to provide a response (Figure 5.1). The server-side program

FIGURE 5.1: Form Processing

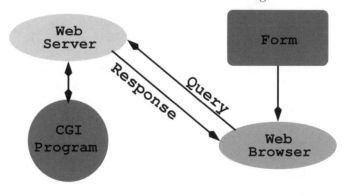

to process incoming HTTP requests can be written in any language as long as it conforms to the *Common Gateway Interface* (CGI). For good reasons, most Web developers prefer PHP for implementing CGI programs. PHP is a language designed for the Web, enjoys native support by most popular Web servers, and offers many powerful features. Furthermore, PHP code can be intermixed within HTML files directly.

This chapter gives you a clear understanding of HTML for forms, enables you to design and style forms, and introduces PHP server-side programming. Examples illustrate how forms work in practice and how to write PHP scripts for form processing. Access to a Web host with PHP support is assumed. The introduction to PHP also gives you an overview and a good foundation for materials on server-side programming in later chapters.

5.1   A Simple Form

We use the HTML `form` element to collect input from users for uploading and processing on the server side.

Different types of *input-control* elements inside can be placed in a form to receive user input for server-side processing. Or they can be placed outside of forms to collect information for JavaScript processing.

A `form` element may contain any flow elements so you can include instructions, explanations, and illustrations together with input controls. For example (**Ex:** `SimpleForm`), the following sample form

```
<form method="post"                              (1)
      action="welcome.php">                      (2)
<pre>
<label for="n">Full Name:</label> <input id="n"
               name="client_name" size="25" />   (3)

<label for="e">Email:</label>      <input id="e"
   type="email" name="client_email" size="25" /> (4)

            <input type="submit" value="Send" /> (5)
</pre>
</form>
```

requests the name (line 3) and email address (line 4) from the user and sends the collected information via an HTTP POST query (line 1) to the program `welcome.php` (line 2) on the server side. The HTTP request `method` for the form can be either `post` or `get`. This form involves three input controls (lines 3, 4, and 5). Each of the first two has a label indicating the requested information. The formdata are sent when the user clicks the **send** button (line 5). Figure 5.2 shows the form being filled out (left) and an attempt to submit the form (right) that resulted in a browser message that a valid email address is needed. HTML5 has many features to make obtaining and verifying user input easier on the browser side.

FIGURE 5.2: A Simple Form

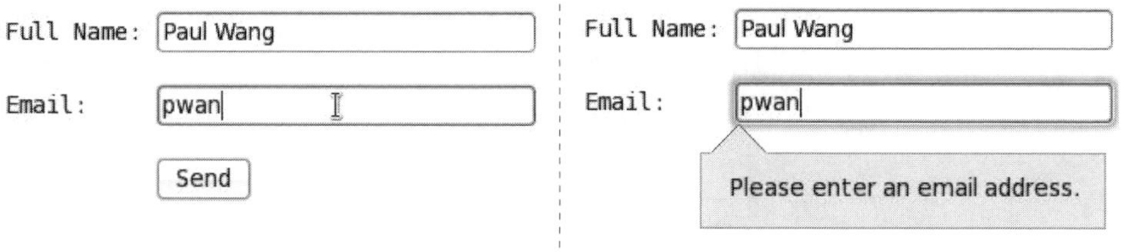

5.2 Basic Input Elements

The HTML fill-out form processing involves three phases: collecting user input, sending HTTP queries, and server-side processing. Let's first look at input collecting with the HTML `form` element and the various input controls available.

A typical `form` element consists of these essential parts:

1. Instructions for the user on what information is sought and how to fill out the form.

2. Blank fields, produced by input collection elements, to be completed by the user.

3. Text or label for each input control to clearly indicate the exact information to be entered.

4. A button or a submit type `input` element, for the user to click and submit the completed form.

5. An HTTP query method, given by the `method` attribute of `form`, for sending the formdata to the server. It is advisable always to use the `post` method even though `get` is the default.

6. A URL, given by the required `action` attribute, for a server-side program to receive and process the collected formdata of the `form` element.

Thus, a `form` is generally coded as follows:

```
<form method="post or get" action="program-URL">
    flow elements each may
        contain input-control elements
</form>
```

Input-control elements include `input`, `textarea`, `button`, `select`, `option`, and more. Input controls are phrasing elements and it is recommended that you enclose them in block-level flow elements inside a form. When a form is submitted, all values collected by input-control elements within the form will be sent to the server-side program.

Input-control elements may also be placed outside a `form` for other purposes such as interactions with client-side JavaScript (Chapter 6). A `form` may not contain another `form`, so no `form` nesting is allowed.

Different kinds of input-control elements are available to make collecting user input easy: short input, long input, multiple choices, selection menus, and file uploading. Let's look at input-control elements.

5.3   Text Input

A short text string is the simplest and most frequent type of user input. An `input` element of type `text` supports such input. For example,

```
<input name="lastname" type="text" size="15" maxlength="25" />
```

displays a one-line input field of size 15 (15 characters wide) with a maximum user input of 25 characters. The `type` attribute may take on other values but `text` is the default. The `name` attribute specifies the *key* for the `input` element (or any other input-control element). User input becomes the *value* of the input control. The received input is submitted as a key-value pair—for example,

```
lastname=Katila
```

to the server-side program. The `value` attribute can be used to specify an initial value for the input.

A `placeholder` attribute can be given to provide an example or hint for what input is expected. If the `value` attribute is not given or is the empty string, the placeholder value is displayed, often in grayed out form, in the input field (Figure 5.3) to guide user input. Here is the code for the example (**Ex: Placeholder**),

```
<label for="so">Social Security No.:</label>          (1)
<input id="so" name="ss" placeholder="xxx-xx-xxxx"
       required="true" />                              (2)
```

FIGURE 5.3: Text Field with Placeholder

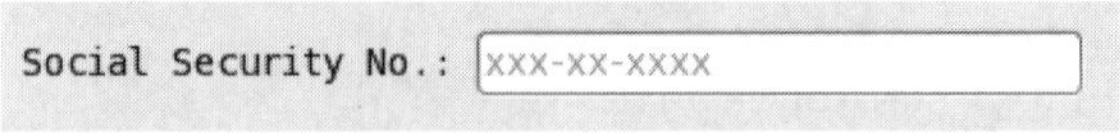

Social Security No.: [xxx-xx-xxxx   ]

Use the `for` attribute of `label` to provide the `id` of the input control it is labeling (line 1). Or instead, enclose the label text and the input element both as child elements of `label` as follows:

```
<label>Social Security No.: <input id="so" name="ss"
placeholder="xxx-xx-xxxx" required="true" /></label>
```

Use the `required` attribute (line 2) to specify an input field that the user must fill in before the form can be submitted. Browsers support this feature by preventing form submission unless all required fields are filled in. It is still good practice to provide clear instructions to the user as to which form fields are required, with red asterisks for example.

The `value` attribute can be set for an `input` element to provide an initial value for that input field.

Multiline text input can be collected with the inline element `textarea`. For example (**Ex: TextArea**),

```
<label for="c">We welcome your comments:</label><br/>
<textarea id="c" name="feedback" rows="4" cols="40">
Tell us what you really think, please.</textarea>
```

can be used to collect user feedback in a four-row by forty-column input field. Text areas are usually resizable by the user. Textual content inside the `textarea` element provides an initial value for the `textarea`.

The user may enter more than four lines, in which case a vertical scroll bar appears. Line breaks, either entered by the user or supplied by the browser due to a line exceeding the column count, are not transmitted to the server side. This is known as *soft wrap* behavior. You can turn it off by adding the `wrap="hard"` attribute for `textarea`.

Also you may use `style="resize: both"` to make a textarea resizable.

5.4   Input in Standard Formats

HTML5 supplies a number of input types for often-used data such as time and date. The input must follow well-defined formats and can be checked by browsers for validity. The following input `type` values can be distinguished:

- `date`, `time`, `datetime`, `datetime-local`—The input string must follow the date and time format specified by the Internet standard (RFC 3339) explained in Section 2.11.1.

- `month`—In the form of *yyyy-mm*.

- `week`—In the form of *yyyy-Www* where *ww* is between 01 and 53.

Additional input formats include

- `email`—An email address in the correct format.

- `url`—An absolute URL or empty string, on one line.

- `number`—A positive or negative integer or floating-point number.

- `range`—A number within a given `min="`*number*`"` and `max="`*number*`"` range with a specific `step`.

- search—A plaintext string for searching, on one line.

- tel—A telephone number, on one line.

- color—A hex color name #*rrggbb*; color names or abbreviations are not allowed.

Additionally, any input field with the required="required" attribute must have a non-empty string value. Figure 5.4 shows an example display

FIGURE 5.4: Sample Input Fields

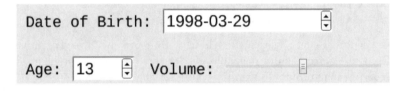

of the following three input elements (**Ex: SampleInput**):

```
<input type="date" name="birth" size="15" />

<input type="number" name="age" size="4"
        min="0" max="200" maxlength="4" />

<input type="range" name="volume" min="0"
                 max="1" step="0.01" />
```

Note we used max and min values to restrict the possible input, which can be further limited by setting step to an integer or a floating-point number. The step specifies the *granularity* or the increment/decrement from one allowed value to the next. In addition to spinboxes (boxes with up-down arrows) and sliders, a browser may display date pickers or other UI widgets to make such input easy for users.

For input elements with formats, user agents should automatically validate the input before submission to the server. Browser support for such checking may not be uniform yet.

5.5 User Choices and Selections

5.5.1 Radio Buttons

A group of *radio buttons* allows the user to choose exactly one from a number of choices. Clicking one choice selects it and deselects all others in the group. Each button is an input element of type radio, and all buttons in the group have the same name.

Figure 5.5 shows the display of the following radio button group (**Ex:** RadioButton):

```
<!-- inside or outside a form -->

<p style="font-size: larger; font-weight: bold;">
   Choose a color:
  <input id="r" type="radio" name="color"
                             value="red" checked="" />   (a)
    <label for="r" style="color: red">Red</label>         (b)
  <input id="g" type="radio" name="color" value="green" />
    <label for="g" style="color: green">Green</label>
  <input id="b" type="radio" name="color" value="blue" />
    <label for="b" style="color: blue">Blue</label>
</p>
```

FIGURE 5.5: Radio Buttons

Choose a color: ● Red ○ Green ○ Blue [Go]

Here we have three radio buttons named `color`. The button with the value `red` is checked (line a) as the initial selection. Selecting a button cancels the previous choice, like pushing buttons on a radio. The data sent to the server-side program are the name-value pair of the selected radio button.

In this example, the label `Red` (line b) is for the input field on the previous line a. By identifying the label, a browser usually allows the user to also click the label to make a selection.

5.5.2 Checkboxes

You can use *checkboxes* to allow users to choose several items from a list of choices. A checkbox is an `input` element of type `checkbox`. Clicking a checkbox selects or deselects it without affecting other checkboxes. Thus, a user may select all, none, or any number of checkboxes.

Figure 5.6 displays checkboxes from the code (**Ex:** CheckBoxes):

```
<!-- inside or outside a form -->
<p style="font-size: larger; font-weight: bold;">
 Sports: <input id="t" type="checkbox" name="tennis" />
    <label for="t">Tennis</label>
 <input id="b" type="checkbox" name="baseball" />
    <label for="b">Baseball</label>
 <input id="w" type="checkbox" name="windsurf" />
    <label for="w">Wind Surfing</label>
</p>
```

FIGURE 5.6: Checkboxes

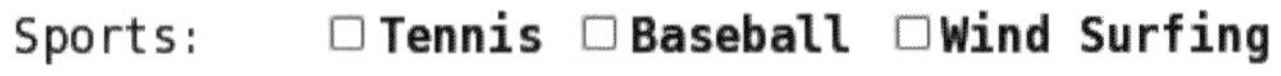

Sports: ☐ **Tennis** ☐ **Baseball** ☐**Wind Surfing**

For checkboxes,

- Any number of items can be initially checked.

- A user may pick any number of choices, including none.

- Each selected item is sent to the server-side program as *name*=on.

A perhaps more convenient alternative implementation of checkboxes uses the same name attribute for each box in a group:

```
<input id="t" type="checkbox" name="sport"
    value="tennis"/> <label for="t">Tennis</label>
<input id="b" type="checkbox" name="sport"
    value="baseball"/> <label for="b">Baseball</label>
<input id="w" type="checkbox" name="sport"
    value="windsurf"/> <label for="w">Wind Surfing</label>
```

In this case, each selected box generates a data string sport=*value*. If such formdata is sent to a PHP program on the server side, we recommend you use name="sport[]" to establish a PHP array $sport in the server-side program to contain all the selected values (Section 5.18).

5.5.3   Pull-Down Menus

When there are many possible choices, a pull-down menu can save space and make the form cleaner and clearer. The select element is an input control for this purpose. You include any number of option elements inside a select element as the choices.

The code for a user to select a state in their address (**Ex:** PullDown) is outlined as follows:

```
<!-- inside or outside a form -->
<label for="st">State:</label
<select id="st" name="state" size="1">
  <option value="0"> Pick One </option>
  <option value="Alabama"> Alabama </option>
  <option value="Alaska"> Alaska </option>
  ...
</select>
```

Figure 5.7 shows the menu before and after it is pulled down. The size attribute specifies how many options are displayed on the menu at one time. If `size` is 1, then the menu initially displays the first option with a pull-down button beside it. Typically, `size` is 1 and the first option is an instruction on how or what to pick. To allow multiple choices, the Boolean attribute `multiple=""` must be given. Otherwise, only one choice is allowed. One or more options can be preselected with the Boolean attribute `selected`. Data sent to the server-side program include one name-value pair for each option chosen.

FIGURE 5.7: A Pull-Down Menu

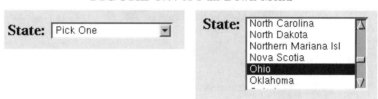

Sometimes options can be divided into different groups to make the selections better organized and easier for the user. `optgroup` can separate options into groups. For example (**Ex:** `OptGroup`), the form

```
<form method="post" action="showdata.php">                      (1)
<select id="menu" name="menu[]" size="8" multiple="multiple">
 <optgroup label="Soup">
   <option value="hot and sour">Hot and Sour Soup</option>
   <option value="egg drop">Egg Drop Soup</option>
   <option value="chicken noodle">Chicken Noodle Soup</option>
 </optgroup><optgroup label="Salad">
   <option value="garden">Garden Salad</option>
   <option value="spinach">Spinach Salad</option>
   <option value="fruit">Fruit Salad</option>
 </optgroup>
</select></p>
</form>
```

results in the display shown in Figure 5.8. The PHP program `showdata.php` (line 1) displays formdata received and can be a good checker for your HTML forms (Section 5.21).

If the `multiple=""` attribute is added to `select`, the user picks the first item with a simple mouse click (left mouse button) and picks each additional item with CTRL-click. Selections are independent of option grouping. It is a good idea to set a sufficient size for menus allowing multiple picks. Some Web surfers may not be aware of this mouse usage, and therefore the `checkbox` may be a better choice for users to pick multiple items.

FIGURE 5.8: Menu Option Grouping

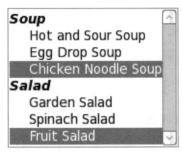

5.6   Form Submission

When a user has finished filling out a form, a simple click on a submit button can trigger a sequence of actions on the part of the browser to submit the completed form. First, necessary validations are performed to make sure required entries have values and values have the correct formats. The user must fix any problems found and submit again. Upon successful validation, the entire set of input, known as the *formdata*, is then encoded and sent to the designated server-side program using the desired HTTP method.

The form attribute `method` indicates the HTTP method `post` or `get`, `action` specifies the server-side program, and `enctype` sets the way data are encoded. Possible `enctype`s are

- `application/x-www-form-urlencoded`—Under this default encoding, the formdata is sent, under either the `post` or `get` method, as a query string consisting of name=value pairs separated by the & character with each name and value URL encoded (Section 5.12.4).

- `multipart/form-data`—Used together with the HTTP `post` method when the form contains file uploading.

- `text/plain`—Working under the `post` method, the name=value pairs are sent, verbatim without encoding, in a string separated by SPACE characters. This enctype is often useful for sending computer codes.

The basic submit button for a form is

```
<input type="submit" value="button-label" />
```

Use a meaningful *button-label* such as `Continue`, `Go`, `Register`, or something similar that relates to the purpose of the form.

A form may have multiple Submit buttons with different values. For example, a membership application form may have a Join button and a Trial button for a regular or trial membership. Formdata include the name-value pair for the clicked `submit` button.

You may customize the look of a Submit button by using an *image input element* (**Ex:** ImageInput):

```
<input type="image" src="url" name="key" alt="..." />
```

Data sent to the server-side program are in the form

key.x=*x0*
key.y=*y0*

where (*x0, y0*) is the position of the mouse cursor in the image when clicked.

As an alternative to the `input`-based Submit button, you can also use the `button` element to create Submit buttons. For example,

```
<button name="submit" value="join">Join the Club</button>
```

displays a Submit button with the label `Join the Club` and sends `submit="join"` to the server-side program. By placing an `img` element inside `button`, you obtain a graphical button of your own design.

A `submit` button, regardless of type, may specify how formdata is submitted, overriding any settings by the containing `form`, with the attributes `formaction` (url of server-side program), `formenctype` (encoding), `formmethod` (post or get), and `formtarget` (valid link target, Section 3.11.1).

It is possible to invoke programmed actions on the client side before submitting a form to the server-side program (Section 6.17).

5.7 File Uploading

So far, user input is collected from the keyboard or the mouse. But it is also possible to receive file uploads from the user's computer. This is done with the `file`-type `input` element (**Ex:** FileUpload).

```
<form method="post" action="upload.php"
      enctype="multipart/form-data">                    (A)
<pre>
<input name="name" placeholder="Name"/> <input         (B)
   type="email" name="email" placeholder="Email"/>

<label for="p">Your photo:</label>  <input id="p"       (C)
   type="file" name="photo" accept="image/jpeg" />

<label for="r">Your report:</label>  <input id="r"      (D)
   type="file" name="report"
   accept="application/pdf" />

<input type="submit" name="submit" value="upload"/>
</pre></form>
```

Figure 5.9 displays the file uploading form requesting a photo (`.jpg` file, line C) and a report (`.pdf` file, line D). The browser displays a `Browse` button for each `file` type input field. Clicking the `Browse` button opens a browser-built-in *file picker* for the user to pick one or more files from the local file system.

FIGURE 5.9: File Uploading

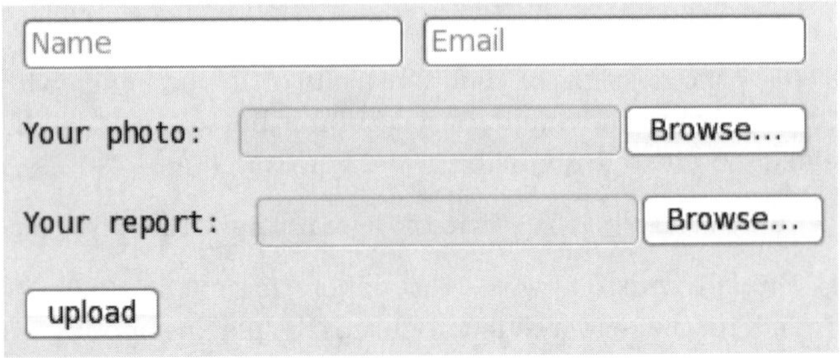

Often, we want to allow users to choose a set of files, pictures at a birthday party for example, to upload. We can enable that by adding the `multiple=""` Boolean option to a file-type `input`.

```
<label>Pictures: <input multiple="" type="file"
  name="picture[]" accept="image/jpeg" /></label>
```

A user then can browse to a pictures folder and pick all desired files to upload at once. To receive multiple files this way, the `name="picture[]"` becomes important for a server-side PHP script to easily pick up all the files.

Several points worth noting about file uploading are

- The query method must be `post`.

- The `enctype="multipart/form-data"` (line A) is needed for a form containing one or more `file` type input.

- The `accept` attribute specifies a comma-separated list of expected valid MIME types for the uploaded file. The accept information is advisory to the browser, and the actual validation of the file type should be performed on the server side when the uploaded file is received. If `accept` is not specified, then no restriction on the file type is indicated.

- Very old browsers may not support form-based file uploading.

- Server-side processing must deal with the `multipart/form-data` data encoding.

See Section 5.22 for server-side processing of form uploading.

5.8   Other `input` Elements

A `hidden` input element is not displayed with the form. But its *key=value* pair is sent along with other formdata to the server-side program. A hidden input field can supply values to customize general-purpose programs on the server side or to provide *session control* where multiple forms are involved in a sequence of user interactions.

For example, a feedback form may use the following hidden input

```
<input type="hidden" name="receiver" value="pwang@cs.kent.edu" />
```

to inform the server-side program where to send the feedback.

A `reset` input element

```
<input type="reset" value="Clear form" />
```

or a `reset` button

```
<button type="reset">Clear All Form Entries</button>
```

displays a button that resets the values of all input fields in a form.

A `password` input element is just like a `text` input element, but the browser will not display the text entered to avoid prying eyes. Formdata containing such personal information must also be secured enroute to the server. This is usually done using *transport layer security* (TLS) via the HTTPS protocol (Section 5.13). When entering a password-protected area of a website, your browser may automatically pop-up an authentication dialog box which is not to be confused with the `password` input in an HTML form.

Any input control can be disabled (displayed with a grayed-out color) with the Boolean attribute `disabled=""`, and can gain input focus automatically after page loading with the Boolean `autofocus=""` attribute.

The `autocomplete="on"` or `"off"` attribute can be given to `form` and input elements to indicate if a browser is to show possible values as the user types based on earlier input values. The default value is `on`.

We already know that an input element does not need to reside inside a form. Such an input element could get input for JavaScript processing and be independent of any forms, or it could be associated with one or more forms via the `form="form_id1 form_id2 ..."` attribute. Browser support for this is not uniform.

When visiting a webpage, a user can press the TAB key to go to the next UI element, such as a link, an input field, or a button, on a page. This feature can make going to the next input field easy when filling out a form.

If for some reason a form wishes to deviate from this *tabbing order*, input controls may be given explicit tabbing positions using the `tabindex="number"` attribute. A smaller positive `tabindex` is visited first. An input control with `tabindex="0"` or no `tabindex` attribute will be visited after any elements with a positive `tabindex`.

An input control (or a link) may also specify an `accesskey="`*char*`"` attribute. It means the user can go to that form entry directly (or activate the link) by typing the character *alt-char*.

To check for correct input, you may give a JavaScript regular expression pattern (Section 6.16.6) to an input element with the `pattern="`*regexp*`"` attribute. This enables the browser to do a validity check based on the pattern provided. For example, Figure 5.10 shows the display of the following (**Ex: PatternCheck**):

```
<label for="c">Country code:</label>
<input type="text" id="c" name="c_code"
       pattern="[A-z]{2}"
       title="Standard two-letter country code" />
```

For any validated input field, it is a good idea to add a `title` attribute to explain the required format.

FIGURE 5.10: Pattern Checking

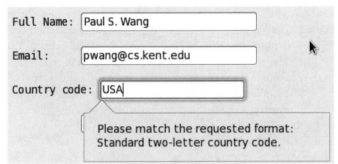

5.9   Layout and Styling of Forms

Forms are critical components of websites. Forms without a good design can be confusing and unpleasant. It is important to design the visual style and layout of a form that is clear, simple, attractive, and easy to follow. The page containing the form must also be visually integrated with the site.

Here are some rules to follow:

- Text to explain the form is placed before and outside the form.

- A form usually uses a single-column format. Related items, such as first name and last name, can be grouped on the same line.

- Entries are labeled consistently, and the `label` tag is used to bind labels to input controls in the HTML code.

- Instructions for individual entries can be placed before, after, or beneath each entry, but they must be consistently placed for all the forms in a site. Placeholders are used where appropriate.

- Required and optional entries are clearly indicated.

- Client-side input validation is employed, and tooltips indicating required input formats are provided.

- Avoid long forms. Group related entries together to divide a long form into smaller, more manageable parts.

- Avoid repeatedly asking for the same information.

The `fieldset` element brackets a set of related input controls and gives it a *legend* (title). For example (**Ex: FieldSet**),

```
<fieldset>
  <legend>Billing</legend>
  ...
</fieldset>
<fieldset>
  <legend>Shipping (if different)</legend>
  ...
</fieldset>
```

groups the input controls and displays the legends, as shown in Figure 5.11.

FIGURE 5.11: Grouping Form Entries

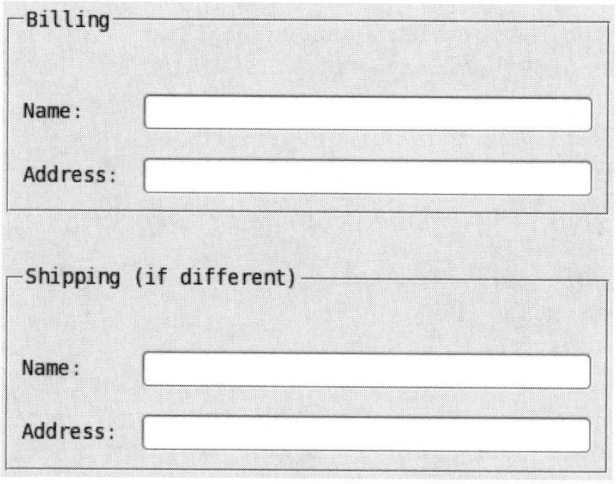

5.9.1   Tabular Form Layout

In practice, forms are often displayed using a table layout to align the labels and input fields in a clear visual grid. Figure 5.12 is an example (**Ex: FormLayout**) of a table layout for a form with several different input fields.

FIGURE 5.12: Table-Aligned Form

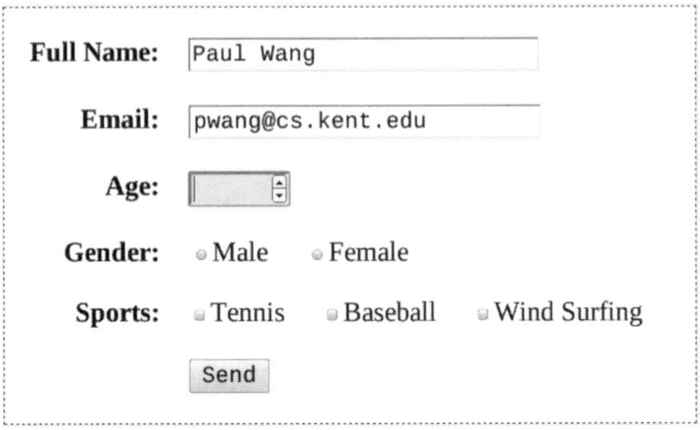

The first entry `Full Name` has the HTML code

```
<form method="post" action="showpost.php">
<div class="entry">
<label for="n">Full Name:</label>
<input id="n" name="client_name" size="25" autofocus="" />
</div>
```

and is displayed as a table row with the style code

```
form
{ display: table; border-spacing: 1em;        /* 1 */
  margin: auto; border: thin dashed black;
  background-color: white;
}

form > div.entry { display: table-row }        /* 2 */

div.entry > label
{ display: table-cell; text-align: right;      /* 3 */
  font-weight: bold;
}

div.entry > input                              /* 4 */
{ display: table-cell; font-family: monospace; }
```

We display this form as a table (line 1) with border spacing. Each (`div.entry`) child of the form is displayed as a table row (line 2). Within each entry, the `label` and the `input` elements are styled as table cells (lines 3 and 4). The Email and Age entries are entirely similar.

The Gender radio buttons have the HTML code

```
<div class="entry"><label>Gender:</label>
<span class="field">
<label><input type="radio" name="sex" value="Male" />
Male</label>
<label><input type="radio" name="sex" value="Female" />
Female</label></span></div>
```

and are styled as follows

```
div.entry > span.field { display: table-cell; }

div.entry > span.field > label
{ padding-right: 0.5em;  margin-right: 0.5em;
  border: thin dashed transparent;
}
```

The checkboxes for Sports are coded and styled the same way. Finally, the Send button at the end of the form uses an empty label, as follows

```
<div class="entry">
<label></label> <input type="submit" value="Send" />
</div></form>
```

Further, we can use :hover and :focus pseudo classes to provide dynamic feedback, with color highlighting (line 5) and area outlines (line 6), as the user navigates through the form (Figure 5.13).

```
div.entry > input:hover, div.entry > input:focus
{ background-color: #def; }                          /* 5 */

div.entry > span.field > label:hover                 /* 6 */
{ border: thin dashed black; }
```

FIGURE 5.13: Form Navigation Feedback

As another example, let's take the shopping cart (**Ex: CartForm**) from Section 9.12 and make it workable as a form and style it nicely with CSS (Figure 5.14).

FIGURE 5.14: Shopping Cart Form

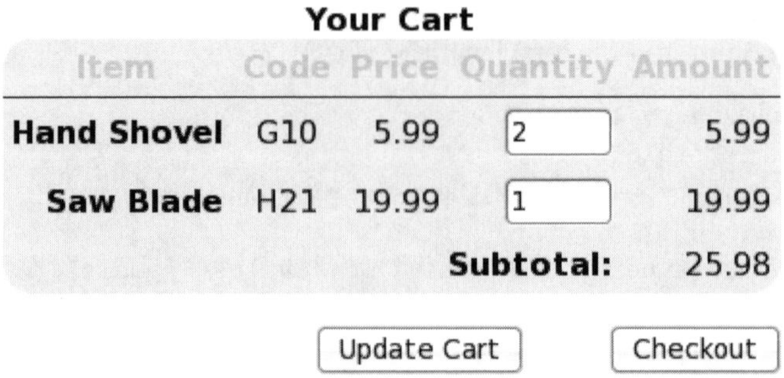

Your Cart

Item	Code	Price	Quantity	Amount
Hand Shovel	G10	5.99	2	5.99
Saw Blade	H21	19.99	1	19.99
			Subtotal:	25.98

Update Cart   Checkout

The form (line A) in this example contains a table (line B) for the cart followed by a paragraph for the two buttons each submits to a different server-side program (lines D and E). The quantity, displayed as a required number type input field (line C), allows user modification.

```
<form method="post" class="shopping">                   (A)
<table id="cart"><caption><b>Your Cart</b></caption>    (B)
<thead><tr><th>Item</th> <th>Code</th> <th>Price</th>
<th>Quantity</th> <th>Amount</th></tr></thead>
<tbody>
<tr>
  <th>Hand Shovel</th>
  <td class="center" >G10</td> <td>5.99</td>
  <td><input type="number" size="4" name="G10_count"  (C)
      value="1" required=""/></td> <td>5.99</td></tr>
<!-- two more rows -->
</tbody></table>
<p class="buttons">
<input type="submit" name="submit" value=" Update Cart " (D)
  formaction="update.php" />       
<input type="submit" name="submit" value=" Checkout "   (E)
  formaction="checkout.php" /></p>
</form>
```

The CSS code used is as follows:

```
form.shopping
{ display: table; margin: auto; }  /* centered form */

form.shopping > p { text-align: right }  /* for buttons */

table#cart  /* cart table */
{ background-color: #f0f0f0; border-radius: 15px;
```

```
  border-collapse: collapse
}

table#cart td, table#cart th { padding: 5px; }

table#cart > thead > tr { color:#fc0;  }

table#cart > thead > tr > th
{ border-bottom: thin solid black  }

table#cart > tbody  { text-align: right }

table#cart > tbody td.center { text-align:center }
```

Of course, the HTML for a shopping cart must be dynamically generated based on user purchases. The complete ready-to-run example (**Ex:** `CartForm`) can be found on the DWP website.

5.10   CSS for Form Presentation

CSS3 introduced several pseudo-class selectors to make it easier to control form input styling resulting in more responsive forms. These include

- `:required` and `:optional`—to style `input` elements with or without the `required` attribute set. It is always good to clearly indicate which form fields are required and which are optional.

- `:disabled` and `:enabled`—to style `input` elements with or without the `disabled` attribute set. For example, the shipping part of a form may be disabled unless the user checks a box indicating that the shipping address is different from the billing address. The box checking can lead to JavaScript actions (Chapter 6) to enable the shipping input fields.

- `:valid` and `:invalid`—to style `input` elements with input that are checked to be valid or invalid. Use these to provide an immediate visual cue as to the correctness of the input.

- `:check`—to style `radio`, `checkbox`, and select `option` elements that are checked/selected by the user.

5.11   Forms and HTTP

To understand formdata submission by a browser and form processing on the server side, it is necessary to have some basic ideas about the HTTP protocol.

On the Web, servers and clients (browsers) follow the Hypertext Transfer Protocol (HTTP) for communication. Hence, Web servers are sometimes called HTTP servers, and browsers are known as HTTP clients, as mentioned in Chapter 1 (Figure 1.5).

The HTTP protocol governs how the Web server and client exchange information. HTTP employs the connection-oriented TCP/IP to support a reliable bidirectional communication channel connecting the client and server. We described HTTP briefly in Section 1.16. Let's now take a closer look.

When a browser requests a page (e.g., the user clicks a link) or sends formdata (e.g., the user clicks the `submit` button), it starts the following sequence of events:

1. *Connection*—The client opens an HTTP connection to a server specified by the URL.

2. *Query*—The client sends an *HTTP request* to access a resource controlled by the server.

3. *Processing*—The server receives and processes the request.

4. *Response*—The server sends an *HTTP response* that can deliver the requested page, results of form processing, or an error message if something is wrong, back to the client.

5. *Transaction finished*—The transaction is finished, and the connection between the client and server may be kept open for a follow-up request, or it may be closed.

The form `action` attribute specifies the URL of the server-side program to receive and process the formdata. The program is usually a CGI program (Section 1.12), but it can be an active page (Section 1.12.1) or a servlet (a special Java program for capable servers). Figure 5.1 shows the form processing data flow.

5.12   HTTP Message Format

An HTTP message can be a request (query) or a response that consists of a sequence of ASCII characters conforming to the required format:

initial line   (different for query and response)
HeaderKey1: value1   (zero or more header fields)
HeaderKey2: value2
HeaderKey3: value3
                       (an empty line separating header and body)
  Optional message body contains query or response data. The amount and type of data in the body are specified in the headers.

The header part consists of one or more *message headers*. Each message header is a key-value pair. HTTP defines what header keys can be used. Each message header ends in RETURN and NEWLINE, but in deference to UNIX systems, it may also end in just NEWLINE.

5.12.1   The Query Line

The initial line identifies the message as a query or a response:

- A query line has three parts separated by spaces: a *query method* name, a server-side path (Universal Resource Identifier, or URI) of the requested resource, and an HTTP version number.

```
GET     /path/to/file/index.html     HTTP/1.1
POST    /cgi-bin/script.cgi          HTTP/1.1
HEAD    /path/to/file/index.html     HTTP/1.1
```

- The GET method requests the specified resource and does not allow a message body.

- The HEAD query requests only the header part of the response for the requested resource.

- The POST method allows a message body consisting of formdata for server-side processing.

5.12.2   The Response Line

A response (or status) line of an HTTP message also has three parts separated by spaces: a version number, a status code, and a textual description of the status. Typical status lines are

```
HTTP/1.1    200    OK
```

for a successful query, or

```
HTTP/1.1    404    Not Found
```

when the requested resource cannot be found.

5.12.3   The POST Query

HTML forms usually specify the post request method resulting in a browser-generated POST query. The query sends the formdata to the program designated by the action URL.

A POST query contains a Content-Type header, a Content-Length header, and a message body consisting of *URL-encoded* formdata. The content-type header is usually

```
Content-Type: application/x-www-form-urlencoded
```

The receiving Web server processes the incoming data by calling the URI-specified form-processing program (typically, a CGI program). The program receives the message body containing the encoded formdata, processes it, and generates a response.

Here is a sample POST query:

```
POST /cgi-bin/register-user HTTP/1.1
HOST: www.SymbolicNet.org
From: jDoe@great.enterprise.com
User-Agent: Mozilla/5.0 Gecko/20100101 Firefox/11.0
Content-Type: application/x-www-form-urlencoded
Content-Length: 132

name=John+Doe&address=678+Main+Street&...
```

5.12.4   Formdata Encoding

Formdata, by default, uses the *form-urlencoded* encoding (media type application/x-www-form-urlencoded). File uploading uses the multipart/form-data media type. Certain formdata, such as computer codes, can be sent verbatim without encoding under the /plain media type.

Formdata is form-urlencoded into a string by the following procedure:

1. The string is in the form of a series of entries separated by &. Each entry is a *name=value* pair.

2. Each SPACE within an entry is replaced by a +.

3. Unsafe characters (Section 1.4.1) in the name and value parts are then encoded. Each unsafe character is replaced by %*hh*, where *hh* is its ASCII code in hexadecimal. For example,

=	%3D or %3d	+	%2B or %2b
&	%26	%	%65

4. Non-ASCII characters are UTF-8 encoded (Section 3.8) into bytes, and then any unsafe byte is %*hh* encoded.

The encoded string can be sent under either the get as the query string or under the post method as request body. A server-side form-processing program must decode the formdata before processing it.

5.12.5   Data Posting via GET Queries

As an alternative to the POST query, formdata can be sent via a GET query by appending the form-urlencoded data at the end of the URL locating the server-side program. In this case, the data sent are known as a *query string* and are joined to the URL by a **?** character:

url_of_server_side_program?*query_string*

The query string is passed to the target program for processing. Many long URLs you see on the Web are caused by attached query strings.

5.12.6   GET versus POST

Clicking or following an http URL generates a GET request. Thus, surfing the Web is basically a sequence of GET requests to various servers. With a query string at the end, a URL can send data to a server for processing. Thus, any link can potentially send data for processing. A form can also elect to use the GET method for submitting the formdata.

So what is the difference between the POST and GET form submission methods? In most cases, there is little difference. Some say if a form does not send user data to be recorded at the server side, the GET method is the right choice. Otherwise, the POST method is the right choice. For example, a form for making payments or membership registration should use the POST method. When in doubt, use the POST method.

Software implementations may put a length limitation on URLs. Thus, a very long query string may not work. File uploading must also use the POST request to send a body using the media type multipart/form-data.

5.13   Formdata Security and HTTPS

Web users care about protecting their sensitive and private information. Every website with forms must display a privacy policy letting users know how the site will protect and use the information collected. It is also important to assure users that the information submitted will not get into the wrong hands.

To protect the formdata during network transmission, *secure HTTP* (HTTPS) can be used for the form pages and their server-side programs. This usually simply means switching from http:// to https:// in the relevant URLs.

While HTTP uses TCP/IP, HTTPS uses *Transport Layer Security* (TLS), a standard derived from SSL (Secure Socket Layer), for transport. TLS can authenticate the originating website, the Web user, as well as perform encryption/decryption to ensure formdata security. Forms that collect any personal or sensitive information must use HTTPS.

5.14   Form Processing Overview

Formdata sent from the client side require customized processing on the server side. Whether via a POST or a GET request, the HTTP query carrying the formdata goes to a Web server. The Web server then relays the formdata to a *target program* (specified by the request URI) for processing. The processing results, as output from the target program, comes back to the Web server, which sends that result back to the client as an HTTP response. The *Web server* here refers to the piece of software, such as Apache, that takes care of HTTP or HTTPS requests on a particular Web host. The target program may be a software module loaded into the server, a Java servlet, an active-page interpreter (ASP, JSP, PHP), or a stand-alone CGI program.

PHP is a very powerful and popular language especially designed for server-side scripting on the Web. The PHP interpreter works especially well with the Apache Web server as a built-in server module. PHP also works well on MS/Windows servers with fastCGI. In this textbook, we will focus on PHP for server-side scripting. The principles and techniques you learn, however, will apply generally.

The Common Gateway Interface (CGI) governs the way a Web server interacts with these external programs. Figure 5.15 illustrates the data flow between a Web server and a PHP program when processing a client request.

FIGURE 5.15: CGI Data Flow

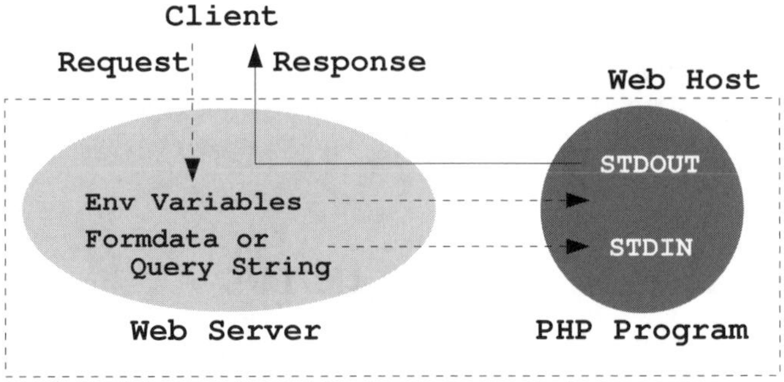

- Any query headers, query string, and other incoming data are sent to the PHP program through a set of predefined environment variables. These are also made available through predefined PHP variables.

- A PHP program receives the GET query string through the environment variable QUERY_STRING or the POST query body through standard input. PHP has built-in support to decode and parse the incoming data and make them available in PHP variables.

- The PHP script may set content-type, content-length, and other

HTTP headers before sending data to standard output that becomes the content of the HTTP response.

Being an active-page technology, PHP scripts are not placed in a cgi-bin, a special folder reserved for CGI programs on the server (Section 10.9.4), but can be placed anywhere in the Web server's document space together with HTML and other files intended for the Web. PHP code can be intermixed in target documents, making it easy to compose and combine static and dynamic parts. PHP files use the `.php` suffix. Many Web servers are also configured to allow PHP files to use the `.html` suffix.

PHP offers additional advantages over regular cgi programs:

- PHP and code libraries are free and widely available.

- PHP is designed for the Web and is a more efficient alternative for CGI. It is most efficient when PHP becomes a resident server module, as in the case of Apache.

- PHP has database, email, ftp, pdf, image, and other support. PHP has a built-in SQLite database and also interfaces well with major database systems, including the free and widely popular MySQL.

- PHP uses familiar C syntax (mostly).

- PHP has a large number of useful functions for the Web.

- PHP offers a command-line interface, making it easy for testing and debugging PHP scripts.

The remainder of the chapter introduces form processing using PHP.

5.15   PHP Scripting Overview

PHP is a high-level, interpreter-based scripting language designed for the Web. PHP code can embedded in HTML or other pages to produce dynamic content.

A PHP script contains active parts enclosed by the bracket `<?php ... ?>` and static parts outside the brackets. The PHP interpreter (Figure 5.16), often inside the Web server, filters a requested PHP file, replacing each bracket with any output generated by the code in that bracket and letting static content pass through untouched as output. The inclusion or not of a particular static part can still be controlled by surrounding logic in php brackets. Within a php bracket, only output-producing statements generate output.

Here is a simple example of a complete webpage with PHP code in it (**Ex: Hello.php**):

```
<html xmlns="http://www.w3.org/1999/xhtml"
      lang="en" xml:lang="en">
```

FIGURE 5.16: PHP Interpreter

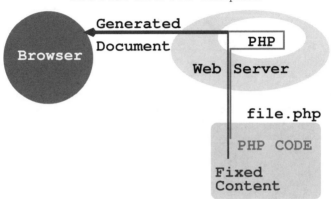

```
<head><meta charset="utf-8"/>
<title>PHP Hello World</title></head><body>
<h1>Hello from PHP</h1>
<p>Hello, it is
<?php
if ( function_exists("date_default_timezone_set") )
    date_default_timezone_set('US/Eastern');          (A)
echo date("l M. d, Y"); ?>,                           (B)
<br />do you know where your project is?</p>
</body></html>
```

A PHP statement is always terminated by the *semicolon*(;). After setting the timezone (line A), we used the PHP operator **echo** to display the current local date computed by the PHP built-in function **date**[1] (line B). The operator **echo** will send to standard output one or more (comma-separated) expressions. After PHP interpretation, line A becomes, for example,

```
<p>Hello, it is Friday Aug. 03, 2012,
```

and the date would of course be the exact date when the page is accessed by a user. Why not visit the DWP website and see **Ex: Hello.php** for yourself?

A PHP page must always produce a correctly formatted HTTP response message. Output from a PHP script constitutes the response body. Response headers are created with the **header** function. For example,

```
header("Content-Type: valid_mime_type");
header("Content-Length: number_of_bytes");
```

---

[1]The date produced by PHP is based on the server location and its time zone setting. To get the client-side date/time, you need to use JavaScript.

For instance, a dynamically generated JavaScript file can use the `application/javascript` or `application/ecmascript` content type.

All `header` calls must be done before any data belonging to the response body is output. If no explicit `Content-Type` header is sent before body output starts, the default

```
Content-Type: text/html
```

is automatically supplied. This makes it convenient for generating HTML pages.

Because no **header** calls are allowed after page output starts, it is important for any PHP page that makes explicit `header` calls to make sure that no characters get sent to the output before all the necessary header calls are made. This also means the page must begin with the character sequence `<?php` and that no white space or unseen characters are placed before it.

Learning PHP involves getting to know the PHP language, built-in objects and functions available, and how to apply PHP in practice. Software releases, available modules, documentation, function references with examples, and more can be found at the PHP official site (`www.php.net`). Our DWP website resource page also provides many links to useful information on PHP.

5.16   An Introduction to PHP

5.16.1   PHP Variables

The name of a variable in a PHP script always carries the `$` prefix. Data from an incoming HTTP request are automatically made available in predefined *super global* variables:

- `$_POST`—for `post` formdata

- `$_GET`—for `get` formdata or query string

- `$_REQUEST`—for either `get` or `post` request data

- `$_SERVER`—for information related to the Web server, HTTP request headers, and the PHP script itself

- `$_ENV`—for CGI defined variables and any form the shell used on particular operating systems

Each of these is an *associative array* (Section 5.18). For example, formdata `item=Hammer&price=4.50` posted by a form is obtained in a PHP script with

```
$product = $_POST['item'];    // $product is Hammer
$cost = $_POST['price'];      // $cost is 4.50
```

This is extremely convenient. Note: Comments using // (to end of line) and /* ... */ are the same as in the C language.

We mentioned in Section 3.12 that a PHP script may include other files, which can also contain PHP code, with **include("path-to-file");**, **include_once**, **require**, or **require_once**. Let's put this idea to use. Say we have scripts front.php and back.php for the leading and trailing parts of pages on our website. The front.php is

```
<!DOCTYPE html>
<html xmlns="http://www.w3.org/1999/xhtml"
      lang="en" xml:lang="en">
<head> <meta charset="utf-8"/>
<title><?php echo $page_title; ?></title>        (1)
</head> <body style="background-color:
        <?php echo $page_background; ?>">          (2)
```

And the back.php is

```
<footer><p style="font-size: small">Copyright
&copy; <?php echo $company; ?>.                  (3)
All rights reserved</p></footer></body></html>
```

Note the PHP code embedded on lines 1, 2, and 3.

A site page template template.php (**Ex: Template**) can become

```
<?php $page_title="Services";
      $page_background="#def";
      require("front.php");       (4)
?>
<p>Page content here </p>
<?php $company="sofpower.com";
      require("back.php");        (5)
?>
```

We set the page title and background variables before including front.php (line 4); then we include back.php (line 5) at the end of the page. Similarly, navbars search boxes, and other common page features are all candidates for inclusion by PHP.

PHP data types include four scalar (one-value) types (boolean, integer, float/double, string), two compound types (array, object), and two special types (resource for I/O steams, etc., and NULL). The NULL type is for the value null (no value). If you unset a variable (**unset($var)**), its value becomes null.

Being a scripting language, PHP does not require (or support) explicit variable type declarations. A variable's type depends on the context within which it is used. PHP converts a variable's type automatically, when appropriate. Explicit type casting is possible with the notation (*type,/*) $var.

5.16.2 PHP Conditionals

For flow control, use the PHP **if** construct:

```php
<?php     // conditionals
   if ( $var )
   { /* $var is not 0, "0", "",
         0.0, FALSE or null */ }
   elseif ($var == 0)
   { /* $var is 0 or "0"    */ }
   else
   { /* otherwise           */ }
?>
```

In place of $var, other test expressions include >, <, ==, >=, <=, !=, === identical (== and same type), and !== (not ===). Predicate functions such as is_string($x), file_exists($file), isset($var), is_array($a), and function_exists($f), which produce a Boolean value (TRUE or FALSE), are also important in tests.

Logical operations on Boolean values are: &&, and, ||, or, !, and xor.

Let's take the navbar in Section 4.11 and make it reusable (**Ex: Navbar.php**). The navbar code in file navbar.php becomes

```php
<nav class="leftnavbar">
<!-- main page -->
<?php if("$page == "index.php") {?>          (I)
<span class="self">Main Page</span>          (II)
<?php } else {?>
<a href="index.php">Main Page</a><?php }?>   (III)
<!-- product page -->
<?php if("$page == "products.php") {?>       (IV)
<span class="self">Products</span>
<?php } else {?>
<a href="products.php">Products</a><?php }?> (V)

...

</nav>
```

The code generates different HTML depending on the value of $page. For the main page, we test for index.php (line I); and for the products page, we test for products.php (line IV). The code generated is a pageid span or a link to the page. Any number of other pages on the navbar can be treated the same way.

A webpage incorporating navbar.php (**Ex: Products**) will use the following code:

```php
<?php $page="products.php";  require_once("navbar.php"); ?>
```

The value for $page is hard-coded here. But it can be computed as well with the code

```
$page=basename($_SERVER['PHP_SELF'])
```

The server-provided value `$_SERVER['PHP_SELF']` gives the full pathname to the PHP script, and the **basename** function picks the simple file name from it. The directory where the PHP script resides can be obtained with **dirname**(`$_SERVER['PHP_SELF']`).

The full example (**Ex: Products**) can be tested at the DWP website.

5.17 Strings in PHP

A PHP string is a sequence of ASCII characters (bytes). You can specify a string by enclosing the characters within single quotes (') or double quotes ("). For example,

```
$name='Paul Wang';
```

```
$first='Paul';  $last="Wang";
$name="$first $last";
```

For single-quote strings, all characters are taken literally except ' and \. To include ' as part of the string, use \'. To include \, use \\.

In double-quote strings, variables and escape sequences such as \n (NEW-LINE), \t (TAB), \r (RETURN), \$ ($), and \" (") are recognized.

Two strings can be concatenated together with the dot operator (.). For example,

```
$me = 'It\'s my name ' . $first . " $last\n";
```

HTTP request data come in the form of strings. PHP offers a large number of string-related functions to make string processing easy. Here are some often useful ones:

- **strlen**(*str*)—Returns the number of characters in *str*.

- **trim**(*str*)—Returns a string by stripping white space (or other characters) from both ends of *str*.

- **substr**(*str*, *i* [, *len*])—Returns a substring of *str* from position *i* (zero-based indexing) to the end or to length *len*.

- **strstr**(*line*, *word*)—Finds first instance of a string *word* in the longer string *line* and returns a substring of *line* that begins with *word*.

- **strtolower**(*s*), **strtoupper**(*s*)—Returns the lowercase, uppercase version of *s*.

- **strcmp**(*str_1*, *str_2*)—Returns a positive, negative, or zero integer if *str_1* is greater than, less than, or equal to *str_2*.

- **md5**(*str*)—Returns the MD5 digest of *str*.

- **urlencode**(*str*)—URLencode *str* and returns a proper query string.

More string functions will be introduced when the need arises. References for PHP string type and string functions are easy to find on the Web. A direct link to any PHP function *fn* is

```
http://php.net/manual/en/function.fn.php
```

5.18 Arrays in PHP

An array groups together a number of related data items into one unit. Each item on the array is an *entry*, or *element* that can hold a scalar value or another array. Arrays are useful in programming in general and very much so in PHP. A PHP array supports numerical indexing (zero based) and string indexing (associative) at the same time. Thus, each PHP array entry is always a key-value pair. The PHP super global variables introduced in Section 5.16.1 are associative arrays. To create a new array, you use the **array** function. Here we create three arrays $a, $b, and $c:

```php
<?php
    $a = array(2, 3, 5, 7);
    // $a is the same as array(0=>2, 1=>3, 2=>5, 3=>7)
    $b = array("first_name" => "Paul", "last_name" => "Wang");
    $c = array(5 => "red", "fox");
?>
```

The first entry of $a is $a[0] with the value 2. The last entry of $a is $a[3] with the value 7. $b["last_name"] is the string "Wang". And $c[6] is the string "fox".

Assigning array values is easy. The assignment $a[3]=5*$a[2] replaces 7 by 25 in $a. While $a[5]=100 adds 100 as $a[5]. It is OK if $a[4] is not defined yet. Similarly, we can do $b['email']="pwang@cs.kent.edu".

You can also add and delete array entries as well as do a host of other operations on arrays. Table 5.1 lists some common array operations. To sort the values in an array, you can use

sort($ar, [*flag*]) (increasing order) **rsort**($ar, [*flag*]) (decreasing order)

The optional *flag* may be SORT_NUMERIC, or SORT_STRING to treat the array entries as numbers or as strings.

TABLE 5.1: PHP Array Functions

Name	Operation
count($ar)	Returns the number of entries in $ar
empty($ar)	Returns true if $ar is empty
unset($ar), unset($ar[$n])	Deletes the array or entry
array_keys($ar)	Returns an array of keys
array_values($ar)	Returns an array of values
array_pop($ar)	Pops and returns last entry, or null
array_push($ar,$e1,$e2,...)	Inserts at end of array
array_shift($ar)	Pops and returns first entry, or null
array_unshift($ar,$e1,...)	Adds at beginning of array
array_reverse($ar)	Returns a reversed array
ksort($ar),krsort	Sorts array keys in increasing or decreasing order

5.19 Getting Started with Form Processing

As our first complete form processing program, let's write a toy version of welcome.php to work with the simple form (Figure 5.2) in Section 5.1. The welcome.php (**Ex: FormAction**) collects the name and email address from incoming formdata and responds with a "Welcome" message (Figure 5.17) or a "Sorry" message if the form is incomplete (Figure 5.18).

FIGURE 5.17: Form Response: Welcome

A Warm Welcome

Hello Paul Wang, it is our great pleasure to welcome you to our site.

We have your email address, pwang@cs.kent.edu, and we will contact you shortly.

FIGURE 5.18: Form Response: Sorry

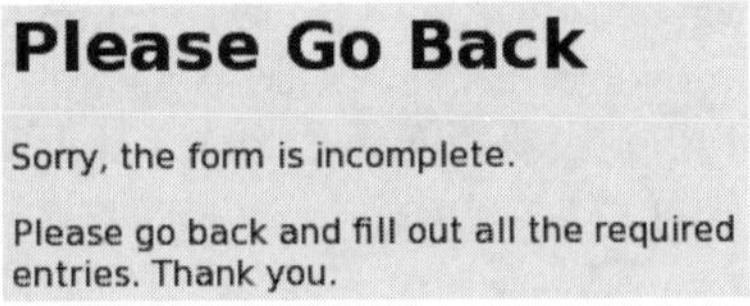

Please Go Back

Sorry, the form is incomplete.

Please go back and fill out all the required entries. Thank you.

The form processing PHP script welcome.php starts with the code

```
<?php
```

```
$title="A Warm Welcome";
if ( empty($_POST['client_name'])        ||
     trim($_POST['client_name'])===""  ||
     empty($_POST['client_email'])       ||
     trim($_POST['client_email'])==="" )  // (1)
{ $error=TRUE;                            // (2)
  $title="Please Go Back";                // (3)
}
?>
```

which checks (line 1) if both name and email are non-empty (filled in). The trim function removes leading and trailing NULL and white space characters from a string. If not, a variable $error is set to true (line 2) and the $title also changes (line 3). These variables are used in generating the response page:

```
<!DOCTYPE html>
<html xmlns="http://www.w3.org/1999/xhtml"
                  lang="en" xml:lang="en">
<head><meta charset="utf-8"/>
<title><?php echo $title; ?></title></head>    (4)
<body style="background-color: #def">
<h1><?php echo $title; ?></h1>                  (5)
<?php if ( isset($error) ) {?>                  (6)
<p>Sorry, the form is incomplete.</p>
<p>Please go back and fill out all the
required entries.  Thank you.</p>
<?php } else { ?>                               (7)
<p>Hello <span style="color: blue">
<?php echo $_POST['client_name']; ?></span>, it
is our great pleasure to welcome you to our site.</p>
<p>We have your email address, <code style="color: blue">
<?php echo $_POST['client_email']; ?></code>,
and we will contact you shortly.</p>
<?php } ?>                                      (8)
</body></html>
```

The variable $title is used in two places (lines 4 and 5). If $error is true (line 6), the "sorry" message is sent (lines 6 through 7). Otherwise (line 7), the "welcome" message is sent (lines 7 through 8).

The complete example (**Ex: FormAction**), form and form processing code, are available for experimentation on the DWP website.

5.20   Form Processing Example: Club Membership

The welcome.php program is very simple. Let's look at a more realistic example (**Ex: JoinClub**) where users join a club of some sort by filling out a form. The form (Figure 5.19) collects the user's full name, email address, age, sex, and favorite sports. The form part of JoinClub.html is included here,

FIGURE 5.19: Joining a Club

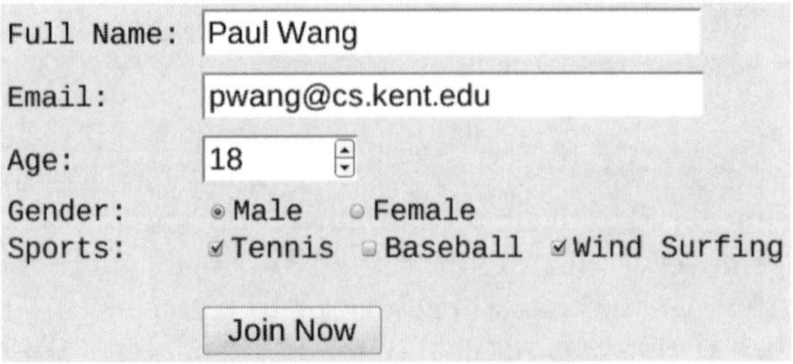

formatted for easier reading:

```
<form method="post" action="joinaction.php">
<label>Full Name: <input required="" name="client_name"
size="25"/></label>

<label>Email: <input required="" type="email"
        name="client_email" size="25" /></label>

<label>Age: <input type="number" title="18-65" name=
"age" required="" size="6" min="18" max="65"/></label>

Gender: <label><input type="radio" name="sex" value=
"Male" />Male </label> <label><input type="radio"
name="sex" value="Female" />Female</label>

Sports: <label><input type="checkbox" name="sport[]"
   value="tennis"/>Tennis </label>
   <label><input type="checkbox" name="sport[]"
   value="baseball"/>Baseball </label>
   <label><input type="checkbox" name="sport[]"
   value="windsurf"/>Wind Surfing</label>

                <input type="submit" value=" Join Now " />
</form>
```

In the form, name, email, and age are required. Age limits apply. And we used the name sport[] to obtain an array of values easily in the form processing program joinclub.php.

The server-side program joinaction.php

1. Checks the incoming formdata (line a)

2. Sends an error message to the user if any input is missing or incorrect (line **b**)

3. Sends the form-collected information by email to the club manager with cc to the user (lines **d** through **e**). It is important to send the correct email **From** field to identify the message as coming from the particular site/organization. Note how we used the PHP constant **PHP_EOL** (end-of-line sequence) to format the correct email headers.

4. Sends a response page to the user

```php
<?php
   if (  empty($_POST['client_name']) ||           // (a)
         empty($_POST['client_email'])||
         empty($_POST['age'])             ||
         ! is_numeric($_POST['age'])
      )
   { $error=TRUE;                                    // (b)
     $title="Please Go Back";
   }
   else
   { require_once("email.php");                      // (c)
     $msg="We have emailed";
     $to="manager@club.com";                         // (d)
     $subject="Club Membership";
     $title="Thanks for Joining Our Club";
     $cc= 'Cc: "' .$_POST['client_name'] .'" <'
              . $_POST['client_email'] . '>';
     $headers = 'From: "Super Club" <service@superclub.com>'
         . PHP_EOL .  $cc . PHP_EOL . 'X-Mailer: PHP-'
         . phpversion() . PHP_EOL;
     if ( ! email_formdata($to,
         $_POST['client_email'], $subject))
     { $msg='<span style="color: red">We
              failed to email</span>';               // (e)
     }
   }
?>
```

The function **empty** (line **a**) returns true if its argument is an undefined variable, not set, is an empty string. But it also returns true for "0" or 0. It also returns true for an empty array. The function is useful for checking required form input where "0" and 0 are impossible or invalid. The call **is_numeric($var)** determines if $var is a number. PHP does not have a built-in function for validating email addresses. But you can find freely available scripts defining an **is_email** Boolean function readily on the Web.

The email.php file (line c, and see Section 5.20.1) defines the function **email_formdata**, which emails the formdata and returns false if emailing fails—in which case, the failed message will be in red (line e).

Having done the processing, the joinclub.php script now begins to send the response page. It first includes a front file rfront.php for the leading part of an HTML page, passing to it page title and background color settings (line f). Then it continues to send the appropriate response page, depending on "form complete" (line h) or not (line g). The matching rback.php ends the form response page (line i).

```
<?php $bg="#def"; require("rfront.php"); ?>     (f)
<h1><?php echo $title; ?></h1>
<?php if ( isset($error) ) {?>                  (g)
<p>Sorry, the form is incomplete.</p>
<p>Please go back and fill out all the
required entries.  Thank you.</p>
<?php } else { ?>                               (h)
<p>Thank you <span style="color: blue">
<?php echo $_POST['client_name']; ?></span>
for joining our club.</p>
<p><?php echo $msg; ?> your request to our
manager,  with a copy to <code style="color:
blue"> <?php echo $_POST['client_email']; ?>
</code>.</p><?php }require("rback.php");  ?>    (i)
```

The success response page can be seen in Figure 5.20. It is important to have appropriate navigation, page layout, and graphics for the response page so it has the same look and feel as the other pages on your site. A common mistake is for the form response page to look bare or be without navigation (a dead-end page).

FIGURE 5.20: Response Page

5.20.1 Sending Email from PHP

The email.php file used in the joinaction.php script contains two user-defined functions: email_formdata and formdata.

```php
<?php
function email_formdata(&$to, &$subject, &$headers)      // (1)
{ if (mail($to, $subject, formdata(), $headers))         // (2)
  { return TRUE;  }
  return FALSE;
}
```

The email_formdata function receives three arguments by reference (line 1) and uses the PHP built-in function **mail** to send email.

mail($to, $subject, $body [, mail headers])

The function **mail** returns false if it fails. The extra header (line 2) we send is a Cc to the applicant.

The email body we send here is the formdata (line 2) in readable form, computed by our **formdata** function.

```php
function formdata()
{  $data="";
   reset($_REQUEST); // resets internal pointer   (3)
   foreach($_REQUEST as $key => $val)             // (4)
   {  if ( is_array($val) )                       // (5)
      {  $data .= ("$key:  " .
           implode(", ", $val) . "\n"); }         // (6)
      else
      {  $data .= "$key:  $val\n";  }             // (7)
   }  // end of foreach loop
   return $data;
}
?>
```

The **formdata** function builds a string $data by looping over the entries in the super global associative array $_REQUEST so it works for GET and POST requests (lines 3 and 4).

For each key-value pair, if the value is a scalar, just append the pair to the end of $data (line 7). Otherwise, if the value is an array (line 5), we call the PHP **implode** function to create a string of comma-separated array values (line 6). Each iteration of the **foreach** loop picks up the next key-value pair, pushing the internal pointer further down the array. At the end of the loop, the internal pointer will be all the way at the end of the array. This is why it is important to call **reset** on any array before you use it in a loop (line 3).

5.20.2   Loops in PHP

You can control iteration in PHP with for, while, and do ... while loops just like in C/C++. The continue and break statements work the same way as well. For example,

```
// displays 0 to 9
for ($i = 0; $i < 10; $i++) { echo "$i\n"; }

$i=30;  // displays 30 to 0
while ( $i >= 0 ) { echo "$i\n"; $i--; }

$i=50;  // displays 50 to 1
do { echo "$i\n"; $i--; } while ($i);
```

PHP also makes it very easy to loop through an array with the each function that returns the next key-value pair of an array when called each time. For example,

```
reset($arr);
while ( list($key, $val) = each($arr) )
{   /* go through array */ }
```

Each PHP array has an internal pointer that keeps track of the position of the next pair. A call to each($arr) also increments the internal pointer in the $ar array. To reset the pointer to the beginning, call reset($arr).

PHP also supports the convenient foreach loop:

```
reset($arr);
foreach($arr as $val) { ... } // loops over array values

reset($arr);
foreach($arr as $key => $val) // loops over key-value pairs
{ ... }
```

Because foreach uses the array internal pointer, it is important to call reset before the loop starts.

5.20.3   Multistep Forms

Forms are used for many purposes on the Web. E-commerce and e-business applications often involve filling out a series of forms. Because HTTP is *stateless*, a prior form transaction has no HTTP supplied connection to the next, and potentially form posts from different clients can be mixed together. This means we must organize the sequence of form interactions from a single user into a session and keep sessions for different users apart.

In other words, a server-side program, PHP or otherwise, must keep track of which subsequent form submission goes with which prior form input. We explain session control in PHP later (Section 8.7).

5.21 HTTP Request Data Exposed

When developing forms and form processing, it is useful to see exactly
what formdata is sent from a particular form. Let's write a PHP script
showdata.php that simply displays the incoming formdata. If you set the
action attribute of a form you are creating to this script, you will be able to
submit the form and see the data it sends. With the input being correct, you
can then develop the desired server-side form processing.

Here is the PHP code in showdata.php.

```php
<?php
if ( ! empty($_REQUEST) ) // if not empty
{   echo "<pre>";
    foreach($_REQUEST as $key => $val)      // (I)
    { if ( is_array($val) )                 // (II)
        echo " $key = ",
            implode(", ", $val), "\n\n";
      else
        echo " $key = $val\n\n";            // (III)
    }
    echo "</pre>";
}
else
    echo "<p>Warning: No input data.</p>";
?>
```

We loop over each key-value pair in the super global array $_REQUEST (line I).
We will display *key = value* if *value* is a scalar (one value, line III). But *value*
could be an array, such as sport[] in our club joining example (Section 5.5.2),
or menu[] in our option group example (Section 5.5.3), in which case, we put
the array entry values into a comma-separated string and display that value
(line II). Figure 5.21 shows a sample display produced by showdata.php.
See **Ex: OptGroup** at the DWP website for a demo of showdata.php. While
$_REQUEST works for both POST and GET requests, the super global $_GET
or $_POST array works for GET or POST requests respectively.

An HTTP request, POST or GET, sends not only query data, but also
HTTP headers. The header information plus server version, paths, and script
locations are sent to a PHP script by the Web server in the super global
$_SERVER.

A script similar to showdata.php can be written to show all the values in
the super global array $_SERVER. The example **Ex: ShowRequest** can be run
at the DWP website to reveal all the values the $_SERVER contains.

FIGURE 5.21: A Sample FormData Display

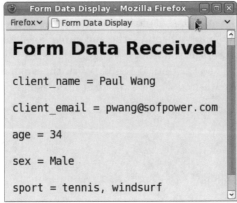

5.22   Processing File Upload

Now, let's write the upload processing code for the file upload form discussed in Section 5.7 that sends two files to the server side:

```
<input type="file" name="photo" accept="image/jpeg" />
<input type="file" name="report"
                    accept="application/pdf" />
```

The image file is sent under the key photo, and the PDF file is sent under the key report.

On the server side, we have the upload.php script to receive the uploaded files.

```
<?php   require_once("uploadfile.php");

if ( ! empty($_REQUEST['name']) &&   // required fields
     ! empty($_REQUEST['email']) )
{ $msg  = upload_file("photo", "image/jpeg", "img_dir");   // (A)
  $msg .= "<br /><br />" . upload_file("report",           // (B)
                    "application/pdf", "report_dir");
}
else
{ $msg='Upload failed.  Please go back and fill out all
        the required form entries.';
}   ?>
```

After checking for the required name and email fields, we call the function **up-load_file**, defined in uploadfile.php, to upload each of the two files (lines A and B) to the target folders img_dir and report_dir, respectively.

The upload results are kept in the $msg string to be displayed for the user in the response page.

```php
<?php $title="File Upload Confirmation";
      $bg="#def"; require("rfront.php"); ?>
<h1>Upload Confirmation for <?php echo $_REQUEST['name'];?>
</h1><p><?Php echo $msg; ?></p>
<?php require("rback.php"); ?>
```

Figure 5.22 shows a sample upload confirmation page.

FIGURE 5.22: File Upload Result

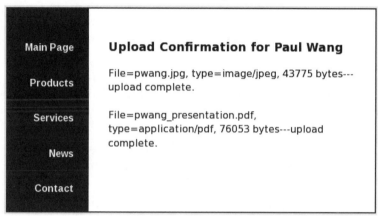

Let's see how **upload_file** is defined in `uploadfile.php`. First the **mime-type** function is defined to obtain the actual MIME type of the file uploaded that must be checked to ensure only allowed types be permitted. The function

```php
<?php
function mimetypeCheck($file, $types)
{   $finfo = finfo_open(FILEINFO_MIME_TYPE);
    $type= finfo_file($finfo, $file);
    finfo_close($finfo);
    return strstr($types, $type);
} ?>
```

checks the MIME type of `$file` against `$types`, a string of one or more allowed types.

From the super global array `$_FILES`, the **upload_file** function obtains the basename of the uploaded file under the key `$formkey` (line 1), the temporary location of the file uploaded (line 2), the file size (line 3), and any associated error (true or false) condition (line 4). The media type of the file is checked (line 3). If there was no error and the type is correct, then `upload_a_file` is called to do the job (line 5). Otherwise, the temporary file is deleted and an error message is returned (line 6).

```php
<?php
```

```
function upload_file($formkey, $check_type, $dir)
{   $file=$_FILES[$formkey]['name'];                     // (1)
    $tmp=$_FILES[$formkey]['tmp_name'];                  // (2)
    $size=$_FILES[$formkey]['size'];                     // (3)
    $error=$_FILES[$formkey]['error'];                   // (4)
    if( !$error && mimetypeCheck($tmp,$check_type) )
    {   return ("<p>" . upload_a_file($file,            // (5)
            $tmp, $size, $dir) ."</p>");
    }
    else
    {   unlink($tmp); return ("<p>upload_file: Sorry    // (6)
                uploading $file failed.</p>");
    }
}   ?>
```

The job of `upload_a_file` is simply to call the PHP built-in function `move_uploaded_file` to move the file from its temporary location to the target folder `$dir`.

```
function upload_a_file($file, $tmp, $size, $dir)
{   if( move_uploaded_file($tmp, "$dir/$file") )
    {   return ("File=" . $file . " (" .  $size
            . " bytes)---upload complete.");
    }
    else
    {   unlink($tmp);
        return ("upload_a_file: uploading $file failed.");
    }
}
```

The Web server must have read, write, and execute access to the destination folder (`$dir`) for **move_uploaded_file** to succeed and for files in the folder to be usable from the Web. See Section 10.7.4 for setting permission for Web files and folders.

The working example (**Ex: FileUpload**), consisting of the HTML form, the `upload.php` action script, and the required `uploadfile.php`, can be found at the DWP website.

5.22.1   Uploading Multiple Files

Uploading many files one at a time is tedious and tiresome for end users. Imagine uploading all your vacation pictures one by one! We can make it much easier by adding the `multiple` attribute to the file `input` element. For example, we can use

```
<label for="p">Your pictures (pick one or more):</label>
<input id="p" size="60" multiple="" type="file"
```

```
name="photo[]" accept="image/jpeg" />
```

to allow the user to pick one or more pictures using the browser-supplied file picker. The file names form a comma-separated list in the input field (Figure 5.23).

FIGURE 5.23: Uploading Multiple Files

The key `name="photo[]"` enables us to pick up the uploaded files as an array from the PHP super global `$_FILES`. Calling

```
upload_multiple_files('photo', 'image/jpeg', $img_dir);
```

will upload all files to the desired folder. The function simply obtains the array of files uploaded (line A), checks (line B), and moves (line C) the files one by one.

```
function upload_multiple_files($formkey,$check_type,$dir)
{  $file_arr=$_FILES[$formkey]['name'];                   // (A)
   $tmp_arr=$_FILES[$formkey]['tmp_name'];
   $size_arr=$_FILES[$formkey]['size'];
   $error_arr=$_FILES[$formkey]['error'];
   $result="<p>"; $len=count($file_arr);
   while ( $len-- > 0 )
   { $file=$file_arr[$len]; $tmp=$tmp_arr[$len];
     $size=$size_arr[$len]; $error=$error_arr[$len];
     if(!$error && mimetypeCheck($tmp,$check_type))  // (B)
     { $result .= upload_a_file($file, $tmp,          // (C)
                   $size, $dir) . "<br />";
     }
     else
     { unlink($tmp); $result .= "upload_multiple_files:
                   uploading $file failed.<br />";
     }
   }
   return $result;
}
```

See **Ex: MultiFile** on the DWP website for the complete example.

5.23 Summary

HTML forms organize a variety of input elements to collect user input. An input field is optional unless designated as `required`. Date, time, URL, email, telephone, and numerical input follow well-defined formats that browsers can validate before submitting the form. The `autofocus` attribute can bring input focus to the first input field of a form automatically. A `button` element, an image `input`, or a submit type `input` can be used to submit a finished form. Browsers will check user input for correctness before allowing a form to submit.

It is important to design and lay out your form for easy and accurate user input.

Formdata is sent in an HTTP POST or GET query to a server-side program for processing. Formdata is normally form-urlencoded unless `enctype` is `text/plain` or `multipart/form-data`.

PHP is a server-side scripting language specifically designed for the Web. Basics of PHP scripting (global and user-defined variables, statements, strings, arrays, conditionals, loops, and simple functions) have been introduced and applied for form processing that should check incoming data for correctness, carry out desired processing, and send a response page back to the user. Super global variables `$_GET`, `$_POST`, `_SERVER`, `$_FILE` make formdata easy to retrieve and use. A PHP script can easily send email, which is often part of form processing.

The response page to form submission must have visual unity with the rest of your site and should provide site navigation links lest it becomes a dead-end page.

The PHP super global `$_FILES` provides data for file uploading. Checking the file for the correct media type is a good idea. Make sure folders to receive uploaded files give read, write, and execute permission to the Web server.

Exercises

5.1. In HTML5, is it possible for an `input` element to be placed outside a form? Or belong to more than one form? Please explain.

5.2. What are possible types of submit buttons for a form? List and explain at least three types.

5.3. Is it possible for a form to have more than one submit button? Is it possible for a submit button to specify where the formdata is sent, regardless of what the form `action` attribute says? Explain.

5.4. Is it correct for a the form `action` attribute to have an empty string (`""`) as value? Why or why not?

5.5. HTML5 forms have automatic client-side validation. Give four example input elements where validation takes place.

5.6. What is the difference between the `post` and the `get` method of form submission?

5.7. What is the purpose of the `name` attribute for form input elements? In what situations should the value of `name` be a simple name or an array name such as `sports[]`?

5.8. What is *form urlencoding*? Please explain in detail.

5.9. Name all the data items sent from a form to a server-side program.

5.10. What PHP super global variables give a PHP script access to formdata, under the `get` or the `post` method. How do you access the individual values? All the key-value pairs without knowing the key names? Loop through all the key-value pairs? Check to see if any given value is present or not?

5.11. What operator in PHP is used to output one or more stings? Give some examples.

5.12. Outline the common tasks a form processing program must usually do.

5.13. If form input has been validated already by the browser, do we still need to check the input on the server side? Why or why not?

5.14. Set up a form with five different products with different prices for choosing. The server-side PHP program will display a response with a table of the picked items, their prices, and a total price. It should be possible to order 0 or more items of the same kind.

5.15. Add receipt by email to the previous exercise.

5.16. Refer to email sending from PHP in Section 5.20. Make the PHP email script more general so it can serve multiple different forms. The idea is to have the form send a hidden `formid` value that can be used to determine the receiver(s) for the generated email. The script can also check the referring page (`$_SERVER['HTTP_REFERER']`) so that only forms on permitted sites can use the script.

5.17. Run the **Ex:** `RequestForm` example and find out the meaning of all the values displayed.

5.18. Modify the `upload.php` file uploading script (Section 5.22) to process multiple pictures and reports from a form.

5.19. Find out if PHP puts any size limit on uploaded files and what can be done for uploading larger files.

5.20. Create a `formdemo.html` to include all types of form input elements in a well-designed layout. Send formdata to `showdata.php`.

Chapter 6

Dynamic User Interface with JavaScript

With an active page technology like PHP, we can add dynamism for webpages on the server side by generating code that is computed or program controlled at the time of page delivery. We have seen some of it in Chapter 5 and we will see much more in later chapters.

On the client side, dynamism for a webpage means a responsive user interface that interacts well with user/browser actions. Such interactions can enhance the functionality and usability of displayed pages and are achieved with JavaScript, a standard scripting language to control browser actions as well as all parts of a webpage.

JavaScript programs associated with a webpage are loaded into a browser together with its HTML and CSS code. In the context of the browser, objects representing windows, the HTML document, elements in the document, attributes, styles, and events are made available for access and manipulation by JavaScript. The HTML5 standard describes the HTML5 *Document Object Model* (DOM), which specifies how different element objects are structured and accessed as well as how they are organized into a DOM tree. The HTML DOM specification provides standardized APIs (Application Programming Interfaces) for document objects and user interface events. All major browsers implement these APIs in JavaScript.

Among many other tasks, you can use JavaScript to

- Monitor user events and specify reactions

- Make computations based on user input and display results

- Change the style and position of displayed elements

- Ask the browser to display informational or warning messages

- Pop up new windows and menus

- Detect browser type, version, and features

- Generate HTML code for parts of the page

- Modify/transform page content

- Check correctness of form input

- Go back and forth on visited pages or load new pages

- Control media loading and playback

- Perform and control CSS transitions and animations

We describe many of these actions in this chapter. Others and more are discussed in later chapters.

We will see how JavaScript is used in combination with other constructs to make a webpage more interactive and responsive. The emphasis is on applying JavaScript in practice rather than studying it as a programming language.

6.1   About JavaScript

Originally developed by Sun Microsystems[1] in the early 1990s for the Netscape Web browser, JavaScript is now the only widely accepted language for client-side scripting. The language is an international standard (ECMA-262), approved by the European Computer Manufacturers Association, in 1997 and later by the ISO in 1998. JavaScript is supported well by all major browsers. We will present JavaScript for adding dynamism and interactivity to webpages.

JavaScript is an object-oriented language, and many constructs are similar to those in C++ or Java. But, JavaScript is not a subset or version of Java. JavaScript programs are embedded in webpages and executed by browsers that provide the *host environment* or execution context. JavaScript code can generate HTML code to be included in the page and perform tasks in reaction to certain events. The host environment supplies document objects, browser objects, and JavaScript built-in objects for script manipulation. These objects can represent the entire document and well-defined parts in it such as windows, buttons, menus, pop-ups, dialog boxes, text areas, anchors, frames, history, cookies, and input/output. Objects provide useful *methods*, functions contained in the objects, for various purposes.

The host environment also provides a means to connect scripting code with events such as focus and mouse actions, page and image loading, form input and submission, error, and abort. Scripting code can also perform computation as the page is loading. Thus, the displayed page is the result of a combination of HTML, CSS, and JavaScript actions.

6.2   Getting Started with JavaScript

As our first JavaScript program, let's display the user agent and current date and time in a simple HTML page (**Ex:** `AgentDate`):

---

[1]Sun was acquired in 2010 and is now part of Oracle America, Inc.

```
<section><h1>Hello JavaScript</h1>
<p>JavaScript programming is fun.</p>
<p>Time = <span class="generated">
<script type="text/javascript">                         // (1)
   var dt = new Date();                               // (2)
   var time = dt.getHours() +":"+ dt.getMinutes();    // (3)
   document.write(time);</script></span></p>          // (4)
<p>UA = <span class="generated">
<script type="text/javascript">
   document.write(navigator.userAgent);               // (5)
</script></span></p></section>
```

Figure 6.1 shows the displayed result in Firefox and in Google Chrome. Note we styled the JavaScript produced content in blue to make it stand out.

FIGURE 6.1: JavaScript-Generated HTML

The `<script>` element (line 1) contains the JavaScript code and the `type` attribute required. Text lines enclosed in the `<script>` element are treated as program code of the given `type` and not to be displayed.

The variable `dt` is set to a `Date` object (line 2) created by the operator `new`, followed by the constructor call `Date()`. A constructor is a function you call to obtain new objects. The `getHours` and `getMinutes` methods of `Date` are used to construct the `time` string (line 3), which is inserted into the page with the `write` method of the `document` object (line 4).

Simple JavaScript statements are terminated by a semicolon (;). Comments are either from `//` to end of line or enclosed by `/*` and `*/`.

In JavaScript, a *string literal* is given inside single or double quotation marks (lines 3 and 5). In string literals, the usual BACKSLASH(\) escape works as well as special characters such as `\n` and `\r`. The operator `+` concatenates strings and objects, such as numbers, to strings.

The built-in object `navigator` stores properties related to the browser as a user agent. The `navigator.userAgent` property (line 5) gives a long and complicated string (Figure 6.1), which is the value for the `User-Agent` header in all HTTP requests made by the browser. Somewhere in this long string you will see a common browser name such as Firefox, Chrome, MSIE, Opera, Safari, iPhone, or android.

This simple example gives an idea of JavaScript programming. Learning JavaScript involves understanding its syntax, semantics, objects and methods, the host environment provided by the browser, events and event handling, and effective applications in practice. Let's first look at how JavaScript code is deployed in a webpage.

6.3   Placing JavaScript in Webpages

Generally, JavaScript can be embedded inside webpages and can be kept in files (with `.js` or `.es` suffix) that are loaded into pages by the browser. A JavaScript program usually consists of various code parts that cooperate to do the job. Where and how to place code in a webpage depend on the purpose of the code.

Inside a webpage, JavaScript can appear either in `<script>` elements or as values of event-handling attributes of HTML elements. Any number of `script` elements can be placed in `head`, `body`, flow, and phrasing elements. The browser executes JavaScript code encountered as it loads the page.

- Code for defining functions or creating data structures is placed in `<script>` elements inside the page's `<head>` element.

- Code for generating HTML to be displayed as part of the page is placed in `<script>` elements inside the `<body>` where the generated HTML code will be inserted.

- Code for actions triggered by events is given as values of *event attributes* of HTML tags. The code is usually a function call or a few short statements. `onfocus`, `onblur`, `onclick`, `onmouseover`, and `onmouseout` are examples of event attributes.

A JavaScript program often works for a set of pages, not just one. In that case, it is better to place the code in a separate file. The file can then be attached to a webpage with

```
<script type="text/javascript" src="file.js"></script>
```

which is usually placed inside the `head` element but can be placed elsewhere on the page. Just before the end of `body` is a good placement because it avoids the JavaScript delaying the page display.

The `src` attribute of `<script>` gives the URL of the program file. Note that we used the conventional `.js` file name suffix.

With file inclusion, any corrections or improvements in the `.js` file will be reflected in multiple pages automatically. Furthermore, a browser will download the program only once, thus reducing download time for these pages.

6.4   Image Rollovers

Have you moused over an image to see a different or enlarged version of it? This rollover effect is widely used by websites and is easy to do in JavaScript.

A basic rollover is simple. Take two images (Figure 6.2): a `webtong.com`

FIGURE 6.2: Image Rollover

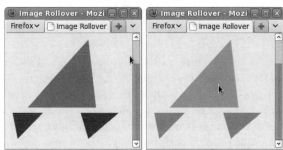

logo `wt1.png` on the left and a white-washed version of it (`wt2.png`) on the right. We want to display the first image, the *base image*, in a page, switch it to the second image, the *mouseover image*, when the mouse is over the image, and then switch back to the base image when the mouse leaves the image.

To achieve the rollover effect, we define actions connected to the `mouseover` and `mouseout` by providing the `onmouseover` and `onmouseout` event handlers for the target image. Here is an example (**Ex: RollImg**):

```
<img onmouseover="this.src='wt2.png'"          (A)
     onmouseout="this.src='wt1.png'"           (B)
     src="wt1.png" id="icon"
     alt="webtong.com logo" />
```

The action for `onmouseover` (line `A`) is a JavaScript expression that sets the source attribute for the image (`this`) to the rollover image `wt2.png`. The `onmouseout` action switches the image back again (line `B`). Note the use of the symbol `this` to refer to the `img` element itself.

In general, you can give one or more JavaScript statements (separated by `;`) inside quotation marks to define the action in response to an event:

on*SomeEvent*="*st1*; *st2*; ... "

To avoid long JavaScript code in an element tag, it is good practice to define an *event-handling function* and simply call that function.

The rollover effect is not limited to a single element. You can easily replace/restore an image when the mouse is over/out some other element. In general, a `mouseover`, a `mouseout`, or any other event from any element can trigger any well-defined JavaScript actions on the page. The dynamic effects are only limited by your design and imagination.

For example, the above-described rollover can be achieved by changing the `opaque` property of the image instead of using a second image:

```
<img src="wt.png" alt="webtong.com logo"
     onmouseover="this.style.opacity=0.4"     (C)
     onmouseout="this.style.opacity=1"  />    (D)
```

We set `this.style.opacity` to 0.4 on mouseover (line C) and back to 1 (line D) on mouseout. In general we can use JavaScript to set the style of any element to any value with the notation

target_obj.`style`.*property=value*

to set the given CSS style *property* of a target element to the given *value*. The *target_obj* is the DOM object for the intended target element. The *property* is a CSS property name. However, if the CSS property name contains a hyphen (`background-color`, for example), you must drop the hyphen and capitalize the next word (`style.backgroundColor`, for example). Rollover using CSS effects can be much faster than displaying another image.

It is also possible to access and set attributes of any element from JavaScript:

target_obj.`hasAttribute`(*attribute_name*) // true or false
target_obj.`getAttribute`(*attribute_name*)
target_obj.`setAttribute`(*attribute_name*, *value*)
target_obj.`removeAttribute`(*attribute_name*)

The `hasAttribute` method detects if the attribute has been explicitly specified. Removing an attribute makes it unspecified.

6.5   Image Preloading

When displaying another image on mouseover or mouse click, the responsiveness depends on the speed at which the image is displayed. Loading from the server side of the needed image is triggered by the `mouseover` or `click` event and can cause a significant delay when it happens for the first time.

To avoid this annoying sluggishness, we can preload any such image immediately after a webpage is loaded by the browser. Loading an image is simple with JavaScript. Create a new `Image` object and set its `src` attribute to the image URL. This causes the image to download across the Internet. Once loaded, an image displays immediately upon request. When called, `loadImage` loads the image given by the `url`:

```
<script type="text/javascript">
function loadImage(url)
{   if (document.images)    // if browser supports images
    {  img = new Image();   // obtains a new image object
       img.src = url;       // downloads the image
       return img;
    }
}
</script>
```

The script is placed inside the **<head>** element.

The **loadImage** function is triggered by the **load** event (line E) on **body**, which occurs immediately after page loading completes:

```
<body onload="loadImage('wt2.png')">          (E)
```

The call (line F) downloads the mouseover image for the rollover (**Ex: RollImg**). Because the mouseout image **wt1.png** has already been loaded as part of page loading, it does not need to be loaded again.

Image preloading is also important for displaying enlarged pictures or slide shows in response to user action.

The **prefetch** meta tag (Section 3.10.2) can be an alternative if it is implemented uniformly by browsers. Another way to preload is to use hidden **img** tags in the page. The JavaScript approach has the advantage of separating the preload code from the page itself.

6.6 Presenting a Slide Show

With JavaScript programmed **onclick** actions for the **click** event we can make a slide show (Figure 6.3). Pictures of delicious dishes are displayed one by one. Users control the display with **next** and **previous** buttons. (**Ex: Slides**).

The HTML for the page can be coded as follows:

```
<body style="background-color: #def"
     onload="changeSlide()">                     (1)
<p style="text-align: center">
<img class="slide" src="" id='display' alt="slide" />  (2)
<br /><br />
<img title="backward" src="backward_s.png"
     onclick="prevSlide()" alt="previous slide" />    (3)
       
<img src="forward_s.png" title="forward"
     onclick="nextSlide()" alt="next slide" />        (4)
</p></body>
```

FIGURE 6.3: JavaScript Slide Show

The centered image with `id='display'` is the slide area where the image `src` is left empty (line 2). Its image source is set with the `onload` event handler of the `body` element (line 1). Clicking the `next` (`previous`) button calls the `nextSlide()` (`prevSlide()`) function (lines 3 and 4).

The `slide.js` file contains the code for loading and changing the images. The `pic` array (Section 6.7) contains the pictures to be displayed and is the only quantity you need to customize for your own slide show. Make sure all images are the same size.

```javascript
var pic = [ "pancake.jpg", "dessert.jpg",
            "meat.jpg", "cake.jpg", "pasta.jpg" ];

var slide = new Array();
var index = 0;              // current slide index

function loadImage(url)
{  if (document.images)
   {    rslt = new Image();
        rslt.src = url; return rslt;
   }
}

// preloading all images
if ( document.images )
   for (var i in pic)                        // (5)
```

```
{ slide.push(loadImage(pic[i])); }
```

The images are preloaded into the `slide` array (line 5). The variable `index` is a global variable that controls which image is shown through a call to `changeSlide()`.

```
function changeSlide()
{
    document.getElementById('display').src = slide[index].src;
}
```

Going forward and backward on the slides is just a matter of incrementing and decrementing `index` in a circular fashion:

```
function prevSlide()
{   if(--index < 0) { index = pic.length-1; }
    changeSlide();
}
```

```
function nextSlide()
{   if( ++index >= pic.length) { index = 0; }
    changeSlide();
}
```

The CSS used for the slide show is

```
img.slide
{ background-color: #060; padding: 40px;
    border: 2px solid white; display: inline-block;
}
```

6.7   JavaScript Arrays

In the slide show (Section 6.6) we used the `pic` and `slide` arrays. JavaScript arrays can be created in the following ways:

```
var a = new Array(arrayLength);
var b = new Array(entry,  ...);
var c = [entry,  ...];
```

where c is a read-only array. The length of an array is recorded in its `length` attribute (e.g., `c.length`). Array elements can be set and retrieved with

```
a[0] = "first";
a[1] = "second";
var value = b[6];
```

The index runs from 0 to length−1. Accessing an undefined array entry gives the value undefined. Assigning to an element beyond the end of the array increases its length. Array methods include

pop()	removes and returns last element
shift()	removes and returns first element, ex: e=a.shift()
unshift(e1,e2, ...)	inserts elements in front and returns new length
push(e1,e2, ...)	inserts elements at end and returns new length
concat(arr2, ...)	returns a new array by joining the array with the given arrays
reverse()	changes the array itself to go backward

6.7.1   The *foreach* Loop

As in PHP and other scripting languages, JavaScript supports associative arrays where the array index can be strings instead of integer indices. For example, we can set up an associative menu array:

```
var menu=Array();
menu['appetizer']="Spring Rolls";
menu['soup']="Miso Soup";
menu['entree']="Steamed Fish";
menu['dessert']="Eight-treasure Rice Pudding";
```

In addition to the usual while and for loops (Section 6.16.5), JavaScript also offers a convenient foreach loop to go through entries in any array *arr*:

```
for (var key in arr)
{ /* do something with arr[key] */ }
```

This works for both indexed and associative arrays because an indexed array can also be considered associative with the keys being the indices 0, 1, 2, and so on. For example,

```
ans=0;
arr=[7,8,9]; for(var k in arr) { ans += arr[k]; }
```

sums the three numbers.

The foreach loop also works on JavaScript objects, such as window and navigator. For example,

```
for (var key in navigator)
{  window.alert( key + "=" + navigator[key]);  }
```

displays the property-value pairs for the built-in object navigator. See Section 6.15 for information on the window.alert method.

6.8   A Conversion Calculator

HTML input controls also provide many opportunities for user interaction. Let's see how `input` elements can be used outside a form to construct an inch-centimeter conversion calculator supported by client-side JavaScript (Figure 6.4).

FIGURE 6.4: A Conversion Calculator

The HTML code for the conversion calculator example (**Ex:** `Convert`) is

```
<head>
<meta charset="utf-8"/>
<title>Unit Conversion Demo</title>
<script type="text/javascript" src="convert.js">
</script></head>
<body style="background-color: #def" onload="init()">        (1)
<section><h1>Inch-Centimeter Converter</h1>
<p><label>in <input id="in" size="10" type="number"
                onfocus="reset()" /></label>                (2)
<input type="button" value="Convert" onclick="convert()" /> (3)
<label><input id="cm" size="10" type="number"
            onfocus="reset()" /> cm</label>                 (4)
</p></section></body>
```

Events from the three `input` elements are connected to event-handling functions. The `inch` (line 2) and `cm` (line 4) input fields will reset themselves when focus is gained. A button click triggers the actual conversion (line 3).

JavaScript is initialized `onload` (line 1) by calling the function `init()`

```
var inf, cmf;

function init()
{    inf = document.getElementById('in');
     cmf = document.getElementById('cm');
}
```

which sets up two global variables `inf` (the inch input element) and `cmf` (the cm input element). These make the convert and reset operations easy to program:

```
function reset()
{   inf.value = "";    cmf.value = "";    }        // (5)

function convert()
{   var i = inf.value.replace(/ /,"");              // (6)
    if ( i )                                        // (7)
    { cmf.value = i * 2.54; return; }               // (8)
    var c = cmf.value.replace(/ /,"");
    if ( c )
    { inf.value = c / 2.54; }                       // (9)
}
```

Both elements are set to contain no input when either one gains focus (line 5).

After entering a number in either field, the user clicks the Convert button (line 3), which calls the convert function to display the conversion results.

The input string from the inch field is obtained (line 6), and any unintentional spaces are removed. If this input is not empty (line 7), the centimeter conversion is computed and displayed (line 8), and the function returns. Here again, you can see how strings and numbers are used interchangeably in JavaScript. If the inch field has no input, then the cm field is processed exactly the same way.

JavaScript allows you to use strings in arithmetic computations (lines 8 and 9). Note: Unlike in Java or C++, the JavaScript / operator always performs a floating-point division and returns a floating-point result.

6.9 Audio/Video Control

HTML5 audio and video elements (Section 3.1 and Section 3.2) provide a media API to allow JavaScript control of various aspects of playback, including play, pause, volume, mute, and playback position settings. By setting the src attribute of the media element to a desired URL, we can load and play any media file.

As an example, let's use our own play, pause, and restart buttons to control a video clip (Figure 6.5): The HTML code is as follows (**Ex**: VideoControl):

```
<video id="myvideo" width="480" height="385"
        poster="nasa.jpg">                                    (A)
<source src="nasa.webm" type="video/webm" />
<source src="nasa.ogv" type="video/ogg" /></video>
<script type="text/javascript">
var me=document.getElementById('myvideo'); </script>  (B)
<p style="width: 480px; text-align: center">
<img src="icons/rewind.png" width="50px"
        onclick="me.currentTime=0.25; me.rewind()" />         (C)
      
```

FIGURE 6.5: JavaScript Control of Media Playback

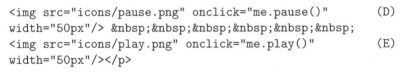

```
<img src="icons/pause.png" onclick="me.pause()"        (D)
width="50px"/>       
<img src="icons/play.png" onclick="me.play()"          (E)
width="50px"/></p>
```

We removed the built-in video player controls and added a still picture poster (line A) for the `video` element. JavaScript code initializes the variable `me` to the target media element object (line B) that is used to play (continue) (line E), pause (line D) or restart (line C) the video playback by rewinding to the 0.25-second mark.

Of course we can set the `me.currentTime` property to any floating-point number representing the playback position in seconds. Also, the `me.volume` property can be set to a number between 0 and 1. The `me.muted` can be set to true/false.

A new audio/video element object can be created with

```
var me=document.createElement('audio');
var me=document.createElement('video');
```

The source URL can be set with `me.src=`*url*, and loading of the source file can be initiated with `me.load()`. The media will begin playback when loading is done if the `me.autoplay` flag is set to true. The media element API provides other properties and events (Section 6.21) for easy JavaScript control and manipulation of audio and video. See also the HTML5 media element documentation for more information.

6.10   Pull-Down Menus

For websites with more than a few pages, a good navigation design is important. One way to organize many links for easy access, yet not crowding the page design, is to use pull-down menus. We will present a solution that can easily be reused in different situations.

Our approach involves menus that are hidden (`display: none`). When the user clicks on a menu label, the associated hidden menu appears below the label, and the user may use the mouse to select a menu item and click on it to follow that link.

Any menu being pulled down puts the navbar of menu labels into the *activated* state. Under the activated state, all menus are said to be active and mouseover any menu label pulls down that menu and pulls up any other menu. Thus, there can be at most one menu pulled down at any given time. To cancel the current menu, simply click some inactive spot anywhere on the page. This pulls up the menu. A menu reappears if the mouse hovers over its label (no clicking needed). The menu will also pull up if the browser window changes size.

To *deactivate* the menus, click on the label of the current menu. Figure 6.6 shows an `Activity` menu below its label on a horizontal navbar.

FIGURE 6.6: A Pull-Down Menu

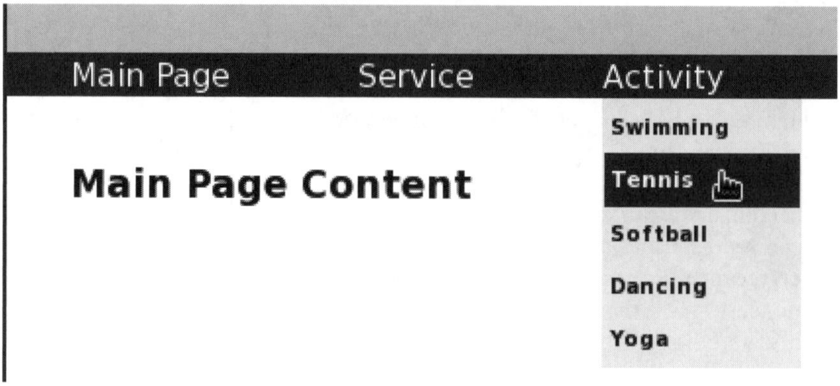

Our pull-down menu implementation involves these components:

- A horizontal navbar where links and menu labels are placed.

- An HTML fragment file (`menu.inc`) containing one or more menus. Each menu is a `nav` element containing child links.

- A style sheet (`menu.css`) for the navbar and the menus.

- A JavaScript file (`menuup.js`) supplying the pull-down, pull-up actions.

To emphasize reusability, we will attach the pull-down menu to a page using the table-styled page layout template discussed in Section 4.10.2. The horizontal navbar code is

```
<header class="banner"> <!-- top banner omitted -->
<nav>
<a href="#">Main Page</a>
<a href="#">Service</a>
<span class="menulabel"                              (A)
    onclick="menuActivate(event,this,'activity',0,0)"
    onmouseover="menuDown(this,'activity',0,0)"
  >Activity</span>
<a href="#">Contact</a></nav>
</header><div id="contentbox" ...> ...                (B)
```

On the navbar, we have three links and a menu label (Activity) that is styled the same way as the links but has JavaScript-defined actions connected to the click and mouseover events (line A). To add another pull-down menu, simply place another such span element on the navbar, thus,

```
menuActivate(event,this,'menu_id', dx, dy)
menuDown(this,'menu_id', dx, dy)
```

The *dx*, *dy* are the x and y offsets for adjusting menu positioning (see below).

Although we show just one menu label here, code in this example works for any number of menus. The very element after the navbar is the `contentbox` `div` that will be used to compute the placement of menus pulled down from the navbar (line B).

The `body` element also has event actions defined:

```
<body id="top" onload="init()"
      onresize="menuUp()" onclick="menuUp()">  (C)
...
<?php require("menu.inc"); ?></body>          (D)
```

A mouse click in any inactive spot in the page, or a window resize event, triggers the `menuUp()` operation (line C). The HTML code for the menus are included with PHP code from the file `menu.inc`. Here is the HTML code for the activity menu:

```
<nav class="menu" id="activity">   <!-- menu id -->
<a href="swimming.html"> Swimming</a>
<a href="tennis.html"> Tennis</a>
<a href="softball.html"> Softball</a>
```

```
<a href="dancing.html"> Dancing</a>
<a href="yoga.html"> Yoga</a>
</nav>
```

The menu has absolute positioning (line E) and is hidden initially (line F). Menu items (links) are styled to respond to mouseover events (lines G and H).

```
nav.menu
{   font-size: x-small; width: 10em;
    background-color: #def;
    color: white; padding: 1px;
    position: absolute; top: 0px; left:0px;     /* (E) */
    display: none;                              /* (F) */
}


nav.menu a:link                                 /* (G) */
{   padding-top: 2px; padding-bottom: 2px;
    color: #000; width: 100%;
    white-space: nowrap; font-weight: bold;
    text-decoration: none; display: block;
}


nav.menu a:hover
{   color: white; background-color: darkblue; } /* (H) */
```

Let's now turn our attention to the all-important JavaScript actions in the file menuup.js. After declaring global variables, we define the init() function, which is called onload and computes the y position of pull-down menus. Our menus are in either the *activated* or *deactivated* state. Only in the active state will menus pull down automatically on mouseover. Clicking on a menu label activates/deactivates the menus:

```
/////    menuup.js    ///////
//    pull-down menus

var menuActive=false;    // true if any menu is activated
var currentMenu=null;    // activated menu
var menu_y;              // y position of menu

function init()
{ menu_y=yPosition(document.getElementById('contentbox')); }

function menuDeactivate()               // pull up menu
{ if ( menuActive )
  { menuUp(); menuActive=false; } // deactivated state
}
```

```
function menuUp()
{ if ( currentMenu )
  { currentMenu.style.display = "none";   // disappears
    currentMenu=null;
  }
}
```

There is at most one menu showing at any given time. The function `menuActivate` activates and deactivates menus. Its arguments are `event` (the click event object), `label` (the menu label element object clicked), `id` (the target menu id string), and `dx` and `dy` (the integer x and y pixel offsets).

If menus are already active, we deactivate them and pull up any displayed menu (line I). Otherwise, we call `menuDown` to pull down the target menu (`id`) (line J) and enter the activated state (line K). Then we need to prevent the click event from propagating (or bubbling) up (lines L and M), because a click event received by `body` will pull up the current menu (line C above).

```
function menuActivate(event, label, id, dx, dy)
{ if ( menuActive ) menuDeactivate();      // (I)
  else
  { menuDown(label, id, dx, dy);           // (J)
    menuActive=true; // activated state    // (K)
    event.cancelBubble = true; // IE       // (L)
    if (event.stopPropagation)
      event.stopPropagation(); // Firefox // (M)
  }
}


function menuDown(label, id, dx, dy)
{ if ( menuActive )
  { down(label, document.getElementById(id), dx, dy); }
}
function down(lb, menu, dx, dy)
{ if ( currentMenu ) menuUp();
  menu.style.top=(menu_y+dy) + "px";           // (N)
  menu.style.left=(xPosition(lb)+dx) + "px"; // (O)
  menu.style.display = "block";                // (P)
  currentMenu=menu;                            // (Q)
}
```

The `menuDown` function will only work in the activated state. It calls the function `down` with the given position offsets `dx` and `dy` (Figure 6.7).

The function `down` pulls up any current menu before pulling down the target `menu`. The y position for the menu is `menu_y` (y of element `contentbox`) plus the offset `dy` (line N). The x position is the x position of the menu label plus the offset `dx` (line O). The offsets make adjusting the positions of individual menus easy, although `dx=0` and `dy=0` will work most of the time. The

FIGURE 6.7: Pull-Down Menu Positioning

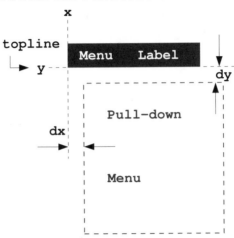

menu is then placed at the desired position and recorded as the current menu (lines P-Q).

The (x,y) position of the upper-left corner of any displayed element can be computed by the functions xPosition and yPosition:

```
function xPosition(nd)
{ var x=0;
  while (nd && nd.offsetParent)
  {  x += nd.offsetLeft; nd = nd.offsetParent; }
  return x;
}
function yPosition(nd)
{ var y=0;
  while (nd && nd.offsetParent)
  {  y += nd.offsetTop; nd = nd.offsetParent; }
  return y;
}
```

Each of these two functions goes up the nd node's *offset parent* chain and sums up the offsets (See DOM Chapter 7). The JavaScript while loop works the same way as in C/C++ or Java.

These position computing functions can be useful in other situations. The working example (**Ex: Menu**) can be found on the DWP website.

6.11 CSS Transitions

A *CSS transition* is the smooth changing of a displayed style from its original starting value to a target ending value in a given time duration. A transition can be defined for any CSS property with a numeric or percentage value. The general form a transition style is

`transition:` *property duration fn,* `...`

where you specify a comma-separated list of one or more properties for transition. Or you can use the keyword `all` for all properties. Each gives a time *duration* and an optional timing function *fn.*

The timing function can be linear where the property moves from its starting value to its final value with a constant speed throughout the transition duration. The linear function would be represented in Figure 6.8 as a straight line linking points P_0 (the starting point) and P_3 (the end point). But you do not have to settle for a linear transition and you can specify the middle points P_1 and P_2 for a Cubic Bézier curve defined function as shown in Figure 6.8.

FIGURE 6.8: Transition Cubic Bézier Timing Function

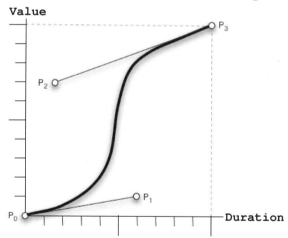

The timing function *fn* can be given in general as

`cubic-bezier(`x_1 `,` y_1 `,` x_2 `,` y_2`)`

giving the coordinates of the points P_1 and P_2. All four values are normalized to lie between 0 and 1. Predefined timing functions are

- `linear`—constant speed, same as `cubic-bezier(0,0,1,1)` a straight line

- `ease`—slow start and end, same as `cubic-bezier(0.25,0.1,0.25,1)`

- `ease-in`—slow start, same as `cubic-bezier(0.42,0,1,1)`

- `ease-out`—slow end, same as `cubic-bezier(0,0,0.58,1)`

- `ease-in-out`—same as `cubic-bezier(0.42,0,0.58,1)`

Let's apply CSS transitions to improve the pull-down menu.

6.12   Menu Fade-In and Fade-Out

The pull-down menu in Section 6.10 appears and disappears abruptly. We can add fade-in and fade-out effects to it with JavaScript and CSS *transitions*.

For menu fade-in and fade-out, we can define a CSS transition on the `opacity` style property of each menu. The menu opacity changes from 0 (completely transparent) to 1 (completely opaque) for pulling down and from 1 to 0 for pulling up. The CSS for a menu now becomes

```
nav.menu
{ font-size: x-small; width: 10em; position: absolute;
  background-color: #def;  color: white; padding: 1px;
  top: 0px; left:0px;
  display: none; opacity: 0;                  /* I */
  -moz-transition: opacity 0.7s ease-in-out;  /* II */
  -webkit-transition: opacity 0.7s ease-in-out;
  -ms-transition: opacity 0.7s ease-in-out;
  -o-transition: opacity 0.7s ease-in-out;
  transition: opacity 0.7s ease-in-out;       /* III */
}
```

The initial `opacity` is 0, which makes it totally transparent. But we still need the `display: none` to make it inaccessible to the user (line I). The transition duration on opacity is set to 0.7 seconds (line II) and the transition-timing function used is `ease-in-out` (lines II and III). The default transition-timing function is `ease`, meaning the target property value varies with a slower start, faster middle, and slower finish. The `ease-in-out` is similar but uses slightly different speeds.

With the CSS in place, we now need to revise the JavaScript for pulling up to

1. Display the totally transparent menu in the correct location (lines IV and V)

2. Transition the `opacity` from 0 to 1 (lines VI and VII)

```
function down(lb, menu, dx, dy)              // (IV)
{  if ( currentMenu ) menuUp();
   var x=xPosition(lb);
```

```
    menu.style.top = (menu_y+dy) + "px";
    menu.style.left = (x+dx) + "px";
    menu.style.display = "block";
    menu.style.opacity = 0;
    currentMenu=menu;                          // (V)
    setTimeout(showdisplay, 50);               // (VI)
}
function showdisplay()                         //  (VII)
{   currentMenu.style.opacity = 1; }
```

The `setTimeout` method of `window`, in the general form,

`window.setTimeout(`*target_function*`, ` *delay*`)`

schedules a call to the given target function to take place *delay* milliseconds later. The target function can be given as a function name as on line **IV** or an anonymous function (Section 6.14). If we call `showdisplay()` directly on line **VI** it does not trigger the CSS transition. We must display the menu with opacity 0 first, then quickly call `showdisplay()` to set opacity to 1. Experimentation shows the 50-millisecond delay is just enough.

The pull-up operation needs to

1. Transition `opacity` from 1 to 0 so the menu is invisible (line **VIII**)

2. Set `display` to `none` so the menu is inaccessible (lines **IX**, **X** and **XI**)

```
function menuUp()
{   if ( currentMenu )
    { currentMenu.style.opacity = 0;           // (VIII)
      up_menu=currentMenu;                     // (IX)
      currentMenu=null;
      setTimeout(nodisplay, 750);              // (X)
    }
}


function nodisplay()
{   up_menu.style.display = "none"; }          // (XI)
```

Note that we recorded the menu element in a global variable `up_menu` in order to use it late in the `nodisplay()` function. Also, our transition to opacity 0 takes 0.7 seconds (line **II**) and the `nodisplay()` call is made very shortly after that (line **X**).

No other changes are needed. Try the working code (**Ex: MenuFade**) online at the DWP website.

To further illustrate CSS transitions, let's use it to animate the pulling up and down of menus.

6.13   Animated Pull-Down Menu

By combining CSS positioning, z-indexing, and transition control, we can animate the pull-up/down effect on menus. It is not complicated. For pulling down, we place the desired menu behind the banner header and transition its `top` value to the final position (Figure 6.9). For pulling up, we do the opposite.

FIGURE 6.9: Animated Menu Pulling

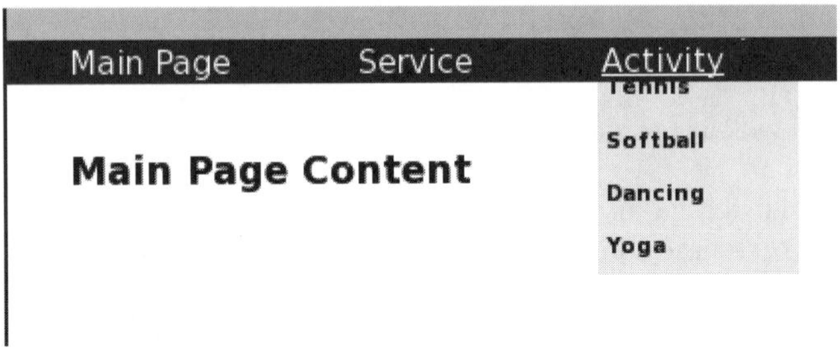

The menu style becomes

```
nav.menu
{  font-size: x-small;
   width: 10em;
   position: absolute;
   background-color: #def;
   color: white; padding: 1px;
   display: block; left:0px; top: -30px;      /* a */
   opacity: 0; z-index: 1;                     /* b */
   -moz-transition: top 0.5s ease-in-out;     /* c */
   -webkit-transition: top 0.5s ease-in-out;
   -ms-transition: top 0.5s ease-in-out;
   -o-transition: top 0.5s ease-in-out;
   transition: top 0.5s ease-in-out;
}

header.banner
{ position: relative; z-index: 20; }          /* d */
```

By setting `top` to a suitable negative value (line **a**), we make sure the menu is hidden behind the `header` (lines **b** and **d**) even when its `opacity` is set to 1 in the JavaScript pull-down action. The transition on the value of `top` supplies the animation (line **c**).

The JavaScript actions for pulling up/down are

```
function down(lb, menu, dx, dy)
```

```
{  if ( currentMenu ) menuUp();
   var x=xPosition(lb);
   menu.style.opacity = 1;              // (e)
   saved_top=menu.style.top;            // (f)
   menu.style.top = (menu_y+dy) + "px"; // (g)
   menu.style.left = (x+dx) + "px";
   currentMenu=menu;
}


function menuUp()
{  if ( currentMenu )
   {  currentMenu.style.top = saved_top;  // (h)
      currentMenu=null;
   }
}
```

To pull down a menu, we set opacity to 1 (line e), save the current top position that hides the menu (line f), then set the new value for top (line g) triggering the transition animation. Because the display value of the menu is never set to none, we do not have the setTimeout complication as in our fade-in case (Section 6.12).

To pull up the current menu, we simply set its top value back to the saved_top value.

Try the working code (**Ex: AnimPull**) online at the DWP website.

6.13.1 Triggering CSS Transitions

Transition-defined animations are useful in many other ways. Even though the CSS rules are static, the transition effects can be triggered by any browser event through JavaScript. For example, we may want to do a transition after a page is loaded (via onload of the body element) or when the user clicks on some button. The JavaScript event handler can simply assign the className attribute for the element to trigger the transitions.

6.14 Sliding Menus for a Left Navbar

We have discussed three versions of the pull-down menu anchored to a horizontal navbar (Section 6.10). It is time to turn our attention to menus for vertical navbars, which are just as popular as horizontal ones, especially with the HD screen ratio.

The basic user interaction with the menu remains the same. A click on a menu label on the vertical navbar expands the hidden menu into view and activates all such menus on the navbar. Another click contracts the menu and deactivates menus. While active, menus expand and contract with mouseover actions. At any time there is at most one expanded menu showing (Figure 6.10).

FIGURE 6.10: Sliding Menus

 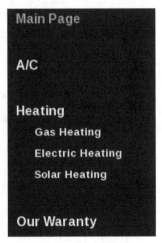

Let's take the left navbar first described in Section 4.11 and put new links plus two menus, one for A/C and the other for heating, on it. The HTML file uses PHP to generate the two menu labels and to include the two menus (lines 1 and 2):

```
<nav class="leftnavbar">
<span class="self">Main Page</span>
<?php $mid="cooling"; $mtext="A/C";        // (1)
      require("label.php"); ?>
<?php $mid="heating"; $mtext="Heating";   // (2)
      require("label.php"); ?>
<a href="#">Our Warranty</a>
<a href="#">Contact Us</a>
</nav>
```

The `label.php` generates the code for a menu with label for the left navbar given its id (`$mid`) and label text (`$mtext`) values (lines 1 and 2). This way menus on the left navbar are coded and styled consistently.

The A/C menu (`cooling.inc`) is

```
<nav class="menu" id="<?php echo $mid ?>">
<a href="central.html"> Central Units</a>
<a href="room.html"> Room Units</a>
<a href="window.html"> Window Units</a>
</nav>
```

The heating menu is similar. Menu-related CSS rules are

```
nav.menu
{ font-size: small;
  background-color: darkblue;                     /* 3 */
  position: absolute; top:-100px; left: -200px;   /* 4 */
  display: block; opacity: 0;                      /* 5 */
  width: 100%; overflow: hidden;                   /* 6 */
  -moz-transition: height 0.7s ease-in-out;       /* 7 */
  -webkit-transition: height 0.7s ease-in-out;
  -ms-transition: height 0.7s ease-in-out;
  -o-transition: height 0.7s ease-in-out;
  transition: height 0.7s ease-in-out;
}
```

The background color of the menu is set to match that of the left navbar (line 3). Initially, the menu is positioned out of the way and hidden (lines 4 and 5). We will expand and contract the height of the menu in response to mouseover and mouse click events on its menu label. Thus, it is critical to set overflow to hidden (line 6). Menu expansion and contraction are animated with CSS transition (line 7).

JavaScript control of the sliding menus is based on

- Computing the full height of each menu

- Switching the menus from their absolutely positioned hiding place back to their static positions, just below their menu labels on the navbar

- Sliding the menus in and out of view by transitioning their height from 0 to full height and back to 0

Menu activation and deactivation are similar to pull-down menus discussed earlier. Because sliding menus stay within the left menu bar, there is no need for clicks inside the body to make them disappear.

The function menuShow expands a menu into view. If the menu is shown for the very first time (line a), its full height is computed and recorded in the global associative array m_h (line b), and the function appear is called to make the menu appear for the first time. Any future expansion of the same menu becomes more straightforward and we use the function show for that (line d).

```
function menuShow(label, id)
{ if ( menuActive )                               // (a)
  { if ( ! m_h[id] )
    { m_h[id]= computedHeight(id);                // (b)
      appear(label, document.getElementById(id),
            m_h[id]);                             // (c)
    }
```

```
      else
        show(label, document.getElementById(id),    // (d)
            m_h[id]);
  }
}
```

To make a menu appear for the first time, we hide any menu already appearing (line e), switch the target menu back to static positioning (line f), setting the opacity to fully opaque, and give it a minimal nonzero height (line h) as a starting point of height transitioning to full height (ht), which is scheduled to take place almost immediately (lines i and j).

```
function appear(lb, menu, ht)
{  if ( currentMenu ) menuHide();                     // (e)
   menu.style.position = "static";                    // (f)
   menu.style.opacity = 1;                             // (g)
   menu.style.height="1px";                            // (h)
   currentMenu=menu;
   setTimeout(function(){ shownow(ht)}, 30);          // (i)
}
function shownow(ht)
{ currentMenu.style.height=ht; }                       // (j)

function show(lb, menu, ht)                            // (k)
{  if ( currentMenu )  menuHide();
   menu.style.height=ht;
   currentMenu=menu;
}
```

We used an anonymous function in setTimeout that provides a *closure*, preserving the value of the local variable ht, in which to call shownow(ht).

After the first time, expanding a menu requires less work (line k). Hiding an expanded menu is just a matter of contracting the height back to zero:

```
function menuHide()
{ if ( currentMenu )
  {  currentMenu.style.height="0px"; currentMenu=null; }
}
```

The critical task of computing the height of each individual menu falls upon the function

```
function computedHeight(el)
{  el = document.getElementById(el);
   var browserName=navigator.appName;
   if(browserName=="Microsoft Internet Explorer")
      return(el.offsetHeight + "px");
```

```
    else
        return(document.defaultView.getComputedStyle(el, "").
            getPropertyValue("height"));
}
```

See the full example (**Ex: SlidingMenu**) at the DWP website.

6.15 Using Windows

The `window` object represents an open window in a browser. Each loaded webpage has its own `window` object that contains all other objects related to the page. If a document contains `iframes`, an additional window object is created for each iframe inside the page.

The `window` object contains all other objects related to the page and many properties, including

- `navigator`—an object representing the browser software

- `document`—the object that represents the structure and content of the webpage (HTML code)

- `location`—an object containing the current URL

- `history`—a sequence of URLs visited previously

- `frames`—an array of windows, each for an iframe in the webpage

Setting `window.location` to another URL loads that target document:

`window.location = ` *someURL*

The window history can be used to go back and forth on the list of visited pages. For example,

```
history.back();          // reloads the previous page
history.forward();       // reloads the next page
history.go(-3);          // goes back three pages
```

Note that `history` is shorthand for `window.history`. Because JavaScript code works in the context of the current window, you can use method and attribute names in the current window directly.

6.15.1 Dialog Windows

The `window` object's `alert`, `confirm`, and `prompt` methods pop up a dialog window. The user usually must respond to the dialog before proceeding.

The `alert` method of `window` displays a warning message. For example

FIGURE 6.11: Alert Window

(**Ex: Alert**), a function that checks the validity of credit card numbers can easily display a warning (Figure 6.11):

```
function checkCC(number)
{  if ( credit card number invalid )
   {   window.alert("The credit card number is invalid.");
       return false;
   }
   else return true;
}
```

Use a prompt box to collect user input interactively. The call

window.prompt(*message, default-string*)

displays a prompt dialog with the given message and the optional default input string in an input box. For example (**Ex: Prompt**),

```
var email = window.prompt('Please enter your email address:');
window.confirm("Email is:  " + email);
```

obtains an email address (Figure 6.12) and displays a confirmation dialog (Figure 6.13) (**Ex: Confirm**). The `prompt` method returns either the input string which can be empty or `null` if the user cancels the dialog. The `confirm` method returns `true` or `false`.

6.15.2   Opening New Windows

To open a new window, use

```
window.open("URL", "window-name", "options")
```

If `window-name` is an existing window, it will be used to display the page identified by the given URL. Otherwise, a new window to display the given page will be made. If the URL is an empty string, a blank window is displayed. For example,

FIGURE 6.12: Prompt Window

FIGURE 6.13: Confirmation Window

```
window.open("http://www.abc.org","Abc")          // (1)
window.open("http://www.abc.org","Abc",          // (2)
            "scrollbars=yes,toolbar=no")
```

opens a new full-size full-featured browser window (line 1) or one without any toolbars (line 2) (Figure 6.14).

If you use pop-up windows for offsite links, then users can close that window to return to your site. To make it less likely for users to wander off to the other site, and to make it clear that it is a side trip, you may decide to omit the toolbars for the new window. Thus, offsite links may be given as follows (**Ex:** PopupWindow):

```
<a href="javascript:window.open('http://www.abc.org')">
```

Alternatively, the function

```
function popWindow(URL, w, h)
{  window.open(URL, "", "toolbar=no" +
                       ",dependent=yes" +
                       ",innerwidth="+ w +
                       ",innerheight="+ h);
```

FIGURE 6.14: A Pop-Up Picture

```
}
```

opens a new window for the given URL, without any toolbars, and with a content area specified by w and h. Such windows are also useful to display a pop-up picture (Figure 6.14), some product data, or other auxiliary information. A website may use pop-up windows for examples, figures, and resource links. Use `width` and `height` instead if you wish to set the outer dimensions of the pop-up window.

The options `screenX` and `screenY` position the upper-left window corner relative to the screen. Some yes-no options for `window.open` are

- `dependent`—Makes the new window a child of the current window so it will close automatically if the current window closes.

- `location`—Displays the `location` entry or not.

- `menubar`—Adds the menu bar or not.

- `resizable`—Allows resizing or not.

- `scrollbars`—Enables scroll bars or not.

- `status`—Adds bottom status bar or not.

- `toolbar`—Includes the toolbar or not.

Opening a new window with no options sets all `yes/no` options to `yes`. But if you supply some options, then most unspecified `yes/no` options are set to `no`. Some features of `window.open()` are still not standardized across all browsers.

Let `win` be the window object for any window; then `win.close()` closes the window and `win.print()` opens a print dialog for printing the window contents.

The `window` object enjoys broad browser support but has not been completely standardized. Other properties and methods of `window` will be explained when used.

6.16 JavaScript Language Basics

Having demonstrated JavaScript for creating dynamic effects, we will now provide a bit more language basics so the power of JavaScript can be put to use in many practical situations. Materials here provide an easy way for programmers to get into serious JavaScript coding.

6.16.1 JavaScript Operators

JavaScript arithmetic operators are the usual

```
+    -    *    /    %    ++    --
```

As we already know, the + operator also concatenates strings. Assignment operators are

```
=    +=    -=    /=    %=
```

JavaScript uses the usual set of comparison operators

```
==    !=    >    >=    <    <=
```

for *comparing strings, numbers,* and *objects,* returning a `true` or `false` Boolean value. The comparison is straightforward if the two operands compared are of the same type. When they are not, the following rules apply:

- When comparing a number to a string, the string is automatically converted to a number.

- When only one of the operands is a Boolean, it is converted to 1 (for `true`) or 0 (for `false`).

- When comparing an object to a string or number, the object is converted to a string (via its `toString` method) or number (via its `valueOf` method). Failure to convert generates a run-time error.

The *strict equality operators* `===` and `!==` can be used if the operands must be of the same type. For more details on comparison operators see, for example,

```
http://developer.mozilla.org/en/JavaScript/Reference/
     Operators/Comparison_Operators
```

You can also use the built-in value `undefined`

```
if ( var == undefined )
```

to see if a variable has been assigned a value.

Logical operators `&&` (and), `||` (or), `!` (not) work on Boolean values.

6.16.2 Built-In Functions

In JavaScript, you can use strings and numbers interchangeably in most situations. You can also explicitly convert a string to a number, and vice versa. The JavaScript built-in functions `Number(` *arg* `)` and `String(` *arg* `)` convert their arguments to a number and a string, respectively. These functions work on many kinds of arguments. For example,

```
var date= new Date();
var n=Number(date);
```

assigns to **n** the number of milliseconds since midnight January 1, 1970 UTC.

The functions `parseInt(` *str* `)` and `parseFloat(` *str* `)` return the integer and floating-point values of the argument *str*, respectively.

The function `encodeURI(` *url* `)` encodes URL special characters (Section 1.4.1) except `/` , `?` : `@` & = `+` `$` # in the given *url*. To also encode these characters use `encodeURIComponent(uri)` instead. The former function is useful for encoding full urls while the latter is for parts in a URL such as the key or value part of a query string pair. Use `decodeURI(` *str* `)` to decode encoded URL.

The function `eval(` *str* `)` executes *str* as a piece of JavaScript code.

6.16.3 JavaScript Strings

In JavaScript, strings are objects and have many useful fields and methods, including

- *str*.`length`—the length of the string

- *str*.`charAt(`*i*`)`—char at position *i*; 0 is the first character

- *str*.`substr(`*i*, *length*`)`—substring starting at index *i* until the end of string or with the optional length

- *str*.`indexOf(`*substr*`)`—first index of *substr* in *str* or −1 if not found

- *str*.`lastIndexOf(`*substr*`)`—last index of *substr* in *str* or −1 if not found

String methods related to pattern matching are described in the next section.

6.16.4   Defining Functions

We have seen many examples of function definition. The general syntax is

function *fn_name*(*arg_1*, ...) {   /* function body */ }

There can be zero or more arguments and they are local to the function. In the function body, you can have any number of variable declarations and statements. Local variables are declared with **var**. For example,

```
var x, y;
var i=0;
```

A variable declared outside of a function is global and accessible from anywhere and in any function.

The **return** operator ends the execution of a function and returns control to the calling point. Any argument to **return** becomes the return value of the function. Here is a simple example:

```
function factorial(n)
{  if (n < 0 )  return NaN;   // NaN, not a number
   if (n < 2 )  return 1;
   return n*factorial(n-1);
}
```

6.16.5   Loops

We have discussed JavaScript arrays in Section 6.7. Loop constructs as in the C language are available in JavaScript. Let **arr** be an array; then

```
for (i=0; i < arr.length; i++)    // for loop
{ /* use arr[i] */  }
```

```
var i=0;
while ( i < arr.length )         // checks before loop body
{ /* do something with arr[i] */
  i++;
}
```

```
do { ... } while( i < arr.length) // checks after loop body
```

```
for (key in arr) { /* use arr[key] */ }  // foreach loop
```

```
for (key in obj) { /* use obj.key */ }   // foreach loop
```

The **break** and **continue** statements work in loops, as you would expect.

6.16.6 Matching Patterns

The concept of pageid has to do with a naming convention for webpages in a site. The idea is to name the file for each page *pageid*.html. The *pageid* can then be used in JavaScript code for page-dependent features.

In JavaScript, a page can compute a `myid` variable from its URL with the `myPageId` function:

```
function myPageId()
{   var str = document.URL;           // (a)
    var re = /([^\/]+)\.html$/;       // (b)
    var found = str.match(re);        // (c)
    if ( found ) return found[1];
    else return null;
}
```

```
var myid = myPageId();
```

The URL of the current page is obtained from the `document.URL` attribute (line a). A *regular expression pattern* (Section 6.16.7) is constructed (line b) to pick the *name* part in the *name*.html suffix of the URL. The `match` method of the `String` object is used to apply the pattern to the string and obtain parts in the string that match the pattern (line c). The matching results are stored in an array and assigned to the variable `found`. If no part of the string matches the pattern, `found` is `null`. Generally, JavaScript converts an empty string, zero, `undefined`, or `null` to `false` and all other values to `true` in a test. If the match is successful, the returned array is organized as follows:

```
found[0]     // the matched string ($0)
found[1]     // the first substring remembered ($1)
found[2]     // the next substring remembered ($2)
...
```

Without assigning a name to the array, the same results can be accessed using the built-in variables $0, ..., $9 as indicated. You request remembering of substrings in a pattern match with parentheses in the regular expression. In the example,

```
var re0 = /[^\/]+\.html$/;
```

matches a sequence of one or more (+) characters, each not a SLASH ([^\/]), followed by .html at the end of the string ($). And it will match the end of strings such as

```
var str ="http://wonderful.com/services/guaranty.html";
```

Thus, the following tests return `true`:

```
str.search(re0)      // returns guaranty, or -1 for no match
re0.test(str_obj)    // returns true, or false for no match
```

The pattern

```
var re = /([^\/]+)\.html$/;
```

adds parentheses to remember the substring preceding `.html` when the pattern matches. For the preceding example, the string remembered is `guaranty`. Thus, either one of the following two statements

```
found = url.match(re);     found = re.exec(url);
```

returns an array where `found[0]` is `guaranty.html` and `found[1]` is `guaranty`.

6.16.7 Patterns

A *pattern* refers to a characterization of a string. Strings that fit the characterization are said to `match` the pattern. For example, "containing `ABC`" is a pattern. Any string that contains the three-character sequence `ABC` matches it. "Ending with `.html`" is another pattern. When you define a pattern, you divide all strings into two disjoint sets: strings that match the pattern and those that do not.

A pattern is expressed using *regular expressions* in the form `/regex/`, where *regex* can be a fixed string such as `/xyz/` to match strings containing the three characters or an expression involving special pattern matching characters. To include a (`/`) character as part of the pattern, you need to escape it with *backslash* `\`.

The character `^` (`$`) when used as the first (last) character of a regular expression matches the beginning (end) of a string or line. The regular expression `\d` matches any digit (0–9). Additional single-character expressions include `\n` (NEWLINE), `\f` (FORM FEED), `\t` (TAB), `\r` (RETURN), `\v` (VERTICAL TAB), `\s` (a white-space character, same as [`\f\n\r\t\v`]), `\b` (word boundary, a white space at the end of a word), `\B` (nonword boundary), and `\cx` (control-x). You compose patterns with additional notations (Table 6.1).

TABLE 6.1: JavaScript Pattern Notations

Notation	Matches
.	Any char except NEWLINE
*	Preceding item zero or more times
+	Preceding item one or more times
?	Preceding item zero or one time
{n} or {m,n}	Preceding item n (or m to n) times
(x)	Item x, captures matching str
x|y	Item x or item y
[abc] or [^abc]	Any listed (or not listed) chars

Here are some examples:

/\d*/	matches zero or more digits
/[A-Z]+/	matches a sequence of one or more uppercase letters
/i\d?/	matches i followed by zero or one digit
/[^\/]+/	matches one or more characters not /
/\/|\\/	matches / or \
/ing\b/	matches ing at the end of a word
/(exp)/	matches a given exp and remembers the match

the usage of patterns, Table 6.2 lists more examples. As you can see, many characters (e.g., ^ and +) have special meaning in a pattern. To match such characters themselves, you must escape them to avoid their special meaning. This is done by preceding such a character with the \ character.

TABLE 6.2: Pattern Examples

Pattern	Matching Strings
/l.ve/	love, live, lxve, l+ve,...
/^http/	http at the beginning of a string
/edu$/	edu at the end of a string
/L+/	L, LL, LLL, LLLL,...
/BK*/	B, BK, BKK,...
/D+HD/	DHD, DDHD, DDDHD,...
/.*G/	Any string ending in G
/G?G/	G or GG
/^$/	Empty string
/[a-z]9/	a9, b9,..., or z9

Let re be a pattern. Then either of the following

```
re.test(str)
str.search(re)
```

performs a quick matching and returns true or false. These are faster than str.match(re) because we are not looking for the matched strings.

Adding the character i to the end of a pattern makes the pattern *case insensitive*. For example,

/#ddeeff/i   matches #ddeeff, #DDEEFF, #dDeEfF,...

You can also use patterns to replace substrings within another string. The replace method of String is used this way:

```
str.replace(re, newStr)
```

For example, we can compute the over image from the out image with

```
outimg = "images/mainout.gif";
overimg = outimg.replace(/out/, "over");
```

Replacement is made for the first occurrence of a match. To replace all occurrences, add the flag **g** to the end of the pattern. For example,

```
line.replace(/ +/g, " ")
```

replaces all sequence of one or more spaces with a single space. Also, we can define these functions:

```
function basename(url)
{  return url.replace( /.*\//, ''); }

function dirname(path)
{  return path.replace(/\/[^\/]*$/, ''); }
```

Often the replacement string is computed from the matched substring. You can use $0, $1, . . . , $9 in the replacement string. And you can use a function call to compute the replacement string.

The **split** method takes a fixed or patterned delimiter and breaks up a string into an array of several substrings at the delimited positions. The following example breaks up a string where a SPACE is found:

```
str = "<p> Chapter Introduction </p>";
arr = str.split(" ");   // delimiter is space
// arr[0] is "<p>",   arr[1] is "Chapter", and so on
```

But there may be extra spaces between the words; hence, this simple split results in many empty substrings in the result array. In such a case, you can use a pattern for splitting:

```
arr = str.split(/ +/);
```

6.16.8   Patterns in Input Fields

You can set the **pattern** attribute for HTML5 input fields to a JavaScript regular expression and the browser will only accept input that matches the pattern. Figure 6.15 shows an example (**Ex: FormPattern**) where we used the following HTML code:

```
<form ...>
<label>Your 5+4 ZIP Code <input name="zip9"
  maxlength="10" placeholder="55555-4444" required=""
  size="10" pattern="[0-9]{5}-[0-9]{4}" /></label>       (A)
   <label>Expiration date
<input name="expiration_date" placeholder="mm/yy"
  required=""  maxlength="5" size="10"
  pattern="(0[1-9]|1[0-2])/[12][1-9]" /></label><br />    (B)
<br /><input type="submit" value="Proceed" /></form>
```

FIGURE 6.15: Input Pattern Checking

where we checked for a 5+4 ZIP code with the pattern [0-9]{5}-[0-9]{4}
and a credit card expiration date with the pattern (0[1-9]|1[0-2])/[12][1-9].
It is important to remember that client-side form validation is only for user
convenience. Actual validation must take place on the server side.

For a full reference to the JavaScript (ECMAScript) language, see documentation on the Web.

6.17   JavaScript Form Checking

When using forms to collect information, it is important to check the correctness of the data entered by Web users. HTML5 input fields provide the
most frequently needed checks, such as required fields, numbers, dates, and so
on, as we have described in Chapter 5. Such supported checks are performed
immediately after the user leaves a particular input field.

Powerful as HTML5 input checks are, there are situations where additional
checks are needed. For that we can write JavaScript checks and attach them
to input fields as event handlers. The form can only be submitted when it
passes all checks.

This does not mean that we will move form input checking from the server
side to the client side because savvy users and Web developers can bypass
client-side checks easily. Hence, client-side checks are primarily to enhance
the user interface and to avoid invalid data being sent to the server side. It
is a good idea to use client-side checks, but they are not replacements for
server-side checks.

We can use custom validation functions by themselves or to supplement
automatic checking by the browser. As an example, let's add a custom check
to the expiration date in the **Ex: FormPattern** example (Section 6.16.7). The
form HTML is as follows:

```
<form method="post"  action="../exc5/showpost.php"
     onsubmit="return(checkForm(this))">              (I)
<label>Your 5+4 ZIP Code <input maxlength="10"
required="" placeholder="55555-4444" name="zip_9"
 pattern="[0-9]{5}-[0-9]{4}" size="10" /></label>
   <label>Expiration date <input
name="expiration_date" placeholder="mm/yy" size="10"
required="" pattern="(0[1-9]|1[0-2])/[12][1-9]"
 maxlength="5" onchange="checkExpire(this)" />      (II)
</label><br /><br />
```

FIGURE 6.16: Custom Client-Side Input Validation

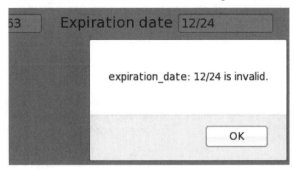

```
<input type="submit" value="Proceed" /></form>
```

We attached our custom check function `checkExpire` to the `onchange` event (line II) that takes place every time the input value changes. And we attached a final checking function `checkForm` to the `onsubmit` event. The `checkForm` will return true or false, which is returned by the `onsubmit` action in line I. This way, the form will only be submitted if `checkForm` returns true.

The JavaScript checking functions are as follows:

```
function checkForm(form)
{  if (checkExpire(form.expiration_date))
      form.submit();
   return false;
}

function checkExpire(field)
{  var arr = field.value.split("/");      // (III)
   var mm=arr[0], yy=arr[1];
   if ( ! (1 <= mm && 12 >= mm  &&
          11 <=yy && 22 >= yy))
   return formErr(field, field.value);
   return true;
}

function formErr(entry, msg)
{  alert(entry.name + ": " + msg + " is invalid.");
   entry.focus();
   entry.select();
   return false;
}
```

Note: We used the `split` method of string to obtain the `mm` and `yy` values to do the checking (line III).

6.18   Smooth Scrolling Text

We now look at a scrolling text animation (**Ex:** Scroll) performed entirely from JavaScript. The user can pause and resume the scrolling at any time. The text box in Figure 6.17 scrolls continually after the page is loaded. Clicking

FIGURE 6.17: Scrolling Text

i do eiusmod tempor incididunt ut labore et dolore magna aliqua

on the scrolling text pauses or resumes the scrolling.

The HTML for the scrolling box is

```
<body style="background-color: #def"
      onload="init_scroll()">                         (1)
<section><h1>Scrolling Text Demo</h1>
<div class="scrollbox" id="sd">
<div onclick="ss(this)" id="st" class="scrolltext">   (2)
<?php include("news.inc"); ?><span id="endMarker">    (3)
  </span></div> (4)
```

Initializations are done by the init_scroll() function (line 1). The function ss pauses and resumes the scrolling (line 2). The text to be scrolled, from the file news.inc, is placed in a div, together with an end marker, inside the scroll box (lines 3 and 4). Scrolling is done by changing the left margin of the scrolltext div.

With styling we create dark blue text (line 5) scrolling in an open-ended box with comfortable left and right margins (line 6).

```
div.scrolltext                           /* 5 */
{   color: darkblue; font-size: medium;
    white-space: nowrap; margin-left: 0px
    font-family: Arial, Helvetica, sans-serif;
}

div.scrollbox                            /* 6 */
{   width: 500px; overflow: hidden;      /* 7 */
    border: 1px solid darkblue;
    border-left: 3em solid transparent;
    border-right: 3em solid transparent;
    padding-top: 0.5em; padding-bottom: 0.5em;
}
```

Some global values are set statically (line 8) and others are set by the function init_scroll: the quantities needed to perform repeated scrolling, including the known width of the scroll box, the left shift advance, and the scrolling speed.

The sd_end is the left end of the visible scrolling range (line 9); sc_offset is the starting position for the scrolling text (line 10).

```
var speed=30, advance=2, // scrolling speed      // (8)
    boxwidth=500,         // scroll box width
    endwidth=50,   // left/right border width
    scrollNode, em, sd_end, sc_offset;

function init_scroll()
{ var scrollDiv = document.getElementById("sd");
  sd_end = xPosition(scrollDiv)+endwidth ;      // (9)
  sc_offset = boxwidth-endwidth;                // (10)
  scrollNode = document.getElementById("st");
  em = document.getElementById("endMarker");
  scrollNode.pauseFlag=false;
  scroll();
}
```

The animation is performed by the function scroll, which shifts scrollNode increasingly to the left (line 11) until the end marker moves beyond the sd_end (line 12) at which time we set it back to its starting position (line 13).

```
function scroll()
{ if ( scrollNode.pauseFlag ) return;
  scrollNode.style.marginLeft = sc_offset+"px";
  sc_offset -= advance;                          // (11)
  if ( xPosition(em)<sd_end )                     // (12)
     sc_offset=boxwidth - endwidth;               // (13)
  setTimeout("scroll()",speed);
}
```

Clicking on the scrolling text triggers a call to ss to pause/resume the scrolling. The pauseFlag property we added to the scrollNode object is checked by scroll before it runs another iteration.

```
function ss(node)
{ if ( node.pauseFlag )
  { node.pauseFlag=false; scroll();  }
  else node.pauseFlag = true;
}
```

From more than one example we see the critical role that window.setTimeout plays in animation. While setTimeout schedules the execution of a function expression after a given time interval, the similar function setInterval sets up a function expression for execution after every delay interval repeatedly:

```
var i_id=window.setInterval(func, delay);
window.clearInterval(i_id);          (cancels further calls)
```

6.19   Animation with CSS

HTML5 and CSS3 aim to make dynamic user interface effects easier to program. Automated form input checking and animated CSS transitions are just two obvious examples. Let's demonstrate this again by implementing the scrolling text example in Section 6.18 with *CSS animation*, which eliminates much of the JavaScript code.

CSS animation supports timed transitions from one set of CSS styles to another. For these animations, you define two components: a set of animation-control styles and a sequence of *keyframes* specifying the starting, intermediate, and final styles of the selected element.

For the scrolling text example, we now add these styles:

```
@-keyframes scroll
{ from   {margin-left: 500px;}      /* starting value keyframe */
  to     {margin-left: -970px;}     /* final value keyframe    */
}

div.scrolltext
{ white-space:nowrap; color:darkgreen; font-weight:bold;
  animation-name: scroll;           /* keyframs reference       */
  animation-duration: 10s;          /* duration, one iteration  */
  animation-iteration-count: infinite; /* no. of iterations     */
  animation-timing-function: linear;   /* transition function */
}
```

Without any JavaScript code, these will cause constant-speed scrolling of the scrolltext `div` repeatedly. To allow click control of pausing and resuming of scrolling, we add the event handler `onclick="ss(this.style)"` on the scrolltext `div` and the `ss` function:

```
function ss(style)
{ if ( style.animationPlayState=="" ||
       style.animationPlayState=="running" )
  { style.animationPlayState="paused";  }    // pausing
  else
     style.animationPlayState="running";      // resuming
}
```

The iteration count can be set to any non-negative number or `infinity`. The timing functions are the same as those for CSS transitions (Section 6.11). The general form for the `@-keyframes` rule is

```
@-keyframes name
{     0%   { styles }
      xy%  { styles }
      . . .
```

```
    100%    { styles }
}
```

where `from` is the same as 0% and `to` 100%.

Try this example (**Ex:** `KeyFrames`) at the DWP website.

6.20 Transform with CSS

CSS `transform` provides ways for us to make two-dimensional geometric transforms on block and inline elements. Such transforms include translate, rotate, scale, and skew. Plus, you can define your own *coordinate transformation matrix* (CTM) for an arbitrary transform.

For example, `translate(50px)` moves an element horizontally to the right by 50px, `rotate(45deg)` rotates the element clockwise 45 degrees (**Ex:** `Transform`), and `scale(2)` scales doubles the size of element.

As an example, let's animate an expanding headline by scaling it up as we increase the letter spacing (Figure 6.18). The HTML is simple enough.

```
<div id="box">
<h1 id="headline">Dynamic Web Programming and HTML5</h1>
</div>
```

FIGURE 6.18: Expanding Headline

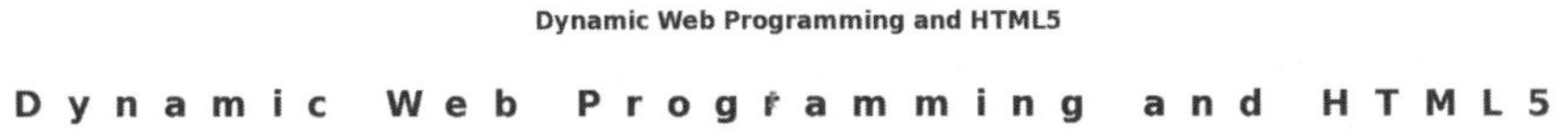

Dynamic Web Programming and HTML5

D y n a m i c   W e b   P r o g r a m m i n g   a n d   H T M L 5

The CSS code combines transforms with transitions to achieve the desired effect. The `div` box provides a centered box to contain the expanding headline (line **a**). The headline is kept on one line and has the initial letter spacing of zero (line **b**). Both the `transform` and `letter-spacing` properties are placed under transition control (lines **c** and **d**).

```
div#box                                /* a */
{   display: table; margin-top: 4em;
    margin-left: auto; margin-right: auto;
}


h1#headline
{   white-space: nowrap;
    letter-spacing: 0px;               /* b */
    transition: transform 1s linear;   /* c */
    transition: letter-spacing 1s linear;   /* d */
}
```

```
h1#headline:hover
{ letter-spacing: 16px; transform: scale(1.5); }
```

On mouseover, the headline expands the letter spacing and scales up the element while keeping it centered on the page. Here we used the pseudo class :hover to activate the animation and transitions. But we can also trigger the transition with JavaScript. For example, we can replace the h1#headline:hover rule by

```
h1#headline.expand
{ letter-spacing: 16px; transform: scale(1.5); }
```

and use

```
document.getElementById('headline').className="expand";
```

to activate the animation and transitions (**Ex: Headline**).

Here are some **transform** functions you can apply immediately:

- **translate**(*x*, *y*)—translates the element by the given lengths in the x and y directions. The *y* is optional.

- **scale**(*xf*, *yf*)—scales the element by the given factors in the x and y directions, keeping the transform origin fixed.

- **rotate**(*angle***deg**)—rotates clockwise around the transform origin.

- **skew**(*xd***deg**, *yd***deg**)—skews the element by the *xd* (*yd*) degrees in the x (y) direction around the transform origin.

The transform origin is a fixed point in the element being transformed. The default transform origin is the center of the element. Or you can set it with

```
transform-origin: x% y%
transform-origin: [left|center|right] || [top|center|bottom]
```

The default is **transform-origin: center center**.

In general, all possible 2D transforms can be represented by a CTM that can be applied to (x,y) coordinates to perform the transform. Such a matrix is in the form

$$\begin{pmatrix} a & c & e \\ b & d & f \\ 0 & 0 & 1 \end{pmatrix}$$

And you can use it in a transform with

```
transform: matrix(a,b,c,d,e,f);
```

where e and f relate to translation and the other four parameters relate to scaling and rotation. More will be said about the CTM in Section 12.3.1 when we cover SVG (Scalable Vector Graphics), which is also part of HTML5.

Browser support for CSS animation or transformation may not be uniform yet. If the above standard notations do not work for a browser, you may need to add a browser prefix (Section 4.5). For example, under Firefox, the JavaScript notation

```
style.animationPlayState
```

becomes

```
style.MozAnimationPlayState
```

6.21   Events and Event Objects

The preceding sections illustrated the application of JavaScript in many useful ways and at the same time provided examples of JavaScript programming together with HTML and CSS usage. One basic technique remains simple: You define event-handling functions and attach them to the appropriate events generated by the browser. A general description of event handling and the available events can help you envision and create many different effects for Web users.

When a particular event takes place, the browser user interface creates an *event object* to represent the particulars of the event (for example, which mouse button and the cursor location) and *dispatches* or *propagates* the event object down the document hierarchy to the final target element that will receive the event. This is known as the *capture phase* of the *event dispatch* (Figure 6.19). After the delivery to the target element (the *target phase*), the event object will travel back up the chain of parent elements toward the root of the hierarchy (the *bubble phase*).

FIGURE 6.19: Event Capturing and Bubbling

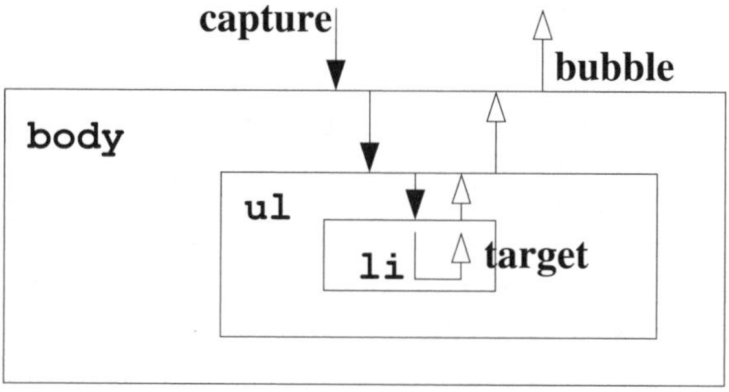

Usually, only the target element will react to an event, but that is not always the case. Event listeners (handlers) can be registered with any HTML element for each of the three phases. Thus, the round-trip affords all elements on the event dispatch path chances to react to the event by registering listeners for any desirable phase. Any event listener along the dispatch path may decide to stop the event from propagating further (Section 6.21.3).

To enable an element to handle a particular type of events, you add an *event listener* (also known as an event handler) for that type of event to the element. Common event types are mouse events and keyboard events. An event listener supplies *callback* code that gets executed upon the particular events. The simplest way is to attach the callback code to an *event handler attribute* for the target element. For example,

```
<img onclick="enlarge(this)" ... />
<input onblurr="validate(this)" ... />
```

And we have seen many such uses before. In general, you attach a handler for *event* by attaching the callback code to the on *event* attribute.

Events and handlers listed by the HTML5 standard include the following:

- Window events (for the `window` object)—Handlers can be defined for the `body` element with event handler attributes including `onload` (window's document loaded), `onunload` (window's document unloaded), `onscroll` (window scrolled), `onresize` (window size changed), `onfocus` (window gained input focus), and `onblur` (window lost focus).

- Mouse events —Handlers can be defined for most HTML5 elements with event handler attributes including `onclick` (a mouse button clicked), `ondblclick` (a mouse button double-clicked), `onmousedown` (a mouse button pressed), `onmouseup` (a mouse button released), `onmousewheel` (mouse wheel rotated), `onmousemove` (mouse moved), `onmouseover` (mouse moved over element), and `onmouseout` (mouse moved out of element).

- Keyboard events—Handlers can be defined for HTML5 elements that may gain input focus with event handler attributes `onkeydown` (any keyboard key is pressed), `onkeypress` (a character key is pressed), and `onkeyup` (a key is released).

- Input control events—Handlers can be defined for HTML5 input elements with event handler attributes `onfocus` (the element gained focus), `onblur` (the element lost focus), `oninput` (when element got user input), `onchange` (content of element changed), `oninvalid` (input field value is invalid), `onsubmit` (form submission requested), and `onselect` (element content is selected).

- Media events—Handlers can be defined for HTML5 media elements (`audio`, `embed`, `img`, `object`, and `video`) with event handler attributes including `onloadstart` (started to load media data), `onprogress` (loading media data), `onloadeddata` (media data finished loading), `onplay` (just before starting to play), `onplaying` (playing started), `onpause` (playing is paused), and `onended` (playing ended). This is part of the *HTML5 media API*.

Additional event handler attributes, such as those for drag-and-drop (Section 7.14), and those for mobile devices and the touch screen (Chapter 13), will be described when they are used.

It is possible to add multiple event listeners to the same event on the same element as long as they are different functions. Such multiple event handlers are called one after another in the order of registration.

6.21.1 Playlists

The HTML5 media elements do not yet support source files of playlist type, such as .m3u. But we can use JavaScript to manage a playlist quite easily. Here is an example (**Ex: Playlist**):

```
<audio onended="playnext(this)"  id="ae"        (I)
   title="Angels of Shanghai--angela with purple bamboo"
   src="angels/angela_with_purple_bamboo.ogg"
   autoplay="" autobuffer="" controls=""></audio>
```

When the first song ends, the event handler `playnext` is called (line I), passing the audio element object `this`.

The `src` attribute is set to the next song (line II) on the `playlist` array and the `title` attribute is set to reflect the song that is playing (line III).

```
const playlist=["angela_with_purple_bamboo.ogg",
          "angels_theme_the_invention_of_love.ogg",
          "butterfly_lovers.ogg", . . .  ];
const dir="angels/";
var index=0;
function playnext(el)
{ if ( index < playlist.length )
  { el.src=dir+playlist[++index];          // (II)
    el.title = "Angels of Shanghai--" +   // (III)
      playlist[index].replace(/.ogg/,"").replace(/_/," ");
  }
}
```

Figure 6.20 shows our playlist in action under Google Chrome.

6.21.2 Using Event Objects

When an event takes place, an *event object* of the appropriate type is created to represent information related to the particular event. This event object `event`

FIGURE 6.20: Playlist

is available for event handlers to extract event-related information such as which key is pressed, which mouse button is clicked, what are the x and y coordinates of the mouse event, and so on.

Consider the slide show example (**Ex: Slides**) in Section 6.6. It would be convenient to be able to use the LEFT- and RIGHT-ARROW keys on the keyboard to go back and forth on the slides. This can easily be done with key events (`KeyEvent`).

Add the key event handler (**Ex: SlideKey**) to the `body` element:

```
onkeydown="keyGo(event)"
```

Note that we transmit the `event` object to the handler function `keyGo`:

```
function keyGo(e)
{ if(e.keyCode == KeyEvent.DOM_VK_RIGHT)      // keyCode 39
     nextSlide();
  else if(e.keyCode == KeyEvent.DOM_VK_LEFT) // keyCode 37
     prevSlide();
}
```

The RIGHT-ARROW goes to the next slide, and the LEFT-ARROW returns to the previous slide. The `keyCode` property of a `KeyEvent` object gives the numerical code of the key involved. The HTML DOM standard defines a set of constants, such as `DOM_VK_LEFT`, available in the `KeyEvent` object. See the DOM level 3 events documentation for a complete list of these keycode constants.

For a `MouseEvent` object `me`, you have the button code (`me.button`), which can be 0 (left button), 1 (middle button), and 2 (right button) for right-hand mouse (reversed for left-hand mouse). The `me.screenX`, `me.screenY` (`me.clientXme.clientY`) give the (X, Y) coordinates of the mouse event location relative to the display screen area (document display area). The `me.ctrlKey`, `me.shiftKey`, `me.altKey`, or `me.metaKey` is set to true if the key was held down for the mouse event.

6.21.3 Adding and Removing Event Listeners

From JavaScript, you can add an event handler or remove it from any given HTML element *el* with

el.addEventListener(*event_type*, *fn*, *capture_phase*)
el.removeEventListener(*event_type*, *fn*, *capture_phase*)

by specifying the event type and the handler function (*fn*) to be added/removed. By default, the listener is registered for the target/bubble phase. Set the optional *capture_phase* to **true** to register a listener for the capture phase.

For example, you can treat the **transitionend** event from CSS transition with the handler function moveOn that can trigger another transition or animation of one kind or another (**Ex: KeyEvent**):

el.addEventListener("transitionend", moveOn);

As an example, let's add JavaScript event listeners to track the progress of the CSS animation in the scrolling text example (Section 6.19). The HTML code

```
<body style="background-color: #def" onload="init()">   (1)
<section><h1>CSS Animation Events Demo</h1>
<div class="scrollbox" id="sd">
<div onclick="ss(this.style)"                            (2)
id="st"> <?php include("news.inc"); ?></div>
</div></section>
<ul id="show"></ul>                                       (3)
</body>
```

adds a **body** onload call to **init()** (line 1) and a **ul** to display our progress messages (line 3).

```
function init()
{ var e = document.getElementById("st");
  e.addEventListener("animationstart", showprogress, false);
  e.addEventListener("animationend", showprogress, false);
  e.addEventListener("animationiteration", showprogress, false);
  e.className="scrolltext";                          // (4)
}
```

The **init** function adds event listeners and puts the **div** element in the CSS class (**scrolltext**) and begins the transform (line 2 and 4). The CSS animation code is

```
div.scrolltext
{     animation-name: scroll;
      animation-duration: 10s;
```

```
    animation-iteration-count: 3;
    animation-timing-function: linear;
}
```

The event handler is

```
function showprogress(e)
{ var l;
  switch(e.type)
  { case "animationstart":
      l="<li>Animation started: elapsed time is "; break;
    case "animationend":
      l="<li>Animation ended: elapsed time is " ; break;
    case "animationiteration":
      l="<li>New iteration began at "; break;
  }
  document.getElementById("show").innerHTML += l +
                          e.elapsedTime +'</li>';
}
```

Running this example (**Ex: KeyEvent**) produces a display shown in Figure 6.21.

While handling an event e, a listener may call e.stopPropagation() to stop triggering listeners not registered in the same element, or call e.stopImmediatePropagation() to stop triggering any other listener altogether.

FIGURE 6.21: Animation Event Listener Display

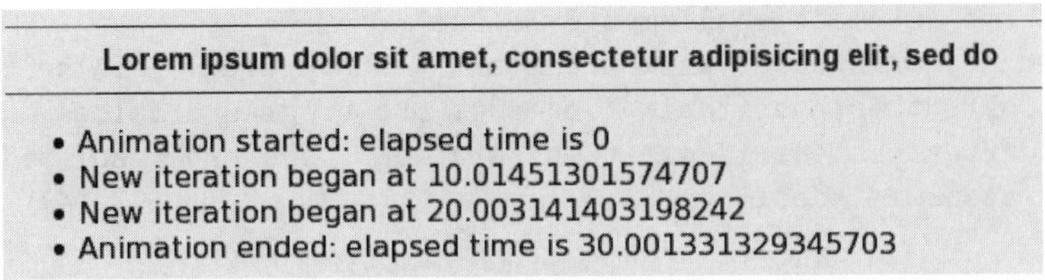

Lorem ipsum dolor sit amet, consectetur adipisicing elit, sed do

- Animation started: elapsed time is 0
- New iteration began at 10.01451301574707
- New iteration began at 20.003141403198242
- Animation ended: elapsed time is 30.001331329345703

Certain types of events, such as drag-and-drop events, have *default actions*. A default action is any supplementary action performed by the browser in addition to event dispatching. Calling the e.preventDefault() method will prevent such actions without stopping the event propagation. See Section 7.14 for an example.

6.22 Testing and Debugging

Serious programming of any kind involves much time on testing and debugging. JavaScript must be tested with a Web browser in the HTML page where it needs to work. If something is wrong, the problem may involve

- JavaScript files not loaded because of a wrong pathname, access permission problems.

- Syntax errors—typos, spelling mistakes, missing quotation marks, strings across multiple lines, missing or unmatched brackets, or incorrect object names or `ids`. JavaScript is case sensitive: `myObject` is not the same as `myobject`. Incorrect syntax is the most common problem.

- Run-time errors—problems that only occur when the script is executing. Accessing a nonexisting object attribute and using a `null` or `undefined` without first checking the quantity are examples.

- Logical errors—mistakes in the solution logic that make the script do the wrong thing or display the wrong information.

- Cross-browser errors —code that works only for certain browsers and not others.

Firefox has a number of tools that help JavaScript testing and debugging. The `tools->Web Developer` option leads to a short list of useful tools. The `Web console` (Figure 6.22) displays errors and warnings for HTML, CSS, and JavaScript (Figure 6.22). The `Scratchpad` allows you to enter and run JavaScript code and inspect the resulting expressions or objects.

FIGURE 6.22: Firefox Web Console

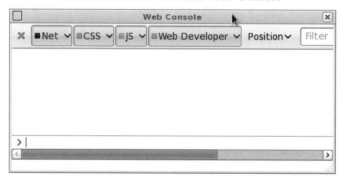

With Firefox, you can select any part of the page display and right click on it to view the HTML associated with that part of the page or to inspect the DOM object for it. The Firebug browser plug-in enables you to edit, debug, and monitor CSS, HTML, and JavaScript live in any webpage. Of course, facilities for developers are not limited to Firefox.

To isolate problems in your code, you can put `window.alert` calls in various key points in your script to trace its progress through the code and display values of variables and function arguments.

You can build your own error reporting window:

```
function errWindow()
{   errWindow = window.open("", "JavaScript Errors",
            "toolbar=0,scrollbar=1,width=400,
            height=300");
}
```

and display your own error messages in it:

```
errWindow.document.writeln( "value of email is " + email );
```

For hard-to-find bugs, you may need to use the Mozilla JavaScript debugger called Venkman. It supports break points, backtracing, stepping, and so on. Venkman offers a visual environment and works on all platforms. After installation, Firebug and Venkman are available under the `Web Developer` option.

6.23   For More Information

An introduction to JavaScript and its application has been provided. For more information on the syntax and semantics of JavaScript, its standardization, events, object model, and tools, see related websites and the resource pages on the DWP website.

Our JavaScript coverage continues in later chapters on HTML5-defined APIs including the `Canvas` element, drag-and-drop, local storage, and DOM. Also covered later will be JavaScript server access known as *Asynchronous JavaScript and XML* (AJAX).

6.24   Summary

JavaScript is the standard language to program browser actions and event handling on the client side. JavaScript code can be placed inside the `script` element, in separate files loaded into the page, and included as values of event attributes of HTML elements.

JavaScript syntax is close to the C language and runs within the context provided by the browser, which supplies built-in objects such as `window`, `navigator`, and `document`. The JavaScript language itself provides constructs, objects, and functions for strings, arrays, pattern matching, and more.

We have provided a good number of examples of using JavaScript in practical situations, including image preloading and display, form checking, audio/video control, pull-down and sliding menus animation, expanding headline animation, pop-up windows, key and mouse events, and so on. In particular, we have seen how JavaScript cooperates with CSS transitions and animations to achieve control over these new CSS features.

Generally, we define the needed JavaScript, and initialize global variables either as the JavaScript loads or with the `onload` event after the webpage is done loading. Then, the JavaScript is ready to react to events to support dynamic effects that enhance the functionality and user experience of our webpages.

Understanding the available events, their propagation (capture, target, bubble), and handling further illuminates how JavaScript code integrates with the user interface.

Exercises

6.1. Describe the ways JavaScript code can be placed in a webpage.

6.2. Which program performs the interpretation and execution of JavaScript code?

6.3. In JavaScript, how is a local variable, a global variable defined? How is an array created?

6.4. For JavaScript, what is the connection between the array and the object?

6.5. How is JavaScript used to send the browser to a new page? Back or forth to a previously visited page?

6.6. Explain the need for image preloading and how that is done.

6.7. Improve the slide show to include a strip of thumbnails that can scroll left and right on mouse click. Clicking on any thumbnail displays that picture in the slide window.

6.8. Refer to the functions `xPosition` and `yPosition` in Section 6.10. While they are good for computing each quantity separately, calling them to get both the x and y positions is not efficient. Write a function `xyPosition` to return both positions at once.

6.9. Explain why a click event on a menu label needs to be captured in **Ex: Menu** (Section 6.10).

6.10. Write a JavaScript function `pageid` that takes the location of the current webpage that ends in *pageid*`.html`, *pageid*`.htm`, or *pageid*`.php`, and extracts the *pageid* part from it.

6.11. In JavaScript, how do you decide if a string contains a certain character or substring?

6.12. Playing background sound can be done with `var me=new Audio("url")` then `me.play()`. Construct a webpage to make this work.

6.13. We have seen `:hover` triggering JavaScript actions. What happens if we use `:active` instead?

6.14. Consider form checking with JavaScript. If a form does pass desired checks, what exactly is needed to stop the form submission so the user can fix the input and submit again?

6.15. Create a form with disabled input fields that are re-enabled through JavaScript. (Hint: See Section 5.10).

6.16. With JavaScript, how do you assign value to the `class` attribute of a particular element? How can this trigger transitions and/or transformations?

6.17. CSS is finalizing the specification for *pointer-events*. Find out what that is and explain in detail.

6.18. How do we detect which mouse button has been depressed? Which key on the keyboard has been depressed? The location of the mouse when a mouse button is depressed?

6.19. Take the audio playlist example (**Ex:** `Playlist`) from Section 6.21.1 and make a working video playlist.

6.20. Write JavaScript to detect page scrolling, when it starts and ends.

Chapter 7

HTML5 DOM and APIs

Webpages as user interfaces can become even more responsive, dynamic, and functional if we give JavaScript maximum control over all aspects of a webpage. An important aspect of the new HTML5 standard is the establishment of APIs giving JavaScript much more control over many aspects of a webpage. More important APIs are listed here; some of these APIs are still under development.

- The *HTML5 DOM* API—for access, traversal, and manipulation of the DOM tree, manipulation of HTML element objects and other node objects on the DOM tree.

- The *Canvas* API—for drawing graphics, manipulating images, making transformations and compositions, performing animations as well as responding to user interface events.

- The *Drag-and-Drop* (DnD) API—for performing DnD operations within the browser context and with external objects.

- The *File* APIs—for accessing files locally on the client side.

- The *Contenteditable* API—for browser-supported rich-text editing.

- The *Selectors* API—for easily locating DOM elements using CSS selector notations.

- The *Offline Cache* API—for updating and controlling browser offline cache.

- THe *Local Storage* API—for client-side persistent storage for Web applications.

The term *Dynamic HTML*, often abbreviated DHTML, refers to techniques combining HTML, CSS, with JavaScript and the DOM API to achieve dynamic effects on document content and presentation. Such effects may alter the content and appearance of any parts of the page. The changes are fast and efficient because they are made by the browser without having to network with any servers.

In this chapter, we focus on the DOM, File, DnD, and Canvas APIs. Practical examples show how they are applied to achieve DHTML. Let's begin with the DOM.

7.1   What Is DOM?

With cooperation from major browser vendors, the W3C has been working on the *Document Object Model* (DOM) as a standard *application programming interface* (API) for scripts to access and manipulate HTML and XML documents. The HTML5 standard contains detailed specification for HTML5 DOM that is a great improvement on past specifications.

Compliant clients, including browsers and other user agents, provide the DOM-specified API to access and modify the document being processed (Figure 7.1). The DOM API gives a logical view of the document structure where

FIGURE 7.1: DOM-Compliant Browser

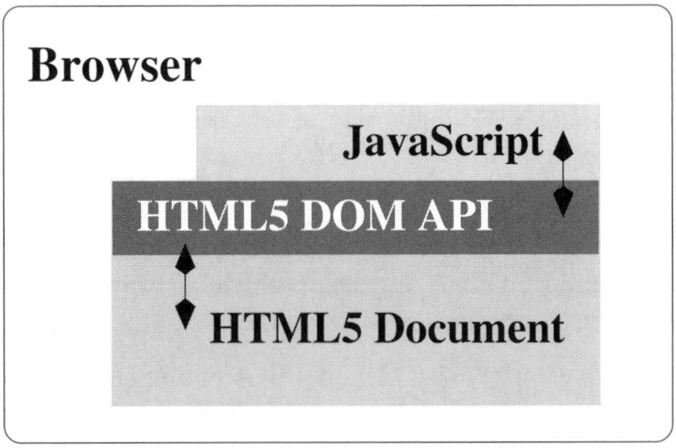

objects represent different parts such as windows, documents, elements, attributes, texts, events, style sheets, style rules, and so on. These *DOM objects* are organized into a tree structure (the DOM tree) to reflect the natural organization of a document. HTML elements used in a webpage are represented by *tree nodes* and organized into a hierarchy. Each webpage has a `document` node at the root of the tree. The `head` and `body` nodes become *child nodes* of the `html` node (Figure 7.2).

From a node on the DOM tree, you can go down to any child node or go up to the parent node. With DOM, a script can add, modify, or delete elements and content by navigating the document structure, modifying or deleting existing nodes, and inserting dynamically built new nodes. Each element object has attributes (called *IDL attributes*) that reflect the values of the element's attributes given in the page's HTML source code (called *content attributes*). Through the IDL attributes we can easily get and set content attributes for any element from JavaScript.

Also attached to the document are its style sheets. Each element node on the DOM tree also contains a *style object* representing the display style for that element. Thus, through the DOM tree, style sheets and individual element styles can also be accessed and manipulated. Therefore, any parts of

FIGURE 7.2: DOM tree structure

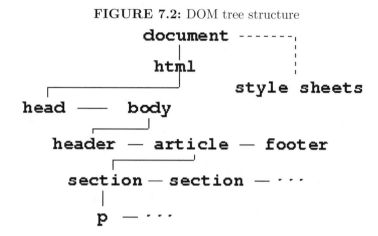

a page can be accessed, changed, deleted, or added, and the script will work for any DOM-compliant client.

7.2 Browser Support of DOM

Historically, the first version of DOM, the so-called DOM level 0, was simply the object model used by Netscape Navigator and perhaps also IE3. Subsequent standardization activities at the W3C led to DOM levels 1, 2, 3, and 4, extending the DOM to XML and beyond JavaScript into an IDL (interface definition language) based interface specification. HTML5 is making DOM an integral part of its specification. HTML5-compliant browsers will make DOM code in JavaScript interoperable.

But it will always be true that a particular browser may have some unique DOM features of its own. It is easy to test if an object, field/property, or method is available in a browser. For example,

```
if ( document.getElementById )
    . . .
```

tests if the `getElementById` method is available in the `document` object. Code and examples in this chapter work in all major browsers.

7.3 HTML5 DOM API Overview

According to the W3C, DOM is a "platform and language neutral interface that allows programs and scripts to dynamically access and update the content, structure and style of documents. The document can be further processed and the results of that processing can be incorporated back into the presented page." The DOM specifies an API (application programming interface) and provides a structural view of the document. The current DOM core draft standard can be found at `dvcs.w3.org/hg/domcore/raw-file/tip/Overview.html`.

DOM, using an interface definition language (IDL), specifies the required objects, together with their methods and attributes, that any implementation must provide to constitute the standard API. An object contains methods and attributes. A method is a function in an object. An attribute, also known as a property or field, is a named value in an object. An attribute value can be a number, string, function, or object. When necessary, we will use the term *IDL attribute* or *DOM attribute* to avoid confusion with HTML content attributes. It is up to compliant browsers (agents) to supply concrete implementation of the required objects, in a particular programming language and environment. DOM specifications are aligned with JavaScript for the obvious reason that all browsers use the language (Figure 7.3).

FIGURE 7.3: The DOM API

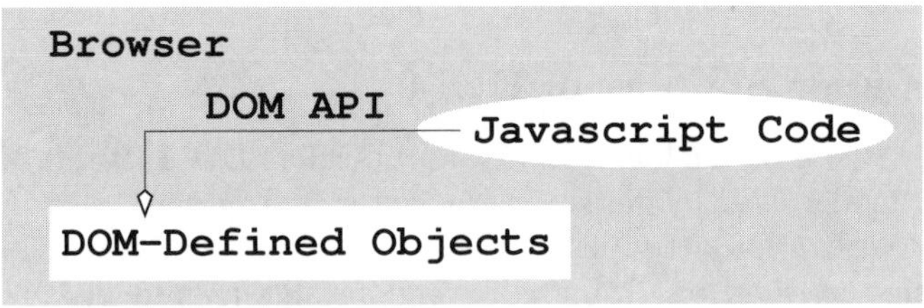

As an interface, each DOM object exposes a set of fields and methods for JavaScript to access and manipulate the underlying data structure that actually implements the document structure. The situation resembles a radio interface exposing a set of knobs and dials to the user. If the radio interface were standardized, then a robot would be able to operate any standard compliant radio.

The DOM tree represents the logical structure of a document. Each tree node is a `Node` object. There are different types of nodes that all *inherit* the basic `Node` interface. Inheritance is an important object-oriented programming (OOP) concept. In OOP, interfaces are organized into a hierarchy where *extended interfaces* inherit methods and fields required by `base interfaces`. The situation is quite like defining various upgraded car models by inheriting and adding to the features of a base model. In DOM, the `Node` interface is at the top of the interface hierarchy, and many types of DOM tree nodes are directly or indirectly derived from `Node`. This means all DOM tree node types must support the properties and methods required by `Node`.

On the DOM tree, some types of nodes are *internal nodes* that may have *child nodes* of various types. *Leaf nodes*, on the other hand, have no child nodes.

For any webpage, the root of the DOM tree is a `Document` node, and it is usually available directly from JavaScript as `document` or `window.document`. The `document` object must also *implement* the `HTMLDocument` interface, giving you access to all the quantities associated with a webpage, such as `URL`, style sheets, scripts, images, media files, links, forms, `body` (the body element node), `head` (the head element node), `title`, `characterSet`, and many others (Section 7.9).

Working with the DOM tree usually involves navigating to a desired target node and manipulating it in some way. We have seen how useful the `document.getElementById(...)` method is for this purpose. You can also use

`document.`*xyz*

to access the node for a `form`, `img`, `iframe`, `object`, or `embed` element that has a `name="`*xyz*`"` or `id="`*xyz*`"` content attribute.[1]

The field `document.documentElement` gives you the child node of type `HTMLHtmlElement` that represents the root `<html>` element of the page (Figure 7.2).

`HTMLElement` (Section 7.7) is the base interface for derived interfaces representing the many different HTML elements.

7.3.1 DOM Tree Nodes

The DOM tree for a webpage consists of different types of nodes (subtypes of `Node`); the three most important types of nodes are as follows.

Document—Root of the DOM tree providing access to pagewide quantities, markup elements, and in most cases, the `<html>` element as a child node (`document.documentElement`).

Element—Internal and certain leaf nodes on the DOM tree representing an HTML or XML markup element. The `HTMLElement` subinterface of `Element` is for HTML elements and provides access to element attributes and child nodes that may represent text and other HTML elements. Because we focus on the use of DOM in DHTML, we will use the terms *element* and *HTML element* interchangeably. The `document.getElementById(`*id*`)` call gives you any element with the given *id*. For non-void elements, the field `innerHTML` gives you the HTML code string inside the element. Setting `innerHTML`[2] changes the HTML

---

[1] Assuming no duplicate `name` or `id` exists.

[2] For IE, the `innerHTML` field is read-only for certain elements, including `table` elements, `html`, `head`, `style`, and `title`.

code in the element. This operation is useful when updating specific parts of a webpage.

The `attributes` field of an element gives you a `NamedNodeMap` of named `Attr` objects. Each `Attr` object provides you the ability to access and set a particular HTML content attribute. The `name` field (a string) of an `Attr` object is read-only, and the `value` field can be set to a desired string.

You can use the notation

element.`attributes`.*xyz*

to access an `Attr` object named *xyz*, if it exists. But it is recommended you use the following to get/set content attributes from JavaScript:

element.`getAttribute`(*'xyz'*)
element.`setAttribute`(*'xyz'*, *value*)

The *element*.`attributes.length` tells you the number of `Attr` objects in the element, and you can also access each attribute object by indexing *element*.`attributes`[*i*]. All DOM indexes are zero-based.

An element provides easy access to its child elements with the fields `children` (an `HTMLCollection`), `firstElementChild`, `lastElementChild`, `previousElementSibling`, `nextElementSibling`, and `childElementCount`. Each child element is also accessible with `children.item`(*i*). Note that a `Text` node is not an element node.

FIGURE 7.4: Paragraph DOM Structure

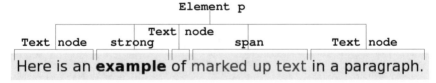

Text—A leaf node containing the text inside a markup element. If there is no markup inside an element's content, the text is contained in a single `Text` object that is the only child of the element. The paragraph

```
<p id="para">Here is an <strong>example</strong> of
<span style="color: blue">marked up text</span>
in a paragraph.</p>
```

is an example (**Ex: TextNodes**) of an element with five child nodes (Figure 7.4).

Using the DOM interface, script-generated results can be placed anywhere on a displayed page by modifying the DOM tree. Figure 7.6 shows the part of the DOM tree that is used to display the results for the calculator.

FIGURE 7.6: Partial DOM Tree for Calculator Example

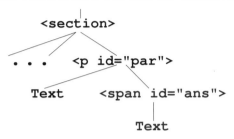

`Document` and `Element` interfaces are important and provide many methods and properties useful in practice. They inherit from the basic `Node` interface, which is presented next.

7.5 Node API

In object-oriented programming, an interface specifies data values (called *fields*[3]) and functions (called *methods*) that are made available to application programs. The `Node` interface is the base of all other node types on the DOM tree and specifies useful fields and methods for them.

7.5.1 Node Fields

Fields provided by a `Node` are read-only and include

- nodeType—A small integer representing the *derived type* of the node. The `Node` interface provides symbolic constants for `nodeType`: ELEMENT_NODE (1), TEXT_NODE (3), PROCESSING_INSTRUCTION_NODE (7), COMMENT_NODE (8), DOCUMENT_NODE (9), DOCUMENT_TYPE_NODE (10), DOCUMENT_FRAGMENT_NODE (11).

  The function `whichType` demonstrates how to determine node type (**Ex: WhichType**):

```
function whichType(nd)                          // (a)
{  if ( nd.nodeType == Node.ELEMENT_NODE )      // (b)
       window.alert("Element Node");
   else if ( nd.nodeType == Node.DOCUMENT_NODE )
```

---

[3]In the official DOM specification, fields are called *IDL attributes*. To better distinguish them from HTML attributes, we use the commonly accepted term *fields* here.

```
        window.alert("Document Node");
     else if ( nd.nodeType == Node.TEXT_NODE )
        window.alert("Text Node");
   ...
}
```

The parameter `nd` is any DOM node whose type is to be determined (line a). The `nd.nodeType` is compared with the type constants defined by the `Node` interface to determine the node type of `nd` (line b).

- `parentNode, firstChild, lastChild, previousSibling, nextSibling`— Related Nodes of a node (Figure 7.7).

FIGURE 7.7: Node Relations

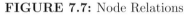

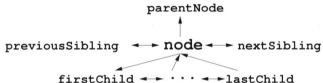

- `nodeName` and `nodeValue`—Strings representing the name and value of a `Node`. The exact meaning of these strings depends on the node type, as shown in Table 7.1. For example, the `nodeValue` of any `Element` or

TABLE 7.1: Meaning of `nodeName` and `nodeValue`

Node Type	nodeName	nodeValue
Element	Tag name	null
Text	#text	Text string
Document	#document	null
Comment	#comment	Comment string
ProcessingInstruction	Target name	Instruction string

Document node is `null`.

- `childNodes`—A `NodeList` of child nodes of the node. Some nodes have children and others do not. For an `Element` node, child nodes represent the HTML elements and text strings contained in that element.

  The `length` field and the `item(i)` method of `NodeList` provide an easy way to visit each node on the node list. For example (**Ex: ChildNodes**), applying the function `visitChildren`:

```
function visitChildren(id)
```

```
{  var nd = document.getElementById(id);
   var ch = nd.childNodes;
   var len = ch.length;          // number of nodes
   for ( i=0; i < len; i++)
   {   nd = ch.item(i);          // node i
       window.alert( nd.nodeName + "   "
                     + nd.nodeValue );
   }
}
```

on the element with `id="par"`

```
<p id="par">Here is <img ...  /><br /> a picture.</p>
```

displays this sequence

```
#text    Here is
IMG
BR
#text    a picture.
```

The `ownerDocument` field of a node leads you to the root of the DOM tree. It is worth emphasizing that the fields of `Node` are read-only. Assignments to them have no effect.

Also, `NodeList` and `NamedNodeMap` objects in the DOM are *live*, meaning changes to the underlying document structure are reflected in all relevant `NodeList`, `NamedNodeMap`, `HTMLCollection` objects. For example, if you get the `childNodes` of an `Element` and subsequently add or remove child nodes, the changes are automatically reflected in the `childNodes` you got before. This behavior is usually supported by returning a reference to the data structure containing the actual child nodes of the `Element`.

7.5.2 Node **Methods**

In addition to fields, the `Node` interface provides many methods inherited by all node types. These fields and methods combine to provide the basis for accessing, navigating, and modifying the DOM tree. Specialized interfaces for derived node types offer additional functionality and convenience.

Among `Node` methods, the following are more frequently used:

- *node*.`hasChildNodes()`—Returns true/false.

- *node*.`appendChild(` *child* `)`—Adds *child* as a new child node of *node*.

- *node*.`removeChild(` *child* `)`—Removes the indicated *child* node from the *node*.

- *node*.insertBefore(*child, target*)—Adds the *child* node just before the specified *target* child of this *node*.

- *node*.replaceChild(*child, target*)—Replaces the *target* child node with the given *child*. If *child* is a DocumentFragment, then all its child nodes are inserted in place of *target*.

- *node*.isSameNode(*n*) and *node*.isEqualNode(*n*)—Returns true if *node* is one and the same or another but equal to the given node *n*.

- *node*.cloneNode() or *node*.cloneNode(true)—Creates a shallow or deep copy of *node*.

Note: If *child* is already in the DOM tree, it is first removed before becoming a new child. Section 7.9 shows how to create a new node.

7.6   DOM Tree Depth-First Traversal

Using the DOM for dynamic effects basically involves accessing nodes and modifying them on the DOM tree. The easiest way to access a target HTML element is to use

```
document.getElementById( id )
```

to obtain the node for the element directly by its *id*.

But it is also possible to reach all parts of the DOM tree by following the parent, child, and sibling relationships. A systematic visit of all parts of the DOM tree, a *traversal*, may be performed *depth-first* or *breadth-first*. In depth-first traversal, you finish visiting the subtree representing the first child before visiting the second child, and so on. In breadth-first traversal, you visit all the child nodes before visiting the grandchild nodes, and so on. These are well-established concepts in computer science.

Let's look at a JavaScript program that performs a depth-first traversal (**Ex: DomDft**) starting from any given node on the DOM tree. The example demonstrates navigating the DOM tree to access information.

```
var result="";

function traversal(node)
{ result = "";                                  // (1)
  dft(node);                                     // (2)
  alert(result);                                 // (3)
}

function dft(node)
{ var children;
```

```
    if (node.nodeType == Node.TEXT_NODE)
        result += node.data;                          // (4)
    else if (node.nodeType == Node.ELEMENT_NODE)      // (5)
    { openTag(node);                                  // (6)
      if ( node.hasChildNodes() )                     // (7)
      {  children = node.childNodes;
         for (var i=0; i < children.length; i++)      // (8)
             dft( children[i] );                      // (9)
         closeTag(node);                              // (10)
      }
    }
}


function closeTag(node)
{ result += "</" + node.tagName + ">\n"; }            // (11)
```

Given any node on the DOM tree, the `traversal` function builds the HTML source code for the node. It initializes the `result` string (line 1), calls the depth-first algorithm `dft` (line 2), and displays the result (line 3).

The `dft` function recursively visits the subtree rooted at the `node` argument. It first checks if `node` is a `Text` node (a leaf) and, if true, adds the text (`node.data`[4]) to `result` (line 4). Otherwise, if `node` is an `Element` node (representing an HTML element), it adds the HTML tag for the node to `result` by calling `openTag` (lines 5 and 6) and, if there are child nodes, recursively visits them (lines 8 and 9) before adding the close tag (line 10 and 11). The subscript notation `children[i]` is shorthand for `children.node(i)`.

```
function openTag(node)
{ result += "<" + node.tagName;
  var at;
  tagAttributes(node.attributes);                     // (12)
  if ( node.hasChildNodes() )
      result += ">\n";                                // (13)
  else
      result += " />\n";                              // (14)
}


function tagAttributes(am)
{ var attr, val;
  for (var i=0; i < am.length; i++)                   // (15)
  {   attr = am[i];   val = attr.value;
      if ( val != undefined && val != null
           && val != "null"  && val != "" )           // (16)
      {   result +=   " " + attr.name + "=\"" +        // (17)
```

---

[4]Alternatively, `node.nodeValue` will also give you the text.

```
                    val + "\"";
        }
    }
}
```

The `openTag` function adds any attributes for the tag (line 12) by calling `tagAttributes`. The open tag is terminated by either `">"` (line 13) for regular elements or `" />"` for void elements (line 14).

The argument `am` of `tagAttributes` is a `NamedNodeMap` of `Attr` nodes. The function goes through each attribute (line 15) and adds each defined attribute (line 16) to the `result` string (line 17). Note the use of the `name` and `value` fields of an `Attr` node. Figure 7.8 shows the first part of the result

FIGURE 7.8: Traversal of DOM Tree

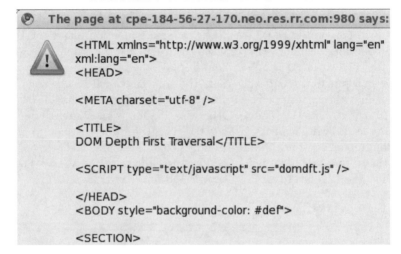

of the depth-first traversal when called on the `document.documentElement` node corresponding to the `<html>` element of the page. The complete example (**Ex:** DomDft) can be tested on the DWP website.

7.7   DOM `HTMLElement` Interface

Derived node types (interfaces extending `Node`) add fields and methods specialized to a particular node type and may provide alternative ways to access some of the same features provided by `Node`. HTML markup elements in a page are represented by nodes extending the base `HTMLElement` derived from `Element` which extends `Node`. For each HTML element, HTML5 provides an interface

HTML*FullTagName*Element

that derives from HTMLElement (Figure 7.9). The complete list of all the

FIGURE 7.9: DOM HTML Interfaces

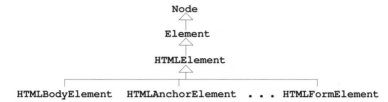

HTML element interfaces can be found in the HTML5 specification.

The HTMLElement interface is rather central for DHTML. Before we systematically discuss the fields and methods of HTMLElement, let's see it in action in an example (**Ex: DomNav**) where we combine navigation and modification of the DOM tree to achieve the kind of visual effects attributable to DHTML.

We can illustrate DOM tree navigation visually by visiting a subtree representing a `<table>` element, for instance. As you traverse the subtree, the part of the table corresponding to the node being visited will be highlighted. A control panel enables you to go up (to the parent node), down (to the first child), left (to the previous sibling), or right (to the next sibling) within the table. The control panel also displays the tag name associated with the current node (Figure 7.10).

FIGURE 7.10: DOM Tree Visual Navigation

row 1, col 1	row 1, col 2	row 1, col 3
row 2, col 1	row 2, col 2	row 2, col 3
row 3, col 1	row 3, col 2	row 3, col 3

```
            up
left    TD     right
          down
```

The HTML code for the table that we will be traversing is shown here in easy-to-read form:

```
<table id="tbl" border="1">
<tbody><tr><td>row 1, col 1</td>
     <td>row 1, col 2</td>
     <td>row 1, col 3</td></tr>
 <tr><td>row 2, col 1</td>
     <td id="center">row 2, col 2</td>
     <td>row 2, col 3</td></tr>
 <tr><td>row 3, col 1</td>
     <td>row 3, col 2</td>
     <td>row 3, col 3</td></tr></tbody></table>
```

In the actual file, we eliminate all line breaks and white spaces between elements to avoid potential extraneous nodes on the DOM tree.

The `init()` function is executed `onload` and sets the stage for the visual navigation:

```
var currentNode, tableNode, nameNode, normal, highlight;

function init()
{   tableNode=document.getElementById("tbl");
    tableNode.normalize();
    highlight="#0ff";
    normal=tableNode.style.backgroundColor;              // (A)
    currentNode=document.getElementById("center");       // (B)
    currentNode.style.backgroundColor  = highlight;      // (C)
    nameNode=document.getElementById("tname").firstChild;
    nameNode.data=currentNode.tagName;                   // (D)
}
```

The JavaScript global variables used are

- `tableNode`—the node for `<table>` that is to be traversed

- `currentNode`—the node for the current traversal position on the `tableNode` subtree

- `nameNode`—the node to display the `tagName` of `currentNode`

- `normal` and `highlight`—the background colors that indicate visually the part of the table being visited

The `init()` function assigns initial values to these variables. The normal background color is set to the background of the table (line **A**). The center cell of the 3-by-3 table is chosen as the starting point of the traversal, and `currentNode` is set (line **B**) and highlighted (line **C**). The text of `nameNode`, at the center of the control panel, is set using the `tagName` field of an `HTMLElement` (line **D**). The `init()` function is called `onload`:

```
<body onload="init()">
```

The control panel (Figure 7.10) for interactive traversal consists of four input buttons and a middle display area. The up button is

```
<input type="button" value="  up  " onclick="up()" />
```

and the display area

```
<span id="tname">tag name</span>                         (E)
```

`id=tname` (line **E**) is used to display the tag name of the current traversal position. The four buttons each trigger a corresponding function that does the obvious. The `up()` function keeps the traversal from leaving the subtree (line **F**).

```
function up()
{  if ( currentNode == tableNode ) return;        // (F)
   toNode(currentNode.parentNode);
}

function down()
{  toNode(currentNode.firstChild); }

function left()
{  toNode(currentNode.previousSibling); }

function right()
{  toNode(currentNode.nextSibling); }
```

Each of these four functions calls `toNode` to visit the new node passed as the argument.

The `toNode` function does the actual work of walking from the current node to the new node given as `nd` (line G). If `nd` is `null` or a leaf node (type `TEXT_NODE`), then nothing is done (line H). If we are leaving an internal node on the subtree, the highlight is removed by calling the `removeAttribute` method of the `HTMLElement` interface (line I). If we are leaving the root `tableNode`, the original background color of the table is restored (line J). The arrival node is then highlighted and set as the current node (lines K and L). Finally, the tag name of the current node is displayed as the text content of `nameNode` (line M):

```
function toNode(nd)                               // (G)
{  if ( nd == null ||
        nd.nodeType == Node.TEXT_NODE )           // (H)
      return false;
   if ( currentNode != tableNode )
      currentNode.style.backgroundColor="";       // (I)
   else
      currentNode.style.backgroundColor=normal;   // (J)
   nd.style.backgroundColor = highlight;          // (K)
   currentNode=nd;                                // (L)
   nameNode.data=currentNode.tagName;             // (M)
   return true;
}
```

The example further illustrates the DOM tree structure, the use of the `style` property of HTML elements, and the `tagName` field. It also shows how DHTML can help the delivery of information, enable in-page user interactions, and enhance understanding.

You can find the complete, ready-to-run version in the example package. You may want to experiment with it and see what it can show about the DOM tree and DHTML.

Exercise 7.11 suggests adding a display of the table subtree to show the current node position on the subtree as the user performs the traversal.

7.8   `HTMLElement` **Fields and Methods**

Every HTML element in a webpage is represented on the DOM tree by a node of type `HTMLElement`. The `HTMLElement` interface extends the `Element` interface, which in turn extends the basic `Node` interface. We list often-used fields and methods available in any node object of type `HTMLElement`:

- `tagName`—A read-only field representing the HTML tag name as a string.

- `style`—A field to a style object representing the style declarations associated with an element. For example, use `style.backgroundColor` to access or set the `background-color` style. Setting a style property to the empty string indicates an inherited or default style. If you change the style of an element by setting its `style` attribute instead, the new `style` attribute replaces all existing style properties on that element, and normally, that is not what you want to do. It is advisable to use the `style` field to set individual style properties you wish to change.

- `className`—A read-write field representing the `class` content attribute of the element.

- `classList`—An object providing convenient methods for accessing and manipulating the list of names in the `class` content attribute: `contains(`*name*`)`, `add(`*name*`)`, `remove(`*name*`)`, and `toggle(`*name*`)` (add if absent, remove if present).

- `innerHTML`—A read-write field representing the HTML source code contained inside this element as a string. By setting the `innerHTML` field of an element, you replace the content of an element. The same field is also supported in the `HTMLDocument` interface.

- `getAttribute(`*attrName*`)`—Returns the value of the given attribute *attrName*. The returned value is a string, an integer, or a `boolean`, depending on the attribute. Specifically, a `CDATA` (character data) value is returned as a string, a `NUMBER` value is returned as an integer, and an on-or-off attribute value is returned as a `boolean`. A value from an allowable list of values (e.g., `left|right|center`) is returned as a string. For an attribute that is unspecified and does not have a default value, the return value is an empty string, zero, or `false`, as appropriate.

- `setAttribute(`*attrName*, *value*`)`—Sets the given attribute to the specified string *value*.

- `removeAttribute`(*attrName*)—Removes any specified value for *attrName* causing any default value to take effect.

- `hasAttribute`(*attrName*)—Returns `true` if *attrName* is specified for this element and `false` otherwise.

- `querySelector`(*selector*) and `querySelectorAll`(*selector*)—Returns first/all descendant elements selected by the given selector, part of the HTML5 Selectors API.

- `focus`()—Causes the input element to get *input focus* so it will receive keyboard input.

- `blur`()—Causes the input element to lose input focus.

When setting values, use lowercase strings for attribute names and most attribute values. When checking strings obtained by `tagName` or `getAttribute`(), be sure to make case-insensitive comparisons to guard against nonuniformity in case conventions. For example,

```
var nd = node1.firstChild;
var re = /table/i;
if ( re.test(nd.tagName) )
{  ...  }
```

tests a `tagName` with the case-insensitive pattern `/table/i`.

HTML input control elements have these additional fields and methods:

- Content attribute fields `name`, `value`, `placeholder`, and so on—Most, if not all, attribute values are directly available as field values by the same name. Field names for multi-word attribute names are capitalized starting with the second word.

- `files`—An array of files entered by the user (for file type input). You can use `files.length`, `files[i].name`, `files[i].getAsText("")` (file contents in UTF-8).

- `validity`—A Boolean for validity checked input.

- `checkValidity`()—Returns Boolean, indicating validity of input.

- `select`()—Selects the current textual content in the input element for user editing or copying.

- `click`()—Causes a click event on the element.

7.9　Document **and** HTMLDocument **Interfaces**

Browsers display webpages in windows. The `window.document` object represents the entire webpage displayed in the window and contains all other elements in the page. The `document` object supports the `Document` interface and also implements the `HTMLDocument` interface (Figure 7.11), providing fields and methods useful for pagewide operations.

FIGURE 7.11: HTML Document Interfaces

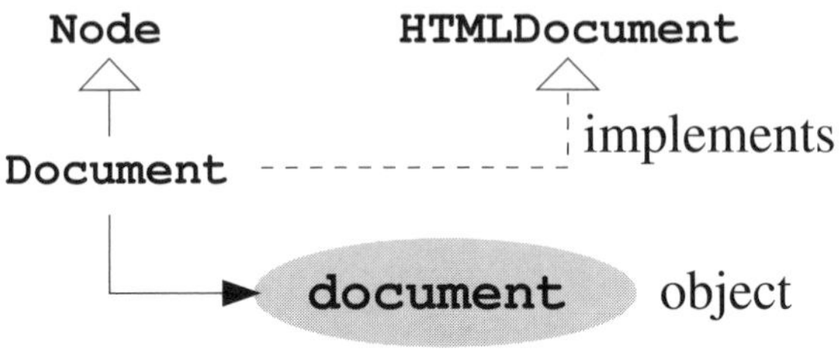

7.9.1　Fields of `document`

A select set of fields available from the `document` object is listed here:

- `documentElement`—The `<html>` element of the page

- `body`—The `<body>` element of the page

- `URL`—A read-only string for the complete URL of the page (also available as `documentURI`)

- `location`—A Location object to obtain (`location.href`) and change (`location=`*url*) or `location.replace(`*url*`)`) the document URL. Replacing removes the current URL from browsing history. The `window.location` returns the `location` of the current document. The `location` object also provides fields for accessing well-defined parts of the URL.

- `title`—The title string specified by `<title>`

- `contentType`—The MIME media type of the document

- `referrer` —The read-only URL of the page leading to this page (empty string if no referrer)

- `domain`—The read-only domain name of the Web server that supplied the page

- `cookie`—A SEMICOLON-separated string of *name*=*value* pairs (the cookies) associated with the page

- applets, forms, images, links—Read-only lists of different elements in the page: <applet> elements, <form> elements, <img> elements, and <a> and <area> elements as href links, respectively; such a list has a length field, an item(*n*) method, and a namedItem(*str*) method, which returns an element with *str* as id or, failing that, as name.

- innerHTML—The HTML code string for the document. Setting this field changes the complete document content.

- activeElement—The element in the document that has input focus or the body element.

- commands—A collection of all a, button, input, option, and command elements that have ids.

- designMode—A Boolean to turn enable/disable editing (Section 8.13) of the document.

7.9.2 Methods of document

Frequently used methods of the document object include

- write(*string*) and writeln(*string*)—Inserts the *string* as document content.

- createElement(*tagName*)—Returns a newly created element object for the <*tagName*> element. By setting attributes and adding child nodes to this element, you can build a DOM structure for any desired HTML element.

- createTextNode(*textString*)—Returns a node of type TEXT_NODE containing the given textString.

- getElementById(*id*)—Returns the unique HTML element with the given *id* string. We have seen this method used often.

- getElementsByTagName(*tag*)—Returns a list of all elements with the given *tag* name in the document.

- getElementsByName(*name*)—Returns a list of all a, applet, button, form, iframe, img, input, map, meta, object, select, and textarea elements elements in the document with the given *name* in the document.

- getElementsByClassName(*class-names*)—Returns a list of all elements in the document with class in the *class-names* (space-separated list of names). The same method invoked on an element gives elements contained in that element.

- `querySelectorAll(`*selector*`)`—Returns a list of all elements selected by the given *selector*. Drop the `All` from the method name to get the first such element. The methods are part of the HTML5 Selectors API.

7.10   Generating New Content

Applying the features discussed in the previous section, let's do more with DHTML by adding new content to a displayed HTML page. The content is computed by JavaScript, built into element nodes, and placed on the DOM tree.

7.10.1   A Session-Recording Calculator

To get started, we can take the interactive calculator example (**Ex:** `DomCalc`) shown in Figure 7.5 and make it more useful by recording the current answer and displaying a history of computation steps. The answer from the previous step can be used in the next step (Figure 7.12).

FIGURE 7.12: Session-Recording Calculator

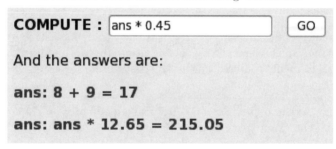

By revising slightly the HTML code for the basic calculator (**Ex:** `DomCalc`), we get the HTML for the session calculator **Ex:** `CalcSession`:

```
<section><h1>Calculator Session</h1>
<p>Simply type in a string such as
<code>32 + 98 * 6</code> and click GO.</p>
<p><strong>COMPUTE : </strong>
<input id="uin" name="uin" maxlength="30" />
  <input value="GO" type="button"
                onclick="comp('uin')" /></p>
<p>And the answers are:</p>
<output id="session" for="uin"
        style="display: block"> </output>          (1)
```

The `<output>` element (for the `uin` button) is where the computation session will be displayed (line 1).

The `init()` function (line 2) is called onload to obtain the `<output>` element and store it in the global variable `session`.

```
var session, ans = 0;   // global variables

function init()
{ session=document.getElementById("session"); }   // (2)

function comp(id)
{ var input = document.getElementById(id);
  var str = input.value;                            // (3)
  ans = eval(str);                                  // (4)
  var ansNode = document.createTextNode("ans: "
                      + str + " = " + ans);         // (5)
  var parNode = document.createElement("p");        // (6)
  parNode.appendChild(ansNode);                     // (7)
  session.appendChild(parNode);                     // (8)
  input.value="";                                   // (9)
}
```

The `comp` function, triggered by the `GO` button, obtains the user input (line 3) and evaluates it. The JavaScript function `eval` (line 4) takes any string and executes it as code. The result obtained is stored in the global variable `ans` (line 4), which can be used in the subsequent step.

To record the computation step, it creates a new text node (line 5), wraps a `<p>` element around it (lines 6 and 7), and appends the element as a new (last) child of the `session` `<output>` (line 8). Finally, the input field is cleared (line 9), ready for the next step. Users may use `ans` in the next computation step to perform a session of steps as shown in Figure 7.12. Further, users may store values for use in subsequent steps by creating their own variables with input strings such as

```
salesTax = 0.08
total = subtotal + subtotal * salesTax
```

7.11   A Smart Form

As another example of dynamically adding and removing page content, let's create a form that adjusts itself to changing user input.

A website in North America, for example, may collect customer address and telephone information without asking for an *international telephone country code*. But if the customer selects a country outside North America, it may be a good idea to require this information as well. In many situations, the information to collect on a form can depend on data already entered on the form. It would be useful to have the form dynamically adjust itself as the user fills out the form.

FIGURE 7.13: Smart Form

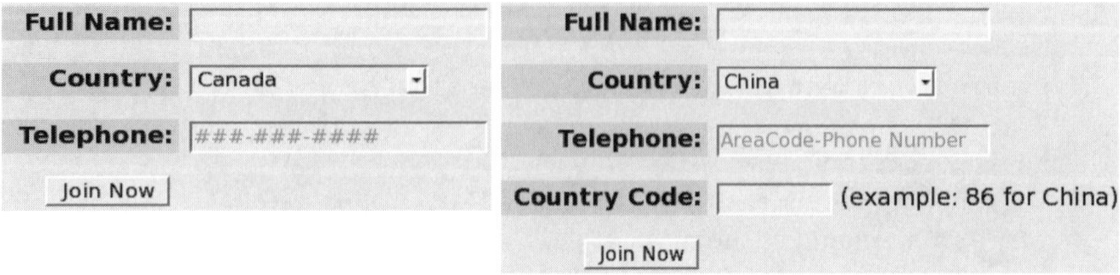

As an example, let's design a smart form (**Ex:** SmartForm) that examines the country setting in the address part of the form and adds or removes (Figure 7.13 an input field for the international telephone code.

Our strategy is straightforward:

1. When the country name is selected, the `onchange` event triggers a call to check the country name.

2. Any country outside North America causes an input field to be added to obtain the telephone country code.

3. If the country is within North America, then any telephone country code input field is removed.

4. Using background color, the form also makes it obvious to the user which input element has input focus.

The `form` code is as follows:

```
<form id="sf" autocomplete="on" method="post"
      action="../../exc5/showpost.php">
<div><label>Full Name: </label><input required=""
   onfocus="highlight(this)"
   onblur="normal(this)" name="fullname" size="20" /></div>
<div><label>Country: </label><select  required="" id="country"
   name="country" size="1" onfocus="highlight(this);"
   onblur="normal(this)" onchange="countryCode(this);">      (A)
    <option value="US" selected="">USA</option>
    <option value="CA">Canada</option>
    <option value="MX">Mexico</option>
    <option value="CN">China</option>
    <option value="RU">Russian Federation</option>
</select></div>
```

```
<div><label>Telephone: </label><input required="" id="phone"
  type="tel" size="20" onfocus="highlight(this)" name="phone"
  onblur="normal(this)"  placeholder="###-###-####"
  pattern="\d{3}-\d{3}-\d{4}" /></div>
<input id="sb" onfocus="highlight(this)"                    (B)
      onblur="normal(this)"
      type="submit" value="Join Now" />
</form>
```

The onchange event of <select> triggers the function countryCode (line A), which can add or remove a form entry for the telephone country code. The new form entry will be a new div element inserted just before the submit button (line B).

The init() function, executed onload, sets global variables and then calls countryCode to add/remove the country code entry (line 1).

```
var base="", high;
var cdiv=null, cc=null, button=null, ph, oph, pattern,
    form, iph="AreaCode-Phone Number", local=true;

function init()
{  high="#9ff";
   ph = document.getElementById("phone");  // phone input field
   oph = ph.getAttribute("placeholder");  // phone placeholder
   pattern = ph.getAttribute("pattern");  // validation pattern
   button = document.getElementById("sb"); // submit button
   form = document.getElementById("sf");   // form element
   countryCode(document.getElementById("country"));      // (1)
}
```

Input focus highlighting is performed by

```
function highlight(nd)  // onfocus
{  base=nd.style.backgroundColor;
   nd.style.backgroundColor=high;
}

function normal(nd)      // onblur
{  nd.style.backgroundColor=base; }
```

The isLocal function checks to see if a country is local to North America (line 2).

```
function isLocal(ct)
{ return(ct=="US" || ct=="CA" || ct==""
                  || ct=="MX"); }              // (2)
```

```
function countryCode(country)
{ if ( isLocal(country.value) )
  {  if ( ! local )                       // (3)
     {  form.removeChild(cdiv);
        ph.setAttribute("placeholder", oph);
        ph.setAttribute("pattern", pattern);
        local=true;                       // (4)
     }
     return;
  }
  if ( ! local ) {  cc.value = ""; return; }
  if ( cdiv == null )                     // (5)
  {  cdiv = document.createElement("div");
     cdiv.appendChild(makeLabel());
     cdiv.appendChild(makeCC());          // (6)
  }
  form.insertBefore(cdiv, button);        // (7)
  ph.setAttribute("placeholder", iph);
  ph.removeAttribute("pattern");
  local=false;                            // (8)
}
```

An onchange event on the <select> element for the country part of an address
triggers a call to countryCode. If the given country is in North America and
local is false, it removes any telephone country code entry from the form,
restores the placeholder instructions and the validation pattern, sets global
variable local to true, and then returns (lines 3 and 4).

If country is outside of North America and local is false, it removes any
previously entered country code string and returns (line 5).

Otherwise (local is true), we need to insert the country code field. This is
done by creating a div element containing a lable and an input element (with
instructions) for the Country Code (lines 5 and 6). The new div is inserted
just before the submit button, the phone input placeholder and validation
pattern are changed, and the local flag is set to false (lines 7 and 8).

The new label is created by makeLabel() (line 9). And the actual input
element (line 10) with instructions (line 11) for the telephone country code is
created by makeCC(), which also sets the global variable cc (Figure 7.14):

FIGURE 7.14: Creating a Form Entry

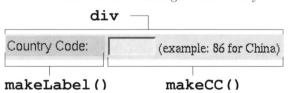

```
function makeLabel()                                        // (9)
{  var t, n;
   n = document.createElement("label");
   t = document.createTextNode("Country Code:");
   n.appendChild(t);
   return n;
}

function makeCC()
{  var t, n;
   cc = document.createElement("input");                    // (10)
   cc.setAttribute("onfocus", "highlight(this)");
   cc.setAttribute("onblur", "normal(this)");
   cc.setAttribute("name", "cc");
   cc.setAttribute("id", "cc");
   cc.setAttribute("size", "7");
   n = document.createElement("span");                      // (11)
   n.appendChild(cc);
   t = document.createTextNode(
               " (example: 86 for China)");
   n.appendChild(t);
   return n;
}
```

Note that these element creation functions use `setAttribute` to set up many attributes so the newly created form entry fits in with the style and dynamic behavior on this smart form. Experiment with **Ex: SmartForm** and see for yourself.

7.12 Client-Side Data Sorting

When displaying many entries in a list or table, the information can be difficult for the user to digest. Websites solve this problem by giving users the ability to sort the information in various ways. The user may want to see the largest dollar amount first, the least expensive item first, inexpensive shipping, or names alphabetically. Often such data sorting is done on the server side. But going back to the server is not necessary if the data entries are not changing and we just wanted to see them in a different order. We can use DHTML techniques to sort the data on the client side. This way we can avoid networking delay, unnecessary load on the server, and make the webpage more responsive to the user.

For example, the shopping cart in Figure 7.15 is in increasing `Amount`. The same shopping cart is shown in Figure 7.16 in increasing unit `Price`.

FIGURE 7.15: Shopping Cart Sorted by Amount

Item	Code	Price	Quantity	Amount
Shovel	G01	14.99	2	29.98
Power Saw	P12	34.99	1	34.99
Hand Shovel	T01	4.99	10	49.90
Hand Saw	H43	24.99	5	124.95

FIGURE 7.16: Shopping Cart Sorted by Price

Item	Code	Price	Quantity	Amount
Hand Shovel	T01	4.99	10	49.90
Shovel	G01	14.99	2	29.98
Hand Saw	H43	24.99	5	124.95
Power Saw	P12	34.99	1	34.99

In this example (**Ex:** DomSort), clicking (double-clicking) a table header cell sorts that column in increasing (decreasing) order.

7.12.1   Sortable Data Organization

The HTML code for the sortable data is organized as a sequence of unordered lists inside a div:

- Items on the first list (line A) serve as column headings that can be clicked or double-clicked to sort the display using data values in a particular column.

```
<ul id="control"><li class="button"                    (A)
    onclick="sortTable(0, 'str', '1');"
    ondblclick="sortTable(0, 'str', '-1');">Item</li>
  . . .
</ul>
```

  The arguments to sortTable are column position (a zero-based index), numerical or alphabetical ordering (num or str), and increasing or decreasing order (1 or -1).

- Following the control list come any number of sortable lists. Each sortable column must contain all numbers or all text strings. Here is a typical data list (five columns):

```
<ul id="aa"><li class="item">Shovel</li>
          <li>G01</li>
          <li class="number">14.99</li>
          <li class="number">2</li>
          <li class="number">29.98</li></ul>
```

The data lists are presented in tabular form with CSS.

```
div#thelist
{ display: table; border: solid;
  border-width: 1px; border-color: black;
  border-collapse: collapse;
}

div#thelist ul
{ display: table-row; background-color:#f0f0f0; }

div#thelist ul li
{ display: table-cell; text-align: center;
  padding: 5px; border: thin solid black;
}

div#thelist ul li.number { text-align: right; }

div#thelist ul#control li
{ display: table-cell; text-align: center;
  font-weight: bold; padding-left: 1em;
  padding-right: 1em; background-color: #fc0;
}

div#thelist ul#control li:hover
{  color: #fc0; background-color: #666; }
```

7.12.2   Data Sorting JavaScript

Now let's look at the JavaScript code for data sorting. As stated in the previous subsection, `onclick` and `ondblclick` events on an active table header trigger calls to the `sortTable` function with appropriate arguments: column position (**c**), numerical or alphabetical ordering (**n**), and increasing or decreasing direction (**d**).

```
var col=null, numerical=false, direction=1;
```

```
function sortTable(c, n, d)
{ if ( col==c && Number(d)==direction ) return;    // (a)
  col=c; direction = Number(d);                     // (b)
  numerical = (n == "num");                         // (c)
  var list = document.getElementById("thelist");    // (d)
  var r = list.childNodes;                          // (e)
  n = r.length;
  var arr = new Array(n-1);
  for(i=1; i < n; i++) arr[i-1]=r.item(i);          // (f)
  quicksort(arr, 0, n-2);                           // (g)
  for(i=0; i < n-1; i++) list.appendChild(arr[i]);  // (h)
}
```

If c is the same as the recorded column position col and d is the same as
the recorded sorting direction (line a), the sorting has already been done, and
sortTable returns immediately. To prepare for sorting, the arguments are
stored in the global variables (lines b and c). The child nodes, except the first
child, of the thelist div are copied into a new array arr (lines d and f).
This array is where the node will be placed in sorted order.

The call to quicksort (line g) sorts the array arr with the quicksort
algorithm, one of the most efficient sorting algorithms known. The elements
on the sorted arr are then appended in sequence as children of div (line h).

Inserting existing nodes from the DOM tree into the DOM tree is very
different from inserting newly created nodes (Section 7.11). An existing node
is first removed from the DOM tree automatically before it is inserted. The
removal is necessary to protect the structural integrity of the DOM tree. This
is why no explicit removal of child nodes from the div is needed before appending the nodes from the sorted array arr.

If you accept the quicksort function as a magical black box that does the
sorting, then we have completed the description of data sorting on the DOM
tree. For those interested, the inner workings of quicksort are presented next.

7.12.3 Quicksort

The basic idea of the quicksort algorithm is simple. First, pick any element of
the array to be sorted as the *partition element* pe. By exchanging the elements,
the array can be arranged so all elements to the right of pe are greater than
or equal to pe, and all elements to the left of pe are less than or equal to pe.
Now the same method is applied to sort each of the smaller arrays on either
side of pe. The recursion is terminated when the length of the array becomes
less than two.

```
function quicksort(arr, l, h)
{    if ( l >= h || l < 0 || h < 0 ) return;    // (1)
     if ( h - l == 1 )                          // (2)
     {    if (compare(arr[l], arr[h]) > 0)      // (3)
```

```
    {  swap(arr, l, h)    }              // (4)
       return;
    }
    var k = partition(arr, l, h);        // (5)
    quicksort(arr, l, k-1);              // (6)
    quicksort(arr, k+1, h);              // (7)
}
```

The `quicksort` function is called with the array to be sorted, the low
index l, and the high index h. It sorts all elements between l and h inclusive.
If the sorting range is empty (line 1), `quicksort` returns immediately. If the
range has just two elements (line 2), they are compared (line 3), and switched
(line 4) if necessary, and `quicksort` returns. For a wider range, `partition` is
called to obtain a partition element and the left and right parts of the array.
Each of these two parts is sorted by calling `quicksort` (lines 6 and 7).

The call `compare(a, b)` compares the arguments and returns a positive,
zero, or negative number depending on whether $a > b$, $a = b$, or $a < b$. The
signs are reversed for sorting in decreasing order.

```
function compare(r1, r2)
{   ke1 =  key(r1, col);                 // (8)
    ke2 =  key(r2, col);                 // (9)
    if ( numerical )                     // (10)
    {   ke1 = Number(ke1);
        ke2 = Number(ke2);
        return direction * (ke1 - ke2);
    }
    return (direction * strCompare(ke1, ke2));   // (11)
}
```

For sorting HTML tables, `compare` is called with DOM nodes r1 and r2
representing two different table rows. The function `key` obtains the string
contents in the designated table cell (lines 8 and 9) and compares them either
numerically as numbers (line 10) or alphabetically as text strings (line 11).

The function `key` extracts the textual content (line 12) of the table cell
from the given row r at the column position c:

```
function key(r, c)
{   var cell = r.firstChild;
    while ( c > 0 )
    {   cell = cell.nextSibling;
        c--;
    }
    return cell.firstChild.nodeValue;    // (12)
}
```

The `strCompare` function compares two text strings a and b by comparing
corresponding characters:

```
function strCompare(a, b)
{   var m = a.length;
    var n = b.length;
    var i = 0;
    if ( m==0 && n==0 ) return 0;
    if ( m==0 ) return -1;
    if ( n==0 ) return 1;
    for ( i=0; i < m && i < n; i++ )
    {   if ( a.charAt(i) < b.charAt(i) ) return -1;
        if ( a.charAt(i) > b.charAt(i) ) return 1;
    }
    return (m - n);
}
```

Swapping two elements on the array is simple:

```
function swap(arr, i, j)
{   var tmp = arr[i];
    arr[i]=arr[j];
    arr[j]=tmp;
}
```

Now we can turn our attention to `partition`, the workhorse in the `quicksort` algorithm. The function is called with an array `arr` and a sorting range, defined by the low index `l` and the high index `h`, which has at least three elements. The function picks the middle element as the `pe` (line 13) and partitions the given range into two parts separated by the `pe`. All elements to the left of `pe` are less than or equal to `pe`, and all elements to the right of `pe` are greater than or equal to `pe`. The index of the `pe` is returned (line 18).

```
function partition(arr, l, h)    // h > l+1
{   var i=l, j=h;
    swap(arr, ((i+j)+(i+j)%2)/2, h);                  // (13)
    var pe = arr[h];
    while (i < j)
    {   while (i < j && compare(arr[i], pe) < 1)  // (14)
        {   i++; }   // from left side
        while (i < j && compare(arr[j], pe) > -1) // (15)
        {   j--; }   // from right side
        if (i < j) {   swap(arr, i++, j); }       // (16)
    }
    if (i != h) swap(arr, i, h);                  // (17)
    return i;                                     // (18)
}
```

Searching from the left (line 14) and right (line 15) end of the range, it looks for an out-of-order pair of elements and swaps them (line 16). When finished, it moves the `pe` back into position (line 17) and returns.

The complete `quicksort` and the table sorting example (**Ex: DomSort**) can be found at the DWP website.

7.13   A Better User Interface for File/Image Upload

It is widely acknowledged that the browser-provided interface for uploading multiple files works fine but leaves much to be desired. Let's improve it by adding a display of the selected files/pictures and a file count to better inform the end user. This will be done using DOM and the *HTML5 File API*.

Consider uploading a set of pictures. We can use the HTML code to get a browser-supplied file picker:

```
<input multiple="" type="file" name="img[]" required=""
    accept="image/jpeg" onchange="thumbnail(this.files)"/>
```

And we can use the `onchange` event to display thumbnails of the pictures chosen to be uploaded, together with a count of the number of pictures (Figure 7.17). When uploading other types of files, we can list the names and sizes

FIGURE 7.17: Improved File Upload Interface

of the files instead.

The picture display HTML is `<div id="thumbnail"> </div>` with the CSS style:

```
<style>
img.tm {width:100px; height:100px; margin:5px}
div#thumbnail
{ border:thin black solid; margin:10px;
    padding-left:10px; padding-right:10px; display: none}
</style>
```

The `thumbnail` function is called `onchange`, and it begins by removing any child in the `filedisplay` div (line 1) and inserting the new file count (line 2).

```
function thumbnail(files)
{ var tn=document.getElementById('filedisplay');
  while ( tn.hasChildNodes() )                          // (1)
  {  tn.removeChild(tn.firstChild);  }
  var h3="<h3>Pictures to upload = "+files.length
      +"</h3>";
  tn.innerHTML=h3;                                       // (2)
  for (var i = 0; i < files.length; i++)                 // (3)
  {  var file = files[i];                                // (4)
     var imageType = /image.*/;
     if (!file.type.match(imageType)) { continue; }      // (5)
     var img = document.createElement("img");            // (6)
     img.classList.add("tm"); tn.appendChild(img);       // (7)
     var reader = new FileReader();                       // (8)
     reader.onload = makeHandler(img);                    // (9)
     reader.readAsDataURL(file);                          // (10)
  }
  tn.style.display="inline-block";
}


function makeHandler(image)                              // (11)
{ return function(event)
  { image.src = event.target.result; };
}
```

Files chosen by the user (one or more) are passed into the function `thumbnail`. Each file is a `File` object, defined in the HTML5 File API, that exposes the file `size`, `type`, `name`, and `lastmodified` date info. For security reasons, `name` is only a simple file name, and no client-side path information is made available to a webpage.

For each chosen file that is an image (lines 3–5), an `img` element is created in the `tm` class and appended to the filedisplay `div` (lines 6 and 7). A `FileReader`, part of the HTML5 File API, is used to read the file and to produce a *data URL* that is then assigned to the `src` attribute of the image element via the `onload` handler for the file reader (lines 8–11). The `onload` handler is made by the function `makeHandler`, which makes the image element available to the event handler returned (line 11).

A data URL is a URL with the data for the resource included in the URL itself. Such data usually use the base64 encoding first introduced in MIME. Data URLs can be useful and speedy in certain operations. Here is a sample data URL:

```
data:image/jpeg;base64,...data...
```

Try this example (**Ex:** `ImageUpload`) at the DWP website.

The file upload user interface improvement here is minimal. Section 7.16 describes how to use drag-and-drop to build a list of files where files can be added and deleted by the user. The files can then be processed by JavaScript in many ways, including uploading them to the server.

7.14 TicTacToe with Drag-and-Drop

Drag-and-Drop (DnD) is an intuitive operation familiar to most users. It can better engage the user and make a webpage more interactive at the same time. The interaction is not limited to inside a webpage either.

While simple and straightforward from a user's viewpoint, DnD is not so simple for programmers. DnD involves a draggable source element, a target dropzone element, a data transfer object, a sequence of drag-and-drop mouse events, and event handlers for those events. The programmer must also take care of preventing any default or other undesirable actions that may be triggered by these events. A DnD operation transfers data provided by the source element to the target dropzone causing a well-defined *DnD effect* to take place. DnD can work within a webpage or may involve a source/target that is outside the webpage. Such an external source/target can be from another GUI application on the client-side operating system or from another webpage.

7.14.1 DnD Life Cycle

From start to finish, a DnD operation goes through a sequence of stages. The DnD life cycle is as follows:

1. DnD begins—The user holds down the mouse button and begins to drag the screen representation of a draggable source element (S). A `dragstart` event is sent to S.

2. DnD in progress—As the user continues the drag, `drag` events are sent to S at regular intervals.

3. Dragging en route—As the dragging passes over elements en route to the target dropzone (T), each element gets a `dragenter`, a `dragover`, and a `dragleave` event. A drop over an element not listening or responding to these events will have no effect and the DnD operation ends.

4. Drag arriving on dropzone—The target T receives and processes the `dragenter` and the `dragover` events, usually also providing visual feedback signaling acceptance of a drop.

5. Drop—The user releases the mouse button over the target T. A `drop` event is sent to T and it retrieves the data delivered by the data transfer

object carried by the event object and processes the drop to achieve the desired effect.

6. DnD ends—After a drop, successful or not, a `dragend` event is sent to S, carrying information about the drop, and the DnD life cycle is complete.

Figure 7.18 illustrates the DnD life cycle and events fired.

FIGURE 7.18: DnD Life Cycle

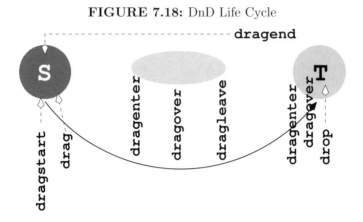

When implementing DnD operations, a Web developer must be fully aware that a Web browser uses DnD for its own purposes, such as dragging and dropping a link to the bookmarks toolbar or dragging selected text/image to the desktop or an application window. Because browsers have internal default handling for DnD-related events, webpage-defined DnD must, in most cases, prevent such default actions from being triggered by stopping event propagation and preventing default actions.

7.14.2 DnD Data Transfer

The actual effect of a DnD is to communicate information between the drag source and the dropzone. At the start of a DnD, the drag source loads data and indicates allowed operation in a data transfer object that is delivered to the dropzone, which receives the information and may indicate success or failure of DnD in the data transfer object to be used by the drop source at the end of DnD.

In all DnD events, the `event.dataTransfer` property gives you that data transfer object which exposes the following properties and methods:

- `files`—An array of files, being dragged, of length `files.length`. Each file is a `File` object (Section 7.13). The array is empty if no files are being dragged.

- `effectAllowed`—Values are `copy` (a copy of the source may be made at dropzone), `move` (the source item may be moved), `link` (a link to the source may be made), `copyLink` (either is allowed), `copyMove`, `linkMove`, `all` (all operations are allowed, the default), and `none` (drop is not allowed).

- `dropEffect`—The actual effect at the dropzone: `copy`, `move`, `link`, `none`.

- `setData(`*type,* *data_str*`)`—Loads a serialized string *data_str* of data of the given MIME *type* to be delivered to the dropzone.

- `getData(`*type*`)`—Returns the data string of the given *type*.

- `clearData([`*type*`])`—Removes data of the given *type* or all data.

7.14.3 DnD in TicTacToe

Let's implement a TicTacToe game where users can drag a game token, either X or O, to a game board with 9 dropzones (Figure 7.19). The draggable token automatically changes after each play and a winner is announced immediately after a winning play. Double-clicking anywhere on the game board starts a new game. Clicking on the draggable token changes it to the other token and also starts a new game.

FIGURE 7.19: TicTacToe Game with DnD

Let's discribe separately the draggable token, which is the DnD source and the game board, which contains nine DnD target dropzones.

7.14.4 Draggable Source

The draggable token is an image:

```
<img style="vertical-align: middle" id="token"
    src="x.png" alt="tictactoe token" draggable="true"  // (A)
    ondragstart="setImgURL(event)"                        // (B)
    ondragend="nextMove(event)"                           // (C)
    onclick="changeToken(); newgame();" />                // (D)
```

The attribute `draggable="true"` (line A) is needed to make any element a
DnD source. Preparing the DnD operation is usually done in the handler
for the `dragstart` event (line B), and finalizing a DnD is performed by the
`dragend` handler (line C). The `click` handler is something, not part of DnD,
to change the play token and to begin a new game (line D).

The event handler `setImgURL` is called with the DnD event object as the
DnD gets underway. The `event.target` property is the DnD source element
(the token image). And we set its opacity to 40% to indicate that the token
is being dragged (line E). The `src` property of an image gives the full URL of
the image file (line F), and that URL is placed in the `dataTransfer` object of
the event under the MIME type `text/uri-list` (line G). The `dataTransfer`
and the DnD events are important parts of the DnD API defined by HTML5.

```
function setImgURL(event)
{  event.target.style.opacity=0.4;                           // (E)
    var img=event.target.src;                                 // (F)
    event.dataTransfer.setData("text/uri-list", img);        // (G)
    event.dataTransfer.effectAllowed = 'copy';               // (H)
}
```

After a drop, the drag source receives the `dragend` event, which is handled by
`nextMove` in this example. If drop is successful (line I), we change the token
for continued play.

```
function nextMove(event)
{ event.stopPropagation(); event.preventDefault();
    event.target.style.opacity=1;
    if ( event.dataTransfer.dropEffect != "none" )          // (I)
    {  changeToken();  }
}
```

```
function changeToken()
{ var img = document.getElementById('token');
    if (img.getAttribute("src")=="x.png" )
        img.src="o.png";
    else img.src="x.png";
}
```

7.14.5   Dropzone

Now let's turn our attention to the game board that is a 3×3 table where each cell is a dropzone with the nearly identical HTML code as follows:

```
<td id="tl" dropzone="copy s:text/uri-list"      (1)
    ondragenter="return false"                   (2)
    ondragover="return feedback(event, this);"   (3)
    ondragleave="this.removeAttribute('class')"  (4)
    ondrop="play(event, this);"> </td>           (5)
```

Of course, each cell has a unique id: tl (top-left), cc (center-center), br (bottom-right), and so on.

In general, to accept a drop, the drop target needs a dropzone attribute (line 1) and handlers for DnD events. The dropzone attribute indicates the data type to accept, for example s:text/uri-list to accept uri text strings, or f:image/jpeg to accept a JPEG image file. The dropzone attribute also indicates the action on the dragged data: copy (data copied), move (data moved), or link (data linked). Browser support for the dropzone attribute is not uniform. In addition to the dropzone attribute, a drop target can handle the dragenter event to accept/deny a drop, and handle the dragover and dragleave events to specify the visual feedback for the user.

Most areas in a webpage or application are not valid dropzones. Thus, the default handling of dragenter and dragover events is to prevent dropping. To allow a drop, use either of the following (line 2):

```
ondragenter="return false"
ondragenter="event.preventDefault()"
```

In our TicTacToe example, we call feedback (line 3) to check if dropping is allowed (line 6) and to provide dropzone visual feedback by highlighting (line 7) the cell. It also important to return false to stop event propagation and cancel default handling (line 8).

```
function feedback(event, node)
{ var id=node.getAttribute("id");
  if ( end || eval(id) > 0)
  // game ended or position occupied
  { event.dataTransfer.dropEffect="none"; // no drop (6)
    return false;  // stop propagation
  }
  node.setAttribute('class','over');          // (7)
  return false;                               // (8)
}
```

Returning false is important (lines 3 and 8) because we need to cancel the default action for dragover, which denies dropping.

In case the drag leaves the cell without dropping, we need to remove the highlighting (line 4). If a drop does occur, we call `play()` to place the dropped TicTacToe token on the target game board cell:

```
function play(event, cell)
{ var id=cell.getAttribute("id");
  event.preventDefault(); event.stopPropagation();
  if ( end || eval(id) > 0 ) return false;          // (9)
  eval(id+(which_tk ? "=1;" : "=2;"));              // (10)
  which_tk= !which_tk;                              // (11)
  tnode=token(event.dataTransfer.                   // (12)
          getData("text/uri-list"));
  if (sp==null) sp=cell.firstChild.cloneNode(true);
  cell.removeChild(cell.firstChild);
  cell.appendChild(tnode);                          // (13)
  cell.removeAttribute('class'); winner();          // (14)
}
```

The `play()` function stops event propagation and default handling, and checks to see if game has ended or the board position is occupied (line 9) before proceeding to place the token. A value of 1 or 2 is recorded in a variable with the same name as the `id` attribute depending on which token is placed (line 10 and 11).

The function `token` is called to create an `img` element using the dropped image src string, and the table cell child is replaced (lines 12 and 13). A clone copy of the replaced child is stored in the variable `sp`, which is used later in restoring the board for a new game. Any hightlighting is removed. Then we check if there is a winner (line 14).

```
function token(img)
{ var t = document.createElement("img");
  t.setAttribute("src", img);
  t.setAttribute("draggable", "false");  // no cheating
  t.setAttribute("width", "35");
  t.setAttribute("height", "40");
  t.style.display = "block"; return t;
}
```

To summarize, the TicTacToe example demonstrates how to create a draggable element and a dropzone, specify allowed drop effect and load data into the `dataTransfer` object, avoid dropping into the same zone twice, prevent moving of dropped tokens, provide visual and program feedback for dropping, receive transferred data and perform intended effects at the dropzone, and prepare for the next drop after a successful play. Also, to allow a drop, the dropzone element's `dragenter` and `dragover` handlers must return `false` to prevent further event propagation and default event handling.

The full example (**Ex:** `TicTacToe`) can be found on the DWP website.

7.15   Tower of Hanoi

To further illustrate DnD, we will implement the well-known *Tower of Hanoi* puzzle in Computer Science.

Legend has it that monks in Hanoi spend their free time moving heavy gold disks to and from three poles made of precious stones. The gold disks are all different in size. Each disk has a round hole at the center to fit the poles. In the beginning all disks are stacked on one pole in a sequence small to large. The task at hand is to move the gold disks one by one from the first pole to the third pole, using the middle pole as a resting place if necessary. There are only three rules to follow:

1. A disk cannot be moved unless it is the top disk on a given pole, and only one disk can be moved at a time.

2. A disk must be moved from one pole to another pole directly. It cannot be set down in some other place.

3. At any time, a bigger disk cannot be placed on top of a smaller disk.

Figure 7.20 shows the first disk being dragged in our implementation of this game where we allow up to eight disks. Each pole is a `div` dropzone (lines **A**,

FIGURE 7.20: Tower of Hanoi Puzzle

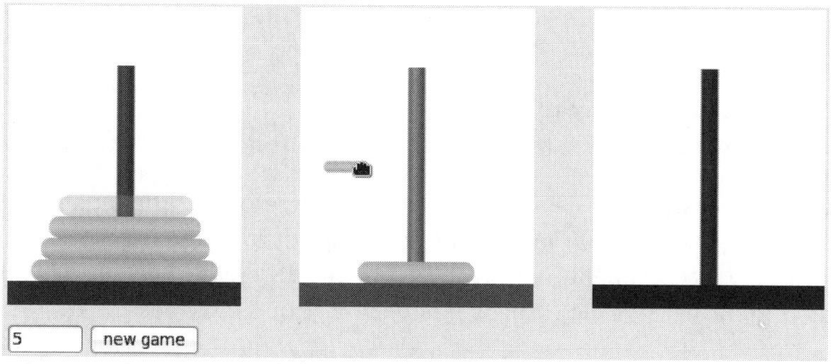

B, and C) and each disk is a potentially draggable `div` with ids `d1` (the largest disk) through `d8` (the smallest disk) that can be child nodes in any dropzone. For his example (**Ex: TowerOfHanoi**), the PHP/HTML code for the three towers is as follows:

```php
<?php $tname="t1"; include("tower.php"); ?>          // (A)
<div class="spacing"></div>
<?php $tname="t2"; include("tower.php"); ?>          // (B)
<div class="spacing"></div>
<?php $tname="t3"; include("tower.php"); ?>          // (C)
</section><section>
```

```
<p><input id="nd" required="" type="number"  min="1"
  max="8" step="1" value="3" size="5" />
<input type="button" value=" new game " onclick=
"newGame(document.getElementById('nd').value)" /></p>
</section>
```

The file `tower.php` is

```
<div dropzone="move s:text/plain"
     ondragenter="return false;"
     ondragover="return feedback(event);"
     ondragleave="this.classList.remove('highlight')"
     ondrop="moveDisk(event, this);"
     id="<?php echo $tname;?>" class="tower"></div>
```

On page load, we call `newGame(3)` to display a three-disk game:

```
function newGame(n)
{ if ( n < 1 || n > 8 ) n=3; removeDisks('t1');
  removeDisks('t2'); removeDisks('t3');     // (D)
  var tower=document.getElementById('t1');  // first tower
  var i;
  for ( i=n; i > 0 ; i--)                   // (E)
  {  tower.appendChild(mkDisk(i));
  allowDrag(tower);                         // (F)
}
```

```
function allowDrag(tower)
{  tower.firstChild.setAttribute("draggable","true");  }
```

A `newGame(n)` call removes all disks from all three poles, appends `n` disks on the first pole (line E), then makes the top-most disk `draggable` (line F).

We used CSS gradient background images to create the visuals for the poles and gold disks. A pole turns golden if a drop is allowed on it.

```
div.tower
{  min-width: 200px; height:  250px;
   display: table-cell;
   vertical-align: bottom; margin: 30px;
   border-top: 50px solid #fff;
}
```

```
div#t1.highlight, div#t2.highlight, div#t3.highlight
{ background-image: linear-gradient(left, #fff 0%,
    #fff 46%, #E9AB17 47%, #FDD017 49%, #FBB117 53%,
    #fff 54%, #fff 100%);
}
```

```
div.spacing
{  min-width: 50px; display: table-cell; }

div#t1
{ border-bottom: 20px solid #700;
  background-image: linear-gradient(left, #fff 0%,
        #fff 46%, #900 47%, #c00 49%, #750000 53%,
        #fff 54%, #fff 100%);
}

/* div#t2 and div#t3 use different colors */

div.disk  /* all disks */
{  height: 18px;
   display: block;
   margin-left: auto;  margin-right: auto;
   border-radius: 9px;
   background-image: linear-gradient(bottom,
      #E9AB17 0%, #FDD017 50%, #FBB117 100%);
}

div#d1 { width: 160px; } div#d2 { width: 145px; }
div#d3 { width: 130px; } div#d4 { width: 115px; }
div#d5 { width: 100px; } div#d6 { width: 85px; }
div#d7 { width: 70px; }  div#d8 { width: 55px; }
```

The preceding gradient code follows the CSS standard. But you may still need to add browser-specific prefixes such as -ms- and so on for certain browsers (Section 4.5).

As a disk begins dragging, its home pole is recorded in the global variable source_tower (line G), and its id is loaded into the dataTransfer object (line H). A custom drag image is also used, with the setDragImage method of the dataTransfer object, for better visual presentation.

```
var source_tower, drag_token=new Image();
drag_token.src="disk3.png";   // drag image

function initMove(event) // ondragstart
{ var disk=event.target; // drag source
  source_tower=disk.parentNode;                       // (G)
  var dt= event.dataTransfer;
  dt.effectAllowed = 'move';
  var id=disk.getAttribute('id');                     // (H)
  dt.setData("text/plain", id);
  dt.setDragImage(drag_token,35,5);
```

```
  disk.style.opacity=0.4;
}


function finishMove(event) // ondragend
{ event.stopPropagation(); event.preventDefault();
  event.target.style.opacity=1;                      // (I)
  if ( event.dataTransfer.dropEffect != "none" )
  { if ( source_tower.hasChildNodes() )
    { allowDrag(source_tower); }                     // (J)
  }
}
```

After a drop, we restore the source disk's opacity (line I) and make sure the
top disk is draggable (line J).

At any dropzone, the `ondragover` handler `feedback` determines if a drop
is allowed (line K) and provides visual feedback (line L) accordingly. In our
case, a pole turns golden, by CSS, to indicate an allowable drop.

```
function feedback(event)
{ var id=event.dataTransfer.getData("text/plain");
  var tower=event.target;
  if ( source_tower == tower || id == null ||
       (tower.hasChildNodes() &&
       id < tower.firstChild.getAttribute("id")) )  // (K)
  { event.dataTransfer.dropEffect = "none";
    return false;
  }
  tower.classList.add("highlight");                  // (L)
  return false;
}
```

The `drop` even is handled by `moveDisk`. If an actual drop took place, the
dropped disk (line M) is added as the first child of the dropzone (line O or P),
and any disk just beneath it is made undraggable (line N). Finally, the pole's
golden color is canceled (line Q).

```
function moveDisk(event, tower)
{ var id=event.dataTransfer.getData("text/plain");
  event.preventDefault(); event.stopPropagation();
  if ( id == null || (tower.hasChildNodes() &&
       id < tower.firstChild.getAttribute("id")) )
  { return false; }  // no drop
  var disk=document.getElementById(id);              // (M)
  disk.style.opacity=1;
  if ( tower.hasChildNodes() )
  { tower.firstChild.setAttribute("draggable",
                              "false");              // (N)
```

```
      tower.insertBefore(disk, tower.firstChild);      // (O)
   }
   else { tower.appendChild(disk); }                   // (P)
   tower.classList.remove("highlight");                // (Q)
}
```

This example applies DOM, DnD API, CSS, and PHP and also shows how they combine and work together. Visit the DWP website to play this game live.

7.16   DnD Local Files

We have been using the browser file picker associated with the file-type `input` element to enter local (client-side) files into the browser.

We can provide a better user interface than the file picker by allowing DnD of local files into an HTML dropzone that displays the file name, type, and size, allowing multiple DnD operations, checks for duplicated files and allowed file types, and support for file removal from the list of files dropped (**Ex: FileDnd**). Figure 7.21 shows mp3 files from a file browser application being drag-and-dropped to a webpage dropzone.

FIGURE 7.21: DnD of Local Files

The resulting list of files can be uploaded to the server side, be played as audio or video, or otherwise processed by the browser.

We focus on the dropzone and the file list management here. In a later section we will see how such list of files can be uploaded, for example.

The dropzone is an `ol` where we change the background to indicate "ready for dropping".

```
<ol id="filelist" ondrop="addFiles(event, this)"
 ondragover="this.style.backgroundColor='#fdd';
             return false;"
 ondragleave="this.style.backgroundColor=drop_bg;
             return false;"></ol>
```

with the following styles.

```
ol#filelist
{  border: thin solid black; height: 80px;
   width: 600px; overflow-x: auto;
   background-color: white;
}
```

```
ol#filelist li { white-space: nowrap;  }
```

The drop handler `addFiles` calls `listFiles` to process the dropped file. The `listFiles` function checks and stores each file in a global array `the_files` (lines 1–3) and appends a child node (a list item) to the dropzone. The item displays the file name, size, and type, and a clickable `delete item` sign (lines 4–6). The dropzone background color is removed, signaling drop completion (line 0).

```
var the_files=new Array(), drop_zone, drop_bg="#fff",
    c_type="image/jpeg,application/pdf,audio/mp3";

function addFiles(event, dz)
{ event.stopPropagation(); event.preventDefault();
  listFiles(event.dataTransfer.files,dz);
  dz.style.backgroundColor=drop_bg;                       // (0)
}

function listFiles(files,dz)
{ var f, pos;
  drop_zone=dz;
  for (var i=0; i < files.length; i++)                    // (1)
  { f=files[i];
    if (! inFileArray(the_files, f) && check_type(f))     // (2)
    { pos=the_files.push(f);                              // (3)
      var item=document.createElement("li");              // (4)
      item.onclick=makeHandler(pos);                      // (5)
      item.innerHTML="<p>"+f.name+" ("+f.size+" bytes, "
        +f.type+")"+"  <img src=
          'delete_small.png' alt='delete' title=
          'delete file' id='cd_" +pos
        + "' style='vertical-align: middle' /></p>";
      dz.appendChild(item);                               // (6)
    }
  }
}
```

Checking for duplicated files is done by the `inFileArray` function. When two files have the same name and size, we assume they are the same file. Allowed file type is checked by the function `check_type`.

```
function inFileArray(arr, file)
{ for (var i=0; i < arr.length; i++)
  { if ( arr[i] !=null &&  arr[i].name == file.name
       && arr[i].size == file.size )
    { window.alert(file.name + " already on the file list.");
      return true;
    }
  }
  return false;
}
function check_type(f)
{  return(c_type.indexOf(f.type) >= 0);  }
```

Clicking on a file list item opens a confirmation dialog to delete the file from the list. The entry on the array `the_files` is set to `null`. Non-null files on the array can then be processed by JavaScript for many different purposes, including file uploading (Section 11.16).

```
function makeHandler(pos)
{ return function()
  { if (window.confirm("deleting file "
            + the_files[pos-1].name + "?"))
    { the_files[pos-1]=null;
      drop_zone.removeChild(this);
    }
  };
}
```

The interface provided by this example is much more user friendly and functional for people who are good with drag-and-drop. But, not everyone is familiar with the intricacies of DnD. The browser built-in file picker does have its advantages. It would be ideal if we added a file picker to work with the same DnD dropzone, allowing users to use both the picker and DnD to add files to the dropzone where unwanted files can also be removed. We can do this easily by combining knowledge from this section and Section 7.13. Readers can do this as an exercise.

7.17   Canvas Element and API

The `canvas` is a new element introduced by HTML5. Similar to the `img` element, the `canvas` element is a phrasing (inline) element providing an area for displaying graphics produced by JavaScript. Unlike the `img` element, `canvas` is not a void element. Here is the general form:

```
<canvas id="someId" width="w" height="h"></canvas>
```

The w and h give the desired width and height (defaults to 300×150) of the area for graphic from JavaScript. All major browsers have support for `canvas`. However, you should place appropriate fall-back content for nonvisual user agents inside the `canvas` element.

HTML5 also defines an API for JavaScript to draw vector graphics and images on `canvas` elements. First, let's look at a simple example.

7.17.1   A First `canvas` Example

To keep the first example (**Ex: HelloCanvas**) simple, we will simply draw a red rectangle on a 200×200 pixel canvas (line B). The `canvas` element is centered on the page by changing its `display` to `block` with automatically computed left and right margins (line C).

```
<body style="background-color: #def" onload=
    "rect(document.getElementById('rect'));">   (A)
<section><h1>Canvas Drawing</h1>
<canvas id="rect" width="200" height="200"        (B)
      style="margin: auto; background: white;
      display: block">                            (C)
Drawing a rectangle</canvas></section></body>
```

The actual drawing is done by the `rect` function called onload (line A).

Each canvas element has its own *rendering context* that controls the context within which graphics rendering is performed in that particular canvas area. To add drawings or images to any canvas `cvs`, obtaining the context object is the very first step. Follow these three major steps to draw graphics in JavaScript:

1. Obtain the rendering context object `var ctx = cvs.getContext('2d');`. Most browsers support the 2D context. Some may offer other contexts.

2. Set the desired property values in the context `ctx`, including color, line/fill style, coordination transformations, fonts, and so on.

3. Draw pixels, lines, shapes (filled or not), or images by calling methods provided by the context object.

The `rect` function here obtains the rendering context object (line D), sets the `fillStyle` to a red color (line E), and calls the `fillRect` method to fill a rectangular area with upper-left corner at (50,50) with a width of 100 and a height of 100 (line F) with the color.

```
function rect(canvas)
{  var ctx=canvas.getContext('2d');    // (D)
   ctx.fillStyle = "rgb(180,0,0)";     // (E)
   ctx.fillRect(50,50,100,100);        // (F)
}
```

To draw (stroke) the borders of a rectangle without filling the area, we can use

```
ctx.strokeStyle = "rgb(180,0,0)";  // color for the stroke
ctx.lineWidth = 20.5;              // line width for the stroke
ctx.strokeRect(50,50,100,100);     // drawing the rectangle
```

Figure 7.22 shows the filled (left) and the stroked (right) rectangles.

FIGURE 7.22: Drawing Rectangles

The default for `strokeStyle` and `fillStyle` is the color black. You can set them to any valid CSS color string, a color gradient, or a pattern, as we will see later. Note that half of the stroke line width is drawn outside the dimension of the rectangle.

In addition to the two functions `strokeRect` and `fillRect`, there is also the `clearRect` that clears all graphics in the give rectangular area.

7.17.2 Drawing Shapes

Of course we can draw all kinds of lines and shapes, in addition to rectangles, on a `canvas`. To draw almost any arbitrary shape, you can define a path outlining the shape and then call either the `fill` or the `stroke` method to draw the shape. Here is how:

1. Call `ctx.beginPath()` to begin defining a new path.

2. Call point position and path construction methods to build up the complete outline desired.

3. Call `ctx.closePath()` to finish the path. This will link the end of the path to the beginning of the path if they are not already the same. In practice, we do not need to call this function explicitly because a call to `stroke()` or `fill()` will automatically close the path for you.

4. Call `stroke()` or `fill()` to do the rendering.

Path-building methods include

- `moveTo(x, y)`—Moves the starting (current) point of the next path segment to the point (x, y).

- `lineTo(x, y)`—Adds to the path a straight line from the current point to the given point, which becomes the current point.

- `arc(x, y, r, θ₀, θ₁, dir)`—Adds to the path a circular arc centered at (x, y) with radius r starting at angle θ_0 and ending at θ_1 (measured from the positive x-axis in a clockwise direction). If *dir* is true, the arc goes clockwise, otherwise counter-clockwise. The current point is moved to the end of the arc.

  For example, either one of the following

  ```
  ctx.arc(100,100,70,Math.PI/4.0,Math.PI*2,true);
  ctx.arc(100,100,70,Math.PI/4.0,0,true);
  ```

  defines the same circular arc of 45 degrees in the fourth quadrant.

- `quadraticCurveTo(x₁, y₁, x, y)` and `bezierCurveTo(x₁, y₁, x₂, y₂, x, y)`—Adds a quadratic or cubic *spline* from the current point to (x, y) using one or two control points specified.

- `rect(x, y, w, h)`—Adds a rectangle to the path. The current point is also moved to $(0,0)$.

As an example (**Ex: Ctc**), let's draw an outer circle, an inscribing equilateral triangle, and an inscribing inner circle in the triangle (Figure 7.23). The canvas is

```
<canvas id="ctc" width="200" height="200"
style="border: thin solid black; margin: auto;
background: white; display: block"></canvas>
```

and the ctc function draws the three figures.

```
function ctc(canvas)
{   var ctx=canvas.getContext('2d');
    var r=70;                          // (I)
    ctx.beginPath();  // draw outer circle
    ctx.arc(100,100,r,0,Math.PI*2,true);
```

FIGURE 7.23: Drawing Figures

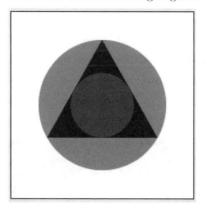

```
ctx.fillStyle = "rgb(0,180,0)";  ctx.fill();

ctx.beginPath();  // draw triangle
ctx.moveTo(100,100-r); // top point of triangle
var dy=Math.sin(Math.PI/6)*r;                 // (II)
var dx=Math.cos(Math.PI/6)*r;                 // (III)
ctx.lineTo(100-dx,dy+100);
ctx.lineTo(100+dx,dy+100);
ctx.fillStyle= "rgb(0,0,180)"; ctx.fill();

ctx.beginPath();  // draw inner circle
ctx.arc(100,100,r/2,0,Math.PI*2,true);        // (IV)
ctx.fillStyle= "rgb(180,0,0)"; ctx.fill();
}
```

The radius of the outer circle is 70 pixels in our example (line I). But the code will work for any radius setting. We used some basic trigonometry to compute the vertices of the inscribing equilateral triangle (lines II and III) and the radius of the inscribing inner circle (line IV).

The order in which the figures are drawn is important. A later figure overwrites existing pixels already on the canvas. Thus, if we draw the outer circle last, then the triangle and inner circle would be wiped out. The canvas context *composite* property

```
ctx.globalCompositeOperation
```

controls how a new drawing interacts with overlapping existing pixels already on the canvas. The default setting is `"source-over"`, which means source (new) pixels overwrite destination (existing) ones. The `"destination-over"` setting is just the opposite. There are a number of other settings governing which parts of the overlapping figures are drawn and how. See the canvas API documentation for details.

7.17.3   Color Gradients and Transforms

Making a few changes to the circle-triangle-circle drawing (Figure 7.23), we can display a simple clock face (Figure 7.24). The simple clock (**Ex:**

FIGURE 7.24: A Clock Face

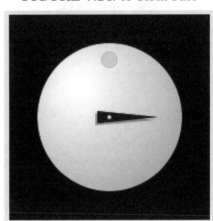

ClockFace) is drawn by the function `clock`:

```
function clock(canvas)
{   var ctx=canvas.getContext('2d');
    ctx.translate(100,100);      // translate coordinates
    drawFace(ctx);
    drawHand(ctx,Math.PI/2);     // points at 3 o'clock
}
```

The clock face is two filled circles each with a color gradient (lines A and B).

```
function drawFace(ctx)
{   var gr=ctx.createRadialGradient(-30,-30,5,-40,-40,140);   // (A)
    gr.addColorStop(0,'#fff'); gr.addColorStop(1,'#c0c0c0');
    ctx.beginPath(); ctx.fillStyle=gr;
    ctx.arc(0,0,72,0,Math.PI*2,true); ctx.fill();

    // Gold Dot
    gr=ctx.createRadialGradient(0,-56,0,0,-56,8);             // (B)
    gr.addColorStop(0,'#ffd700');
    gr.addColorStop(0.9,'#ffd700');
    gr.addColorStop(1,'#daa520');
```

```
    ctx.beginPath(); ctx.fillStyle=gr;
    ctx.arc(0,-56,8,0,Math.PI*2,true); ctx.fill();
}
```

As the example illustrates, to draw with a color gradient, you fiirst create a gradient object (radial or linear), add color stops, and then assign the gradient object to either the fill or the stroke style property.

```
createLinearGradient(x₁,y₁,x₂,y₂)          (point 1 to 2)
createRadialGradient(x₁,y₁,r₁,x₂,y₂,r₂)        (circle 1 to 2)
```

A single clock hand (line E) with shadow (line D), pointing at the direction specified by the t (radian) argument (line C) is drawn by the function drawHand:

```
function drawHand(ctx, t)
{   ctx.fillStyle="#009"; ctx.rotate(t);        // (C)

    ctx.fillStyle="rgba(80,80,80,0.4)";        // (D)
    ctx.beginPath();    // hand shadow
    ctx.moveTo(3,-45); ctx.lineTo(-4,14);
    ctx.lineTo(8,14); ctx.fill();

    ctx.fillStyle="#009";                      // (E)
    ctx.beginPath();    // hand
    ctx.moveTo(0,-48); ctx.lineTo(-6,12);
    ctx.lineTo(6,12); ctx.fill();

    ctx.beginPath();    // pin dot            // (F)
    ctx.arc(0,0,2,0,Math.PI*2,true);
    ctx.fillStyle="#ffd700"; ctx.fill();
}
```

A small golden dot (line F) pins down the hand.

Canvas 2D exposes these transform methods:

- ctx.translate(x, y)—Moves the origin to (x, y).

- ctx.rotate(θ)—Rotates the coordinate system clockwise θ radians.

- ctx.scale(h, v)—Scales up/down by a floating-point factor of h horizontally and v vertically. A negative factor reverses the direction of the x- or y-axis.

- ctx.setTransform(a, b, c, d, e, f)—Sets the coordinate transform matrix (CTM) to the matrix given (Section 12.3.1).

- ctx.transform(a, b, c, d, e, f)—Multiplies the current CTM by the matrix given (Section 12.3.1).

We have used translate and rotate in drawing the clock.

7.17.4   Canvas Animation

To demonstrate canvas animation, let's simply make the clock face (Figure 7.24) tick. Here is our strategy:

1. Display the clock face without the hand in a canvas (clockface, line 2).

2. Overlap another canvas of the same size (clockhand, line 3) on top of the clock face and draw the clock hand in it.

3. A click event on the hand canvas starts/stops the ticking (line 4). Each tick involves erasing the old hand and drawing it in a new position $\pi/30$ radians from the old position.

The HTML code is as follows:

```
<body style="background-color: #def" onload=
    "clockFace(document.getElementById('clockface'));
    handInit(document.getElementById('clockhand'));">   (1)
<section><h1>Canvas Clock Animation</h1>
<canvas id="clockface" width="200" height="200"          (2)
    style="background-color: #006; margin: auto;
    display: block"></canvas>
<canvas id="clockhand" width="200" height="200"          (3)
    style="margin: auto; margin-top: -200px;
        display: block" onclick="ss()"></canvas>        (4)
</section></body></html>
```

On page load (line 1), we draw the clock face in `clockface` and a clock hand pointing north.

```
var pi=Math.PI, sec=0, handCtx, stop=true;

function clockFace(canvas)
{  var ctx=canvas.getContext('2d');
   ctx.translate(100,100); drawFace(ctx);
}

function handInit(canvas)
{  handCtx=canvas.getContext('2d');
   handCtx.save();                      // (5)
   handCtx.translate(100,100);          // (6)
   drawHand(handCtx,pi*sec/30.0);       // (7)
}
```

The `handInit` function saves the state of its original graphics context (line 5) for later use, before moving the origin to (100,100) (line 6). A north-pointing hand is then drawn (line 7).

The click event handler `ss` toggles the global `stop` flag and calls `tick()` to set the clock in motion (line 8).

```
function ss()
{  stop = !stop; if ( ! stop ) tick(); }  // (8)

function tick()
{  if ( stop ) return;                    // (9)
   handCtx.restore(); handCtx.save();     // (10)
   handCtx.clearRect(0, 0, 200, 200);     // (11)
   handCtx.translate(100,100)             // (12)
   sec = ++sec % 60;                      // (13)
   drawHand(handCtx,pi*sec/30.0);         // (14)
   setTimeout(tick, 1000);                // (15)
}
```

The function `tick` is the heart of this animation. It checks the `stop` flag before continuing (line 9). The first task in the animation is to restore the saved state of the graphics context by calling `ctx.restore()`, then save it again for the next time (line 10). It proceeds to translate the coordinates (line 11), increment the second count (line 12), and calls `drawHand` with the correct radians position (line 13). Finally, it calls `setTimeout` to call `tick()` again after 1 second (line 15).

Restoring the state gets rid of any transform, color, and other settings used in the previous tick so new drawings can begin with a fresh (saved) state. To see the clock animation in motion try the complete example (**Ex:** `ClockAnim`) at the DWP website.

7.17.5 Drawing Images and Text

Image drawing on a canvas is a simple three-step process:

1. Create a new JavaScript image object (`img=new Image();`).

2. Set the image source (`img.src="`*url*`";`).

3. Draw the image with the `ctx.drawImage` method.

```
drawImage(img, x, y)    // positions it at (x,y)
drawImage(img, x, y, w, h)  //  scales to width and height
drawImage(img, x0, y0, w0, h0, x, y, w, h)
```

The third version draws a `w0` by `h0` slice of the image at (`x0,y0`) on the canvas at point (`x, y`). Note: The `img` object can also be some in-page canvas element.

To render text on a canvas, you have the methods `ctx.fillText("`*str*`"`, `x,y`) and `ctx.strokeText("`*str*`"`, `x,y`). Before calling either, you can set the `ctx.font`, fill style, stroke style, and transform properties. See the HTML5 canvas standard and other online documentation for more details.

7.18 For More Information

While we have presented many basic and important topics on HTML5 DOM and APIs, the coverage is by no means complete. Some APIs such as editable elements, client-side file access, offline cache, and client-side local storage will be discussed in later chapters.

The HTML5 DOM and API specifications may still be evolving. The DWP website resources page contains links to full documentations and specifications.

7.19 Summary

DHTML is a technique that combines JavaScript, CSS, and DOM. Part of the HTML5 standard, the DOM API is used to access and modify the DOM tree, which is a data structure representing a loaded HTML5 document in a browser. DHTML with DOM results in powerful and cross-platform code.

Browsers support the DOM-specified API and provide the required objects and methods for JavaScript programming. The `window.document` object, implementing both the `Document` interface and the `HTMLDocument` interface, gives you the root node of a webpage. Each element on this DOM tree corresponds to an HTML element in the source document and implements the `HTMLElement` interface derived from `Element`, which extends `Node`. The `document.getElementById` method is handy for obtaining the DOM object representing any HTML element with an assigned `id`. Starting from any node *e* on the DOM tree, you can follow the child (*e*.`childNodes`, *e*.`firstChild`), parent (*e*.`parentNode`), and sibling (*e*.`nextSibling`, *e*.`previousSibling`) relations to traverse the entire tree or any parts of it. You can also access and modify element attributes (*e*.`getAttribute`(*a*), *e*.`setAttribute`(*a*, *value*)) and styles (*e*.`style`.*property*).

The DOM API allows you to systematically access, modify, delete, and augment the DOM tree, resulting in altered page display:

e.`removeChild`(*node*)
e.`appendChild`(*node*)
e.`replaceChild`(*new*, *old*)
e.`insertBefore`(*new*, *node*)

New tree nodes can be created in JavaScript with

`document.createElement`(*tag*)
`document.createTextNode`(*str*)

Drag-and-drop (DnD) involves a source (draggable) and a destination (dropzone), and a data transfer object carried by the various events related to DnD. The DnD API can make webpages more interactive.

The Canvas API gives you full scripting control over 2D vector graphics generation, display, and animation. Image and font drawing are also supported.

When you combine event handling, including those generated by `window.setTimeout()`, and style manipulations, many interesting and effective dynamic effects can be achieved for your webpages.

Exercises

7.1. Explain in your own words what DHTML is. What are three important enabling technologies for standard-based DHTML?

7.2. What is DOM? The DOM tree? The most basic `Node` interface? Name important types of nodes on the DOM tree and describe the DOM tree for a webpage in detail.

7.3. Write the JavaScript code for obtaining the DOM node for an HTML element with a given *id* and for determining its node type.

7.4. What is the `nodeName` and `nodeValue` for an `HTMLElement`?

7.5. Explain the fields and methods in the `HTMLDocument` interface. Which object available by JavaScript supports this interface?

7.6. Describe the difference between a shallow and a deep copy of a node.

7.7. How does JavaScript modify the presentation style of an element on the DOM tree?

7.8. Explain the concept of Web service and the access of Web services with DHTML.

7.9. Modify the **Ex:** `DomCalc` in Section 7.10.1 and present a display in the form *string* = *result*.

7.10. Consider **Ex:** `DomDft` in Section 7.6. The traversal does not take comments into account. Modify the `dft` function to remedy this and test the traversal on pages that contain comments.

7.11. Consider the visual navigation of a DOM tree (Section 7.7). Take **Ex:** `DomNav` and add a tree display of the table. As the user navigates the table, also show the current position on the DOM tree display.

7.12. Modify the tic-tac-toe program (Section 7.14) to support *"I changed my mind"* by allowing the latest move on the game board to be dragged to another board position, but only if the opponent has not made a move yet. Of course, earlier moves cannot be taken back.

7.13. Improve the tic-tac-toe program in Section 7.14 by adding the ability to play with the computer. (Hint: Add move generation.)

7.14. Study the Tower of Hanoi puzzle program (Section 7.15) and make a solution demonstration for up to seven disks. The user picks the number of disks and the program displays an optimal sequence of moves by moving the disks from Tower A to Tower C. The program also displays the move count as the solution makes progress.

7.15. Construct a pocket calculator. Layout the LCD window and calculator buttons with a `table` and simulate the functions of the common calculator with `onclick` events on the buttons.

7.16. Do the exercise suggested at the end of Section 7.16.

7.17. Use DnD to implement the game of Halma.

7.18. Add a minute hand and an hour hand to the clock animation example **Ex: ClockAnim** in Section 7.17.4.

7.19. Consider the clock animation example **Ex: ClockAnim** in Section 7.17.4. Instead of using the `ctx.clearRect` method, we can clear the canvas using other ways:

(1) `ctx.width=ctx.width;`
(2) `ctx.globalCompositeOperation="copy";`

Revise the example to use each technique and discuss the pros and cons of all three techniques.

7.20. (**Interactive Panoramic View**) Take the 360-degree panoramic picture of Yosemite National Park (`yosemite.jpg`, Figure 7.25) supplied on the DWP website and use a Canvas to make it a continuous loop,

FIGURE 7.25: Yosemite Panoramic

moving either right or left with the mouse. This technique can be used in many different situations to provide the user with a good view of some target scene.

7.21. (**Scrolling Thumbnails**) Write your own JavaScript code (do not use libraries such as JQuery) to display a thumbnail strip similar to Figure 7.26

Requirements are:

FIGURE 7.26: Sample Thumbnail Strip

- The thumbnail strip scrolls left/right when the user clicks on one of the arrow icons.

- Each scrolling operation will smoothly move a number of off-strip pictures into view, then stop (a part of an old picture should still be in view).

- The scrolling stops automatically after the last picture, on either end, moves fully into view.

- On mouseover, a thumbnail displays the original-size picture in an area above the thumbnail strip, nicely positioned and formatted. On click, a thumbnail makes the picture the current picture.

- The current picture in the thumbnail strip is highlighted by border or background (your design).

- Your JavaScript code is easily customizable with parameters: width, height, background color of the strip, size of the thumbnails, the array of the image URLs, the arrow icons, and the scrolling speed.

- Your solution must contain at least twelve pictures of your choice.

- Add a caption to each picture and display the caption as tooltip on the thumbnails and as an actual caption for the current picture.

7.22. Add JavaScript to the **Scrolling Thumbnails** exercise (7.21) to allow a user to pick pictures and send their names in a request form to the site manager.

Chapter 8

Server-Side Programming with PHP

Having discussed in Chapter 5 some important basics of PHP and simple form processing, we are now ready to go in-depth and put PHP to use in practical situations as well as present additional aspects of PHP. We will see best practices of PHP in

- Dynamic page templates

- Integrated form delivery and processing

- File and directory manipulation

- Session control

- User authentication and login support

- Security code verification

- Delivering different content types

- Image manipulation and processing

- Content editing and management

- Error reporting and debugging

As we discuss these applications, we will also introduce important PHP language constructs such as function definitions, reference parameters, passing function names as arguments, I/O and networking, and OOP in PHP.

Let's get started.

8.1   Page Templates

A good design gives a website a consistent and unified look and feel throughout all its pages. Usually, templates are constructed for pages at different levels, the site entry, main page, first-level pages, second-level pages, and so on. The templates supply layout, graphical design, logo, navigation, styling, and other statically designed aspects of pages. Contents can then be loaded into the templates to realize each page.

The easiest way to implement page templates is to use an active page technology such as PHP to incorporate the reusable templates into pages on-the-fly. The approach separates page content from the template parts and allows each to be changed without affecting the other. Furthermore, program logic can be embedded in pages and templates to increase their reusability by making them as flexible as possible.

Let's illustrate page templates with a simple example (**Ex: ClubForm**). Figure 8.1 shows a page that combines the following parts.

- `rfront.php`—A template for the leading part of an HTML5 page, including document type, `head`, page title, style files, JavaScript files, and a navbar.

- `rback.php`—A template for the trailing part of a page.

- `vbar.php`—A left-side navigation bar.

- `clubform.php`—The join club page that combines a form and other content with the above templates.

FIGURE 8.1: A Template-Produced Page

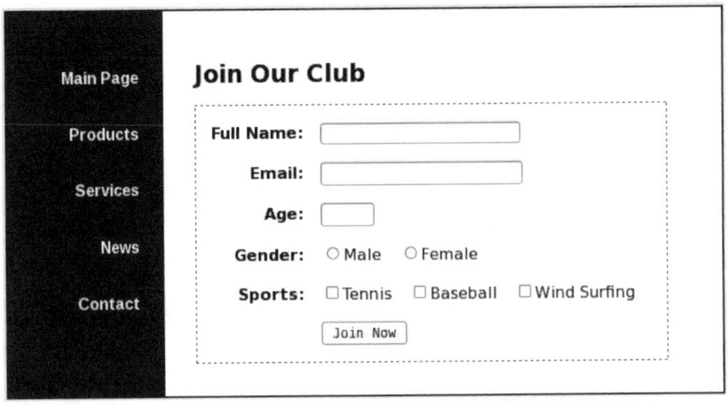

The three-part structure of `clubform.php` is as follows:

```
<?php $title="Join Our Club";                           (a)
      $css=array("basic.css", "form.css");              (b)
      require("rfront.php");
?>


page content: heading and form code goes here


<?php require("rback.php"); ?>
```

The parameterized values (lines a and b) are used by `rfront.php`, which adds the desired page title (line c), any CSS, JavaScript, and other head elements (line d), as well as code from given files (line e). A template may choose to place Javascript code at the end of a page rather than in the `head`.

```
<?php  /////   rfront.php    /////
require_once("tempfns.php"); ?>
<!DOCTYPE html>
<html xmlns="http://www.w3.org/1999/xhtml"
                  lang="en" xml:lang="en">
<head><meta charset="utf-8"/>
<title><?php echo $title; ?></title>              (c)
<?php addCss($css); addJs($js); addAny($any);     (d)
      addFromFile($srcfile); ?>                    (e)
</head><body>
<div id="centerpage"><section id="main">
<?php $page=basename($_SERVER['PHP_SELF']);
      require_once("navbar.php"); ?>              (f)
<article id="content">
```

Then, the $page value is set and the left navbar navbar.php is included
(line f).

```
<?php  /////   navbar.php    ///// ?>
<nav class="leftnavbar">
<?php if($page == "index.php") {?>
<span class="self">Main Page</span><?php } else {?>  (g)
<a href="index.php">Main Page</a><?php }?>

<?php if($page == "products.php") {?>
<span class="self">Products</span> <?php } else {?>  (h)
<a href="products.php">Products</a><?php }?>

<?php if($page == "services.php") { ?>
<span class="self">Services</span><?php } else {?>
<a href="service.php">Services</a><?php }?>

<?php if($page == "news.php") { ?>
<span class="self">News</span><?php } else {?>
<a href="news.php">News</a><?php }?>

<?php if($page == "contact.php") { ?>
<span class="self">Contact</span><?php } else {?>
<a href="contact.php">Contact</a><?php }?>
</nav>
```

The navbar.php produces the desired left-side navbar that has the correct
page identification feature by using the class="self" style (lines g and h, for
example).

The PHP functions used on line e are defined in tempfns.php.

```
<?php  /////   tempfns.php    /////
```

```
function addCss(&$arr)  // addsstyle sheet
{ if ( count($arr) )
  { reset($arr);
    foreach($arr as $file)
    { echo '<link rel="stylesheet" type="text/css" href="'
         . $file .  "\" />\n";
    }
  }
}

function addJs(&$arr)  // adds JS file
{ if ( count($arr) )
  { reset($arr);
    foreach($arr as $file)
    {  echo '<script type="text/javascript" src="'
         . $file .  "\"></script>\n";
    }
  }
}

function addAny(&$arr)   // adds arbitrary strings
{ if ( count($arr) )
  { reset($arr);
    foreach($arr as $str) {  echo "$str\n"; }
  }
}

function addFromFile(&$arr) // adds code from files
{ if ( count($arr) )
  { reset($arr);
    foreach($arr as $file)
    {  require($file);  }
  }
}
?>
```

The function `addFromFile` is useful to include meta information such as page description, keywords, favicon, and search engine related information. The pass-by-reference formal parameter `&$arr` makes these functions more efficient, as we will see next.

8.2   PHP Functions

One advantage of PHP is its rich set of built-in functions. To avoid reinventing the wheel, check before writing a function of your own.

To define a function *xyz*, use the declaration

```
function xyz($arg1, ...)        // header
{   ...  }                      // body
```

PHP function names are case-insensitive. Also, it is not allowed to redeclare a function. Thus, having `addFromFile` already, declaring the `addfromfile` function would cause a fatal error.

A function may receive zero or more arguments. The formal parameters are local variables to the function. Variables introduced in the function body are local to the function.

To use a global variable `$gvar` in a function, declare `global` in the function body:

```
global gvar;
global v1, v2, v3, ...;
```

or use the notation `$GLOBALS["gvar"]`.

PHP built-in super global variables (`$GLOBALS`, `$_SERVER`, `$_GET`, `$_POST`, `$_FILES`, `$_COOKIE`, `$_SESSION`, `$_REQUEST`, `$_ENV`) are directly accessible from any scope. There is no need to declare them `global` in functions or methods.

Normally, function arguments are passed by value. A pass-by-reference formal parameter can be declared with the notation `&$arg1`. Constants cannot be passed by reference, only variables can. Passing string and array variables by reference can make your function call much faster. Functions defined in `tempfns.php` (end of Section 8.1) are examples.

In a function call, make sure to supply all the arguments. Missing arguments cause the corresponding formal parameters to become **undefined** variables. Trailing formal parameters may have default values. For example,

```
function address($st, $city="Kent", $state="OH")
 { ... }
address("Kent Road", "Stow");    // sample call
address("Main Street");          // sample call
```

A by-reference version of `address` is

```
function address(&$st, &$city="Kent", &$state="OH")
{ ... }
$a="Main Street";   address($a);
```

Use `return($expr);` to exit a function and return the value of $expr. If a function is terminated by `return;` or by running out of statements, then `NULL` is returned. To return a reference from a function, prefix the function name with the symbol **&**.

It is allowed to pass a function name, `strcmp` say, to a formal parameter, `$fn` for example. Then, in the called function, the expression

```
$fn($s1, $s2)
```

can be used to call `strcmp`.

Passing too few arguments in a function call produces a warning message. But passing too many arguments is allowed. In a function, you can always access incoming arguments with

```
func_num_args();    // returns the number of incoming args
func_get_arg($i);   // returns the ith argument, 0-based indexing
func_get_args();    // returns array of all incoming args
```

Thus, the `sum` function can compute the total of any number of arguments:

```
function sum()
{ $ans=0;
  foreach(func_get_args() as $a)
  {  $ans += $a;  }
  return $ans;
}
```

PHP has many built-in functions. We have seen some string (Section 5.17) and array (Section 5.18) functions. Other functions will be discussed as the need arises.

Part of learning PHP is to get to know the commonly useful functions and to know how to quickly find others. The resources page on the DWP website has quick links to PHP functions and documentation.

8.3   Form Generation and Processing

In Chapter 5, we began our discussion on forms and form processing. We saw the different form input fields, how they can be used, and how to write PHP scripts to process incoming formdata.

Here we demonstrate a *best-practices* approach for form generation and processing that provides the following features:

- Integrating form delivery with form processing.

- Supporting both HTTP POST and HTTP GET methods.

- Automating formdata validation using supplied checking functions.

- Generating error-fixing forms (*eff*) based on detected input errors. An eff has all the input data pre-filled and the error entries clearly marked, together with instructions for correcting the problems.

- Displaying forms and processing results using supplied page templates.

- Applying any given CSS sheet for form layout and style.

8.3.1 Form Control Page

To illustrate this form generation and processing approach, let's apply it to the join club form. The implementation is general, and the code is applicable and reusable for many forms.

The control page `joinclub.php` begins with customizable settings. You name the file for the form (line 1) and the file for form processing (line 2).

```php
<?php /////    joinclub.php    /////
$the_form="clubform.php";                // (1)
$the_action="joinaction.php";            // (2)

$check_list=array(                       // (3)
   "client_name"=>"require",
   "email"=>"require",
   "age"=>"age_check");

function age_check($age)                 // (4)
{  if ( !empty($age) && is_numeric($age)
       && $age >= 18 && $age <= 65)
   { return "OK";  }
   else
   { return "must be 18 to 65"; }
}
```

You then set up the global `$check_list` for the required and other input (line 3), supplying the functions needed for validating all input that needs checking (line 4).

At the end of `joinclub.php` comes the fixed code (do not change) for form display and form processing (lines 5–8).

```php
require_once("formfns.php");                 // (5)
if (($_SERVER['REQUEST_METHOD']=="GET"       // (6)
    && count($_GET)==0 ) ||
    ! formdata_check($_REQUEST) )            // (7)
{ require($the_form); } // display form
else
{ require($the_action); }// process form     // (8)
?>
```

We display the form upon a GET request with no query string (line 6) or upon failure of form validation (line 7). Otherwise, the validated form input is processed by loading the given action script (line 8).

Formdata validation is performed by the function `formdata_check` (defined in `formfns.php`). The function uses the global `$check_list` array to validate each input field given by the reference parameter `&$fd`. Rach required input field must be set and non-blank (line 9). Any error messages are recorded in the global array `$formErr` that will be used in generating the eff. The function returns 1 if no errors are found and 0 otherwise.

```
function formdata_check(&$fd)
{   global $formErr, $check_list;
    $pass=1; reset($check_list);
    foreach($check_list as $key=>$val)
    { if ( $val == "require" )
      { if(isset($fd[$key]) && trim($str)!=="") // (9)
        { $pass=0;
          $formErr[$key]="please fill in";
        }
      }
      else
      { if (($s=$val($fd[$key])) != "OK" )
        { $formErr[$key]=$s; $pass=0; }
      }
    }
    return $pass;
}
```

8.3.2   Form Generation

The `clubform.php` demonstrates an efficient way to use a single PHP file to generate the initial form and subsequent eff in case of form errors. The code also takes advantage of a nice table layout defined in `form.css` (line A) and incorporates any site template and overall styling (line B)

```
<?php $title="Join Our Club";
      $css=array("basic.css","form.css");                    (A)
      require("rfront.php"); ?>                               (B)
<h2>Join Our Club</h2>
<form method="post" action="">
<div class="entry">
 <?php $id="client_name";
      genLabel("$id","Full Name:"); ?>                        (C)
 <span class="field"><input required="" size="25"            (D)
 <?php text("$id"); ?> autofocus="" />                        (E)
 <?php genErr("$id");?></span></div>                          (F)

<!-- more form entries -->
</form><?php require("rback.php"); ?>
```

Each form entry consists of a label (line C) plus an *input field span* (line D) that contains an `input` element followed by optional instruction and/or error texts. We lay out the label and the input field span as two cells of a table row.

Form code generation is done by calling functions such as `genLabel`, `text`, and `genErr`. Each is passed the *input key*, which is the value of the `name` attribute of the `input` element (lines C–F).

The **Full Name**, **Email**, and **Age** entries have the same structure.

The input field span for the **Gender** entry contains radio buttons (lines G and H). And the input field span for the **Sports** entry deploys checkboxes (lines I and J).

```
<div class="entry">
<?php $id="sex"; genLabel("$id","Gender:"); ?>
<span class="field">
<label><input <?php radio("$id","Male");?>                (G)
type="radio" />Male</label><label><input type="radio"
<?php radio("$id","Female");?> />Female</label>          (H)
<?php genErr("$id");?></span></div>


<div class="entry">
<?php $id="sport"; genLabel("$id","Sports:"); ?>
<span class="field">
<label><input <?php check("$id","tennis"); ?>           (I)
type="checkbox" />Tennis</label><label><input
type="checkbox" <?php check("$id","baseball"); ?>/>
Baseball</label><label><input type="checkbox"
<?php check("$id","windsurf"); ?>/>Wind Surfing
</label><?php genErr("$id");?></span></div>              (J)


<div class="entry"><label></label>                       (K)
<input type="submit" value=" Join Now " /></div>
```

The **Join Now** submit button, which needs no label, concludes the form entries (line K).

The form code generation functions, kept in `formfns.php`, help create labels (normal or error), error messages, as well as text, radio, and checkbox input fields.

```
///// code producing functions in formfns.php
function genLabel($key, $label)  // for labels
{ global $formErr;
  echo "<label for=\"$key\"";
  if ( !empty($formErr[$key]) )
  { echo ' class="fixthis" '; }
  echo ">" . $label . "</label>";
}

function text($key)   // for text type input
{ if ( !empty($_REQUEST[$key]) )
    { echo 'name="' . $key . '" id="' . $key .
```

```
        ' " value="' .$_REQUEST[$key] . '" ';
    }
    else
    {  echo 'name="' .$key. '" id="' .$key. '" '; }
}

function radio($key, $val)    // for radio button
{  echo 'name="' . $key . '" value="' .$val. '" ';
   if ( !empty($_REQUEST[$key]) )
   {  if (  $val == $_REQUEST[$key] )
      {  echo ' checked=""'; }
   }
}

function check($key, $val)    // for checkboxes
{  echo 'name="' .$key. '[]" value="' .$val. '" ';
   if ( !empty($_REQUEST[$key]) &&
        is_array($arr=$_REQUEST[$key]))
   {  if ( in_array($val, $arr) )
      {  echo ' checked="" ';  }
   }
}

function genErr($key)               // for error msgs
{ global $formErr;
  if ( !empty($formErr[$key]) )
  {  echo '<br /><span class="fixthis">'
         . $formErr[$key] . "</span>";  }
}
?>
```

Figure 8.2 shows a sample error-fixing form where the Age field is highlighted and supplied with an error message.

8.3.3　Form Processing

When a submitted form passes input validation, the designated $the_action PHP script is called to process the data (line 8). In this example, it is the joinaction.php script that sends the membership application to the club manager by email with Cc to the applicant (line I).

```
<?php  /////   joinaction.php   /////
  require_once("email.php");
  $to='"Mr. Manager" <manager@club.com>';
  $subject="Club Membership";
  $title="Thanks for Joining Our Club";
```

FIGURE 8.2: An Error-Fixing Form

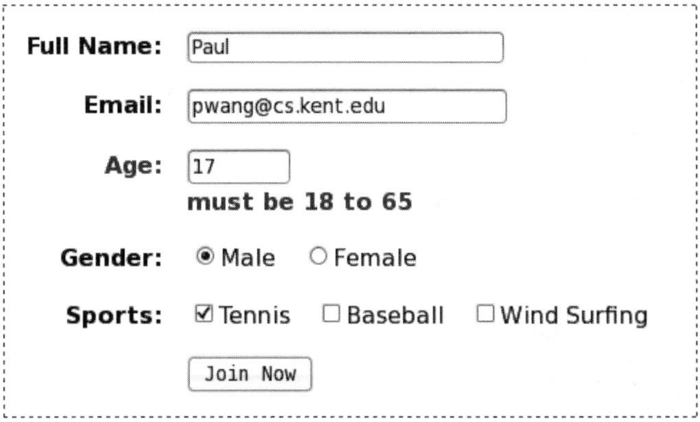

```
$cc= 'Cc: "' .$_POST['client_name'] .'" <'
        . $_POST['email'] . '>';
$headers = 'From: "Super Club" <service@superclub.com>'
      . PHP_EOL . $cc . PHP_EOL . 'X-Mailer: PHP-'
      . phpversion() . PHP_EOL;
$msg="We have emailed";
if ( ! email_formdata($to, $subject, $headers) )
{   $msg='<span style="color: red">We failed to email</span>';
}
$title=$subject; $css=array("basic.css", "form.css");
require("rfront.php");
?>
```

```
<h2><?php echo $title; ?></h2>
<p>Thank you <span style="color: blue">
<?php echo $_POST['client_name']; ?></span>
for joining our club.</p> <p><?php echo $msg; ?>
your application to our manager,  with a copy to
<code style="color: blue">
<?php echo $_POST['email']; ?></code>.</p>
<?php require("rback.php"); ?>
```

A confirmation response page from the joinaction.php script can be seen in Figure 8.3.

This example illustrates a general form generation and processing practice that is efficient, flexible, and reusable. The full example (**Ex: FormBP**) can be found at the DWP website.

FIGURE 8.3: Form Processing Result Page

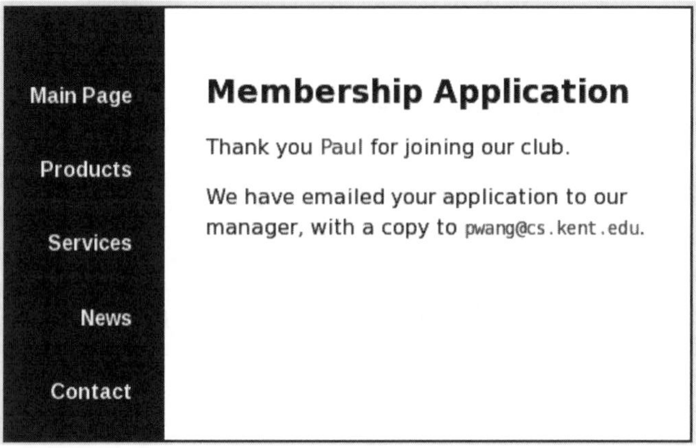

8.4 File I/O in PHP

We have seen some file operations in file uploading (Section 5.22). PHP provides a complete set of file and directory functions enabling you to easily access and manipulate files and folders on the local file system.

Common file system functions include

- File tests—file_exists, is_dir, is_file, is_link, is_readable, is_writable, is_executable, filesize($file)

- File manipulation—copy (copies a file), unlink (deletes a file), rename (renames a file), chgrp (changes file group), chmod (changes file mod), chown (changes file owner)

- File I/O—fopen (opens file for I/O, returns handle), fclose (closes handle), feof (tests eof), fflush (flushes buffer), fread (reads from handle), fwrite (writes to handle), readfile (sends file to output)

- Directory handling—mkdir (creates dir), rmdir (removes dir), chdir (changes dir), closedir (closes dir handle), getcwd (returns current working dir), opendir (opens dir, returns handle), readdir (reads from dir), rewinddir (rewinds dir handle)

- File status— fileatime (last access time of file), filectime (inode change time of file), filegroup (file group), filemtime (file modification time), fileowner (file owner), fileperms (file permissions), filesize (file size), filetype (file type)

We also have basename(*path*) (extracts the basename from the given path), dirname(*path*) (extracts the directory name). We have seen the usefulness of include(*file*), include_once(*file*), require(*file*), and require_once(*file*) to include other PHP files in a PHP script. The function file(*name*) returns the lines of the given file into an array while the function file_get_contents(*name*) returns the file contents as a string.

PHP file I/O work not only with file names, but also URLs to access remote files.

8.4.1   Listing Files in Directories

File/folder access and manipulation are clearly important for various Web operations such as file upload/download, slide shows, audio and video playlists, file templates, and content management, just to name a few.

Often, a folder can be set aside for contents uploadable by users or the site administrators. Such contents can be text files, code files, images, audio, video and so on. We can then use PHP script to deploy the contents in webpages and to allow registered users to add, delete, or modify the files.

The function dir_list (**Ex: DirList**) produces a sorted array of files, in a given folder, specified by $path, that satisfy some criteria specified by $test.

```php
function dir_list(&$path, $test)
{ if ( !($dir_handle=@opendir($path)) )      // opens dir (A)
  { trigger_error("Unable to open $path");
    return null;
  }
  $theFiles = array();
  while ($file = readdir($dir_handle))       // gets next file
  { if ( $test($file) )                      // if passes test
    array_push($theFiles, "$path/$file"); // puts on array
  }
  closedir($dir_handle);
  sort($theFiles);  reset($theFiles);        // sorts array
  return $theFiles;
}
```

Say we are interested in image files with .png and .jpg suffixes. We can pass is_img as the $test in a call to dir_list.

```php
function is_img($file)
{  return ( preg_match ("/\.jpg/", $file)
      || preg_match ("/\.png/", $file) );
}
```

See the next section for the meaning of the @ sign on line **A**.

8.5   PHP Error Reporting and Debugging

Just as other languages, PHP scripts may run into errors. When a PHP-containing page is being loaded by the server, syntax errors may occur. During execution, a script may also encounter other run-time errors.

To debug a newly constructed page containing PHP, it is a good idea to run it through the PHP interpreter from the command line to check syntax:

php *some_page*.php

Having fixed any syntax problems with command-line testing, we can check the page from the Web to see if there are any run-time errors. Common run-time problems include undefined variables or array entries, file and folder access problems, and setting headers after standard output already took place.

If you get a blank or partial page, it usually means the PHP code ran into a problem not encountered from the command line. It can often be a file access problem from the Web.

PHP helps debugging by reporting errors. You can control whether errors, and what types of errors, are reported. To control error reporting, you may use the function

```
error_reporting($n)   // sets error reporting level
error_reporting()     // gets error reporting level
```

The first version also returns the previous error reporting level. PHP defines integer constants (bit patterns) for various error reporting levels (see the PHP manual). But the most useful constant for debugging is `E_ALL`, which ensures the reporting of all errors. For debugging, putting

```
error_reporting(E_ALL);       // reports all errors
ini_set("display_errors", 1); // enables error display
```

at the beginning of your PHP code will cause any and all error messages to be displayed. User error messages reported via the function `trigger_error` are subject to error reporting control. The function `ini_set` allows you to set PHP configuration options, such as `display_errors`, at run-time.

For a production page, you may want instead to use

```
error_reporting(0);           // reports no errors
ini_set("display_errors", 0); // disables error display
```

When making a function call, prepending @ to the function name (line A) sets the error reporting level to 0 for the duration of the function call.

Instead of displaying an error message, which can be undesirable in many situations, you can use the function `error_log` to send the message to a log file for later diagnosis:

```
error_log(msg) // Adds to default log file
```

```
error_log(msg, 1, email_addr) // Sends email
error_log(msg, 3, file_path) Appends to file
```

See the `php.ini` file (Section 10.23) for the error log location and other PHP configuration option settings.

8.6 Login with Basic Authentication

It is often the case that certain parts of a website will have restricted access. Such protected resources are not open to everyone on the Web and are reserved for specific users. For example, online merchants and businesses may require login to access areas for members or employees only.

To arrange for user login, a website usually needs to perform user *registration* and *authentication*. Registration involves recording personal information about individual users and assigning each a login userid and password. Authentication involves checking client supplied userid and password against stored data on the server side.

For security reasons, passwords are only stored in encrypted form produced by one-way encoding functions. Such encryption makes it almost impossible to recover a password from its encrypted form. This is why a forgotten password cannot be retrieved and must be reset.

The command-line tool **htpasswd**, supplied by the Apache Web server, enables you to easily store userids and passwords in a password file. Such a password file contains lines in the form

userid:*encrypted-password*

To create a new password file with one user in it, run the command

htpasswd -c *password_file userid*

You will be prompted for a password. The same command, with no -c option, adds a new user to an existing password file. Use the -D option to delete a user from the file.

For encrypting passwords, the **htpasswd** command uses **crypt** on Linux/UNIX systems or MD5 on Windows, by default. Other encryption options exist.

8.6.1 HTTP Basic Authentication

To authenticate is to check and verify the identity of the incoming person/program. Authentication is the basis for granting or denying access to protected resources.

The HTTP protocol provides two ways for the Web server and browser to cooperate in performing access authentication: *Basic Authentication* and *Digest Authentication*. The former is simple but not secure because userid and password are sent essentially unencrypted. The latter is more complicated but only slightly more secure.

In practice, it is best to switch from HTTP to HTTPS (HTTP over TLS, RFC2818). Under HTTPS, all communication between the Web server and client is encrypted and secured at the transport layer. Thus, using the simple Basic Authentication under HTTPS is a recommended approach.

FIGURE 8.4: HTTP Basic Authentication

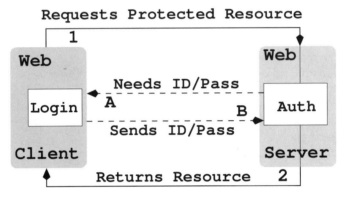

Figure 8.4 shows the client-server interactions for HTTP Basic Authentication.

1. Request for protected resource—A Web client sends an HTTP request to a server for a resource that is protected.

2. Authentication—The Web server checks the userid and password in the message *authorization header* sent by the client:

   `Authorization: Basic` *auth-string*

   where the *auth-string* is a base64 encoded string of *userid:password*. If the userid and password are correct and allowed to access the requested resource, then the resource is returned (Figure 8.4, step 2). If there is no such header or the userid/password is incorrect, the server sends back an *authorization challenge* (Figure 8.4, step A)

   `HTTP/1.0 401 Unauthorized`
   `WWW-Authenticate: Basic realm="`*realm_name*`"`

   and perhaps also the HTML for a *cancel page*. A *realm* is a collection of resources under the same access protection. The realm name can be any string (case insensitive) but should be a clear indication of where the user is logging in.

3. Authorization Response—Upon receiving the authorization challenge, the Web client (browser) usually displays a login dialog indicating the realm for which the user credentials are sought. If the user cancels the login dialog, the supplied *cancel page* or an empty page is displayed. Otherwise, the client re-sends the resource request, adding the user credentials in an authorization header (Figure 8.4, step B).

On the server side, a realm usually consists of all files and subfolders in a designated folder. But, it is also possible to protect just a single file. For a Web client, the realm name in combination with the root URL of the target server defines the protection space. User credentials are cached for each protection space and automatically sent when requesting any resource located in/under the realm's folder. Closing the client/browser usually clears the authentication cache.

To arrange for access protection, a server administrator simply needs to add access restrictions for the target files or folders in the Web server configuration files (Section 10.11). But we can also use PHP to achieve such protection without modifying Web server configuration files.

8.6.2 Basic Authentication via PHP

Here we describe an approach to use PHP for Basic Authentication under HTTPS.

For any page to be protected, simply place this line

```
<?php require_once("login.php"); ?>
```

at the beginning of the page and customize the login.php file:

```
<?php /////    login.php    /////
    require_once("../auth.php");
    $AuthRealm = "dwp auth example";    // realm
    $AuthUserFile = "../dwp_passwd";    // password file
    $cancelPage="../cancel.html";       // cancel page
    $groupName="DWPBook";               // group name
    $groupFile="../dwp_group";          // group file
    $user=authenticate($AuthRealm, $AuthUserFile, $cancelPage,
                    $groupName, $groupFile);
?>
```

The heavy lifting is done by the auth.php library script, which requires no modification in most cases. Let's look at the functions defined in auth.php.

```
function authenticate(&$realm, &$pwFile,
        $cancelPage="", $group="", $gFile="")
{ https();  // insists on HTTPS                    // (1)
  if(!empty($_SERVER['PHP_AUTH_USER']))            // (2)
```

```
      && ($user=$_SERVER['PHP_AUTH_USER'])
      && groupCheck($user, $group, $gFile)        // (3)
      && passwdCheck($user, $pwFile)              // (4)
    )  return $user; // login success
  authResponse($realm, $cancelPage);            // (5)
}

function https()
{ if (! isset($_SERVER['HTTPS']) || $_SERVER['HTTPS']!="on")
  { header("Location: https://" . $_SERVER['SERVER_NAME']
          . $_SERVER['REQUEST_URI']);
    exit;   // back to browser
  }
}
```

The `authenticate` function is called to request/check user credentials. It requires the realm and password file arguments. The cancel page, group name (string), and group file name (string) are optional.

It first calls `https()` to switch to HTTPS if that is not already in use (line 1). The three conditions that must hold for login success are: userid supplied (line 2), user name in the required group (line 3), and the password checks out (line 4). The userid is returned upon login success or we send an authorization challenge by calling `authResponse` (line 5).

We assume a group file consists of one line for each group in the form

group_name: *userid_1 userid_2 userid_3* ...

The function `groupCheck` tests if a give user is in the indicated group:

```
function groupCheck(&$user, &$group, &$gFile)
{ return ($group == "" || preg_grep(
    "/^($group).*[: ]($user)[ ]*.*$/", file($gFile)));
}
```

The PHP `preg_grep` function applies a regular expression to each string in an array. The array here is one produced by calling the `file` function on the given group file. PHP supports *Perl compatible regular expressions* (PCRE), which is very similar to patterns used in JavaScript.

The `passwdCheck` function extracts the relevant line for the given user from the password file (line 6) and isolates the **crypt**-encrypted password and the salt (line 7). The incoming password is **crypt**-encrypted (using the same salt) and the result compared with that in the password file (lines 8 and 9).

```
function passwdCheck(&$user, &$pwFile)
{ if (!($authUserLine = array_shift(preg_grep
      ("/$user:.*$/", file($pwFile)))))         // (6)
```

```
        return false;
    preg_match("/$user:((..).*)$/",
                $authUserLine, $matches);
    $thepw = $matches[1]; $salt = $matches[2];      // (7)
    $pw = crypt($_SERVER['PHP_AUTH_PW'], $salt);    // (8)
    return ($pw == $thepw);                         // (9)
}
```

Failure to log in successfully triggers a call to

```
function authResponse(&$realm, &$cancelPage)
{ // headers to request login dialog
    header("WWW-Authenticate: Basic realm=\"$realm\"");
    header('HTTP/1.0 401 Unauthorized');
    // HTML page to display if user cancels login
    if (!empty($cancelPage)) require($cancelPage);
    exit;   // back to browser
}
```

that issues an authorization challenge back to the client. Fully working login examples (**Ex: BasicAuth**) using this approach can be found at the DWP website[1].

8.6.3 Basic Authentication Logout

The HTTP Basic Authentication does not provide a way to log out from a realm. Once a Web client caches the credentials for that realm, it continues to send the credentials automatically upon each access to a resource in the realm folder. In order to log out, a user usually must close the browser, which can be inconvenient.

But, from PHP we can invalidate the browser's authentication cache for a particular realm with a logout page in the realm folder coded this way:

```
<?php require_once("logout.php"); ?>
<!DOCTYPE html>
<html xmlns="http://www.w3.org/1999/xhtml"
      lang="en" xml:lang="en">
<head><meta charset="utf-8"/>
<title>Logout Confirmation</title></head>
<body style="margin: 50px">
<h2>Logout Success</h2>
<p>You are logged out.</p>
</body></html>
```

where the logout.php file is

---

[1] userid=**open**, password=**sesame**

```
<?php  require("../auth.php");
$AuthRealm = "dwp auth example";
https();
header("WWW-Authenticate: Basic realm=\"$AuthRealm\"");
header('HTTP/1.0 401 Unauthorized');
?>
```

The idea is to force another browser login even though we may already have received the correct credentials. This normally will cause the browser to invalidate its cached credentials for the particular realm. On the client side, the user who is logging out would cancel the authentication dialog to complete the logout.

8.7   Session Control

8.7.1   What Is a Session?

The Web uses the HTTP protocol, which is *stateless*, meaning that each request-response transaction is self-contained and independent. The protocol itself offers no *state* or memory across transactions. In other words, there is no automatic association of one request from a user to the next.

We can illustrate the situation by a crazy pizza ordering scenario. A person calls the pizza place, says "Large pizza," and then hangs up! That is request number one. Then, the person calls the pizza place again and says, "extra cheese," and hangs up again! That is request number two. If the pizza place is serving only this customer and there are no other calls coming in, then perhaps it can work. But, of course the pizza place may have any number of other calls (requests) coming in and all of them are hanging up like that. Just imagine how confused the pizza place can become!

This explains why some websites display lengthy forms to collect all information from you at once. However, it is not possible or practical to always obtain all information with one form submission. Just think about the multiple steps (requests) you need to go through for online shopping (Figure 8.5). Thus, we must find a way to connect one request from a user to the next from that same user. The sequence of requests from each individual user form a separate *session*. *Session control* involves creating, saving, resuming, and terminating sessions.

8.7.2   Session Control Techniques

The *hidden form fields* approach connects one request to the next by placing all the information collected from a request into hidden form fields in the response page containing the form for the next request. When the response page is submitted back, it carries all the data from the hidden form fields as

FIGURE 8.5: A Typical Session

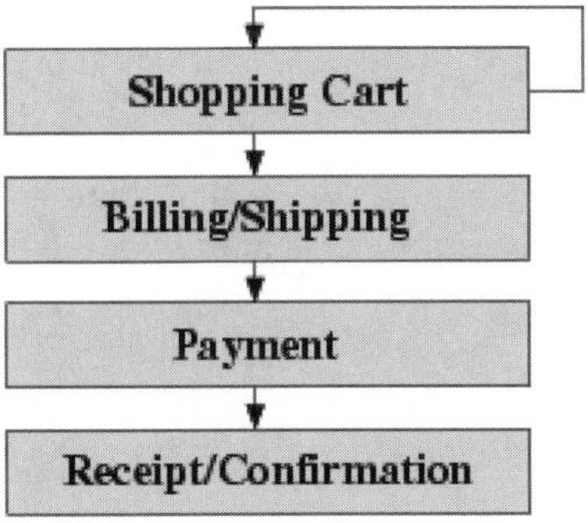

well as new information from the user. This technique is not secure because the hidden values can be changed before the second submission.

The *URL rewriting* approach puts the known values as a query string for the next form action URL. This way, again information from the same user can be accumulated through multiple requests, but the query string is also subject to change before the next submission. Also, browsers may have restrictions on how long a URL can become.

The third, and recommended, approach is to use *cookies*. A cookie is a piece of state information a server places in a user's browser for safekeeping, via the `Set-cookie` HTTP header. The browser will send back the stored cookie information, via the HTTP `Cookie` header, to the same server on every subsequent request to that server. The server may delete a cookie by setting its value to the empty string or attach an expiration time to it. The cookie approach of session control is more secure than the previous two.

8.7.3 PHP Session Support

PHP supports sessions by giving each session a unique *session identifier* (the `PHPSESSID` session id), storing session states in the server host's local file system, and retrieving the session state automatically based on the session id. If browser cookie support is enabled, a `PHPSESSID` cookie is used to handle the session id. Otherwise, PHP switches to URL rewriting automatically.

PHP provides a session API that makes session control very simple and easy for the programmer:

- `session_start()`—A call to this function either creates a new session with a blank session state or resumes an existing one by recovering the stored session state. Each participating page in a session calls this function to *join the session* before any page output begins. It is a good idea to call `session_start()` at the very beginning of such a page.

- $_SESSION—A super global associative array for storing the session state. For example,

```
$_SESSION['userid']=$user
```

stores the value of **$user** in the session state as **userid**. To remove a value from the session state, use **unset**. For example,

```
unset($_SESSION['userid']);
```

The entire **$_SESSION** array is saved when a page finishes and recovered later when any page calls **session_start()** to resume a session under the same session id received via a **PHPSESSID** cookie, or a **PHPSESSID** GET/POST value.

- Session id—By default, a randomly generated session id is used for a new session. Or you can set a desired session id with a call to **session_id($str)** just before creating a new session. The call **session_id()** always returns the session id for the current session. Here is an example of setting your own session id:

```
session_id(md5(mktime() . rand() .
           $_SERVER['REMOTE_ADDR']));
```

Usually the **PHPSESSID** is set at the start of a session and will not change until the session ends.

- Ending a session—After calling **session_start()**, you can call your own **session_end** to terminate a session:

```
function session_end()
{  foreach ($_SESSION as $key=>$val)     // (A)
   {  unset($_SESSION[$key]);  }          // (B)
   session_destroy();                     // (C)
}
```

The function wipes out the session state (lines **A** and **B**), and releases all session resources (line **C**). To end any current session (line **D**) and start a new one (line **E**), use the sequence

```
session_end();                            // (D)
session_start();                          // (E)
```

FIGURE 8.6: Crazy Pizza Shop

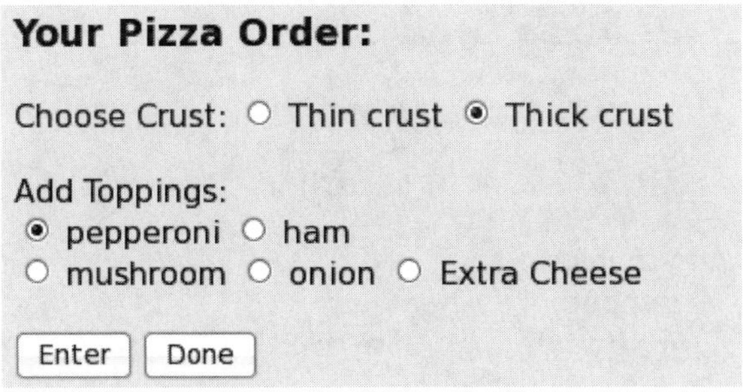

Your Pizza Order:

Choose Crust: ○ Thin crust ● Thick crust

Add Toppings:
● pepperoni ○ ham
○ mushroom ○ onion ○ Extra Cheese

[Enter] [Done]

8.8   Crazy Pizza Shop

To further illustrate session control, let's try pizza ordering in a sequence of requests. Figure 8.6 shows an order form where you can pick a crust and one topping at a time (crazy!).

The page is produced by `pizza.php`, which begins by checking if there is no form submission (line A), in which case a new session is created (lines B and C), session variables are initialized (line D), and the order form is displayed (line E).

```php
<?php require_once("sessionfns.php");

if( empty($_POST['submit']) )          // (A)
{ session_id(md5(mktime() . rand().    // (B)
         $_SERVER['REMOTE_ADDR']));
  session_start();                     // (C)
  $_SESSION['crust']="";               // (D)
  $_SESSION['toppings']="";
  include("head.inc");
  require("orderform.php");            // (E)
}
elseif ($_POST['submit']==" Enter " ) // (F)
{ session_start(); // resumes session
  processOrder();                      // (G)
  include("head.inc");
  require("orderform.php");
}
```

Otherwise, if the order is continuing (line F), the session is resumed, the order formdata is processed (line G), and the order form, with the current order status, is displayed (Figure 8.7).

The user may add another topping and/or change the crust and press **Enter** again, leading to another display (Figure 8.8). If the order is done

FIGURE 8.7: Pizza Order Step 2

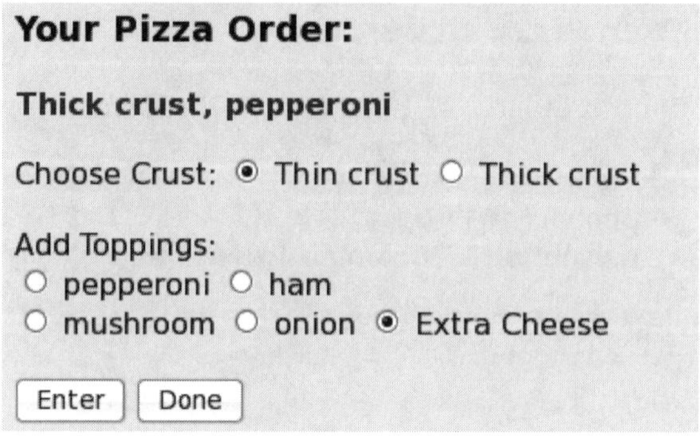

Your Pizza Order:

Thick crust, pepperoni

Choose Crust: ● Thin crust ○ Thick crust

Add Toppings:
○ pepperoni ○ ham
○ mushroom ○ onion ● Extra Cheese

[Enter] [Done]

(line H), the session is terminated (line I). Order confirmation and a link for another order are displayed.

```
elseif( $_POST['submit']==" Done ")    // (H)
{ session_start(); session_end();      // (I)
  include("head.inc");
  echo '<p>Your order is complete.
       Thanks for your business.</p>';
  echo '<p><a href="">Another Order</a></p>';
} ?>
</body></html>
```

FIGURE 8.8: Pizza Order Step 3

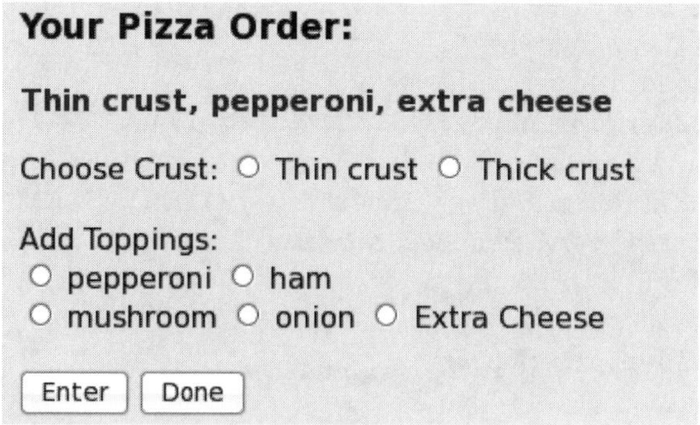

Your Pizza Order:

Thin crust, pepperoni, extra cheese

Choose Crust: ○ Thin crust ○ Thick crust

Add Toppings:
○ pepperoni ○ ham
○ mushroom ○ onion ○ Extra Cheese

[Enter] [Done]

Order processing is performed by

```
function processOrder()
{ $cr = $_POST["crust"];
  if ( $cr ) { $_SESSION['crust'] = $cr; }
```

```
$top = $_SESSION['toppings'];
$it = $_POST["top"];
if ( $it && ! strstr($top,$it) )
{ if ( $top )
  { $_SESSION['toppings'] = "$top, $it"; }
  else { $_SESSION['toppings'] = $it; }
}
}
```

which updates the crust and toppings session variables with the formdata posted. The session example here involves repeated access to a single PHP page `pizza.php`. But a session can easily involve multiple cooperating pages. Experimenting with the complete example (**Ex: Pizza**) on the DWP website can help you understand session control.

Moving away from this make-believe example, the next section shows a practical application of session control.

8.9 Security Code Verification

For security reasons, websites often need to double-check the identity of a visitor. This is especially true if a user accesses a bank or other account when traveling or from an unknown computer.

In addition to requiring login, a website can also send a *security code* to the registered email address of the user and can ask the user to enter the code in a form for verification. Such a verification procedure involves several steps in a session:

1. Generating a verification code, sending it by email, and storing it in the session state.

2. Displaying an HTML form requesting the verification code to be entered by the user.

3. Receiving the formdata and checking the incoming code with that stored in the session state.

4. Saving the verification result in the session state, and redirecting to a designated next-step page, depending on the verification result.

Our PHP implementation `verify.php` (**Ex: Verify**) can be used in any session that requires such a verification. For example, the code

```
<?php session_start();   // starts or resumes a session
    $_SESSION['email']="pwang@cs.kent.edu";
    $_SESSION['pass']="target.php";
    $_SESSION['fail']="sorry.php";
    require_once("verify.php"); ?>
```

sets up the email address and the next-step pages for passing/failing the verification before loading `verify.php`.

The script `verify.php` first sends a generated code and a verification form. Later it receives the form submission and verifies the code entered by the user.

The `verify.php` script assumes that `session_start()` has been called already and begins to verify any posted code.

```php
<?php   /////   verify.php   /////
if ( !empty($_POST['vcode'] ) ) //  verification phase
{ $_SESSION['vresult']=
    ($_POST['vcode'] == $_SESSION['code'] );          // (1)
  if ($_SESSION['vresult'])
  { header("Location: " . $_SESSION['pass']); }    // (2)
  else
  { header("Location: " . $_SESSION['fail']); }    // (3)
  unset($_SESSION['code']);
  unset($_SESSION['pass']);
  unset($_SESSION['fail']); exit;                   // (4)
}
if ( empty($_SESSION['email']) ||                   // (5)
     empty($_SESSION['pass']) ||
     empty($_SESSION['fail']) ) { exit; } ?>
```

The incoming code (`vcode`) is compared with that in the session state (line 1) and the result saved in the session state. Depending on the result, the user is directed to the correct locations (lines 2 and 3). The script ends after unsetting values no longer needed (line 4).

When `$_POST['vcode']` is empty, the script checks the required email, pass, and fail session values before proceeding (line 5) to send email and the verification form.

After sending the usual front part of an HTML5 page, the script generates a random code if one has not been supplied in the session state (line 6), and emails the code to the indicated address (line 7).

```html
<!DOCTYPE html>
<html xmlns="http://www.w3.org/1999/xhtml"
   lang="en" xml:lang="en">
<head><meta charset="utf-8"/>
<title>Verification Code</title>
</head><body style="background-color: #def">
<h2>Verification Code</h2>
<?php
if (empty($_SESSION['code']))                    // (6)
{  $_SESSION['code']=rand();   }
if(mail($_SESSION['email'],                      // (7)
       "Verification Code",
```

```
        "Your Verification Code: " .
        $_SESSION['code'] .  "\n\n") )
{ /* send verification form */  }
else { /* send email problem error */ }
?>
```

If email is successful, the verification form (Figure 8.9) is sent. Otherwise, an

FIGURE 8.9: Code Verification Form

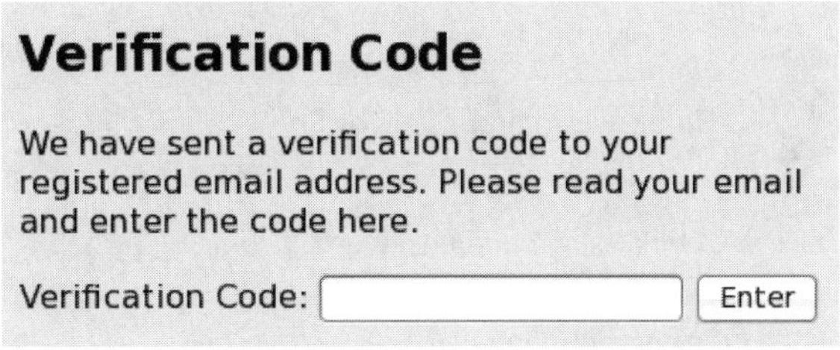

appropriate error page is sent. The form code is as follows:

```
<form method="post" action="">
<label>Verification Code:
<input required=""  name="vcode" /></label>
<input type="submit" value=" Enter " />
</form>
```

In our example, we direct both pass and fail to the page `next.php`, which extracts the verification result and ends the session at the very beginning of the page.

```
<?php session_start();
  $result=$_SESSION['vresult'];
  require("sessionfns.php");  // defines session_end()
  session_end();
?>
```

The PHP function `rand()` (line 6) produces a nine-digit decimal number. The `md5` function, called on the current time and the user's IP address perhaps, can be used to generate a 32-character hexadecimal number. There are other possibilities.

8.10   Login Sessions

Websites often have areas and services reserved for members who are registered or have accounts with the site/organization. Each member often have his/her own login userid and password to gain access to member-only or account-specific information or services.

Such a website normally needs to provide all of the login-related functionalities as indicated by Figure 8.10. Therefore, user authentication is only one piece of the login puzzle.

FIGURE 8.10: Login Logic Flow

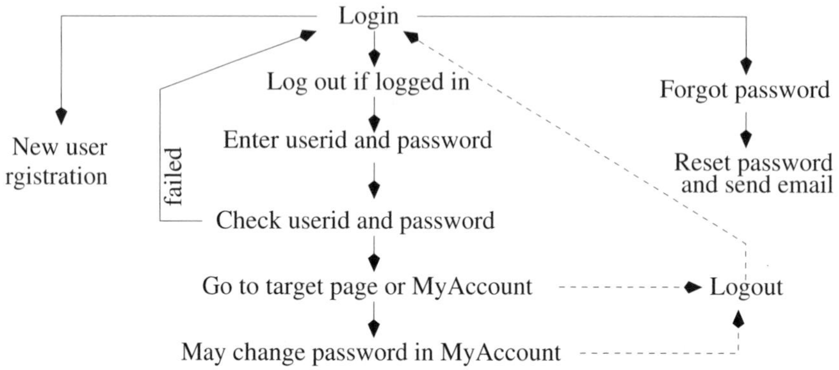

We have seen login with Basic Authentication under HTTPS in Section 8.6. But we can handle login without relying on HTTP supported authentication by using PHP to display our own login page and by using session control.

A login session usually consists of a login page, a logout page, and other *participating pages* accessible only after login and before logout.

We follow these basic ideas to handle the login session:

- The login page uses an HTML form to collect the userid, password, and perhaps other information for login. The login page always terminates any existing login session and starts a new session. We also switch from HTTP to HTTPS to secure the login and the session.

- Login form processing records the userid and other related information in the session state upon successful login. The presence of such state information indicates the login status. The login page then redirects to a designated *target page* to continue the login session.

- Each participating page resumes the session and checks the login status ($_SESSION['userid'], for example). If no login yet, or if the page is reached *out of nowhere*, the page redirects to the login page. Otherwise, the page is reached in sequence within the login session and proceeds with its business and often will display the next logical step (usually another participating page).

- A login navigation link leads to the login page. As soon as login is successful, the login link changes to a logout link.

- To log out, we terminate the login session, change the logout navigation link back to login, and redirect the user to an appropriate non-participating page, such as the website entry page. Logout also may switch from HTTPS back to HTTP.

- If a user visits the login page while already logged in, the login session is almost aways terminated (i.e., logout) before the login page is shown again.

8.10.1 Login Session with PHP

We now present a PHP implementation of login consisting of these pages: `login.php`, `memberarea.php`, `register.php`, `logout.php`, and `forgotpw.php`. The `login.php` page displays a login page (Figure 8.11) and processes the login form.

FIGURE 8.11: Login Page

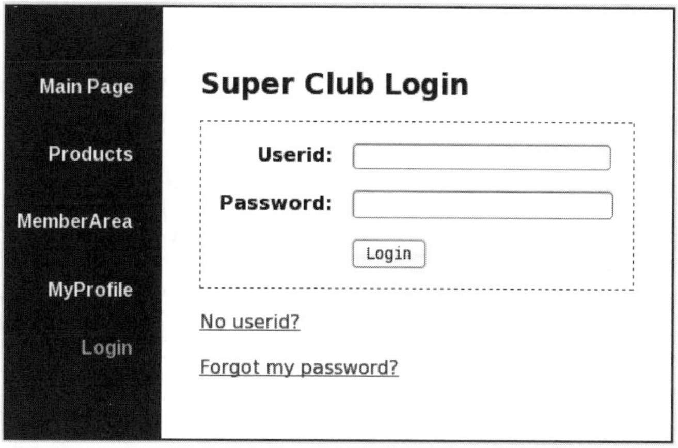

The `login.php` script begins by starting/resuming a session (line 1) and loading login-related functions (line 2). For a login post request, the session value `target` must also be set; the script then checks the userid and password (line 4). If login is successful, the userid is recorded in the session state (line 5) and the user is redirected to the desired target page (line 7).

```php
<?php session_start(); $realm="Super Club";    // (1)
      require_once("loginfns.php");            // (2)
if( !empty($_POST['submit'])                   // (3)
    && $_POST['submit']=="Login"
    && !empty($_SESSION['target']) )
{ if( check_pass($_POST['uid'],$_POST['pass'], // (4)
            $PASSWORD_FILE) )
```

```
  { $_SESSION['user']=$_POST['uid'];              // (5)
    login_fn();                                    // (6)
    header("Location: ".$_SESSION['target']);      // (7)
    exit;
  }
}
```

On line 4, check_pass is called to check the userid and password supplied against data stored in a login file designated by the global variable
$PASSWORD_FILE.

```
function check_pass($id,$pass,$file)
{ if (!($authUserLine = check_user($id,$file)))
     return false;  // no entry for user
  preg_match("/$id:(.*)$/", $authUserLine, $matches);
  $ps = $matches[1];
  return (crypt($pass, $ps) == $ps);              // (8)
}
```

```
function check_user($id,$file)
{ return array_shift(preg_grep("/$id:.*$/",
                     file($file)));
}
```

The PHP built-in function crypt is used to match the password (line 8). The
same crypt function is also used to encrypt the stored password in the first
place (Section 8.10.4).

On line 6 we call

```
function login_fn() {  touch("login.mark");  }
```

to leave a marker in the file system. A page in the login session can check
$_SESSION['user'] to determine the login status. A page outside the login
session can check the login.mark.

If login fails, we display the login form with a failure message (lines 9 and
10).

```
  else  // login failed
  { $title="Login Failed";                        // (9)
    $css=array("basic.css","form.css");
    require("rfront.php");
    echo "<h2>Incorrect Userid/Password</h2>";
    echo "<p>Please try again.</p>";
    require_once("loginform.php");                 // (10)
  }
}
```

Otherwise, when not receiving a login request, `login.php` switches to HTTPS and starts a new session (line 11). A target page (or a default) is recorded in the session state (line 12) and a login page is displayed (lines 13 and 14).

```
else // new login
{ require_once("sessionfns.php");
  https(); new_session(); logout_fn();        // (11)
  if( !empty($_REQUEST['target']) )
  { $_SESSION['target']=$_REQUEST['target']; }  // (12)
  else
  { $_SESSION['target']='memberarea.php'; }
  $title="Login";                             // (13)
  $css=array("basic.css","form.css");
  require_once("rfront.php");
  require_once("loginform.php");              // (14)
}
require_once("rback.php");
```

We have seen the functions `https()` in Section 8.6.2. The function `new_session()` ends any current session (Section 8.7) and starts a new one:

```
function new_session()
{ session_end(); session_start();  }
```

The `logout_fn` removes any login marker:

```
function logout_fn()
{  if ( file_exists("login.mark") )
      unlink("login.mark");
}
```

In our implementation, we use the login marker to conditionally display the Login/Logout navigation link:

```
<?php
function loginLabel()
{ echo file_exists("login.lock") ? "Logout" : "Login"; }
function loginLink()
{ echo file_exists("login.lock") ? "logout.php" : "login.php"; }
?>
```

8.10.2   A Login-Protected Page

The collection of all pages requiring login forms the protected realm. The `memberarea.php` is a typical participating page in a login session. Each login-protected page resumes the session (line a) and checks the login status (line b). If the user has not yet logged in, it redirects to the login page (line c), which

FIGURE 8.12: Protected Page Access

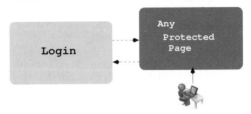

will direct the user back after login (Figure 8.12). The page content is displayed only for a logged-in user.

```php
<?php session_start();                    // (a)
if ( empty($_SESSION['user']) )           // (b)
{  header("Location: login.php?target="
      . $_SERVER['PHP_SELF']);            // (c)
}
$title="Member Area";
$css=array("basic.css","form.css");
require("rfront.php");
?>
<h2>Welcome back
<?php echo $_SESSION['user']; ?></h2>
<p>Here is your member-only area.</p>
<?php require("rback.php"); ?>
```

Often, after login, an ordered sequence of steps (an online purchase, for example) must be performed. In such a case, each page must also enforce the proper order in the progression of steps. Any out-of-sequence access, which normally should not occur, will have to be dealt with, usually by redirecting to the first step or ending the session altogether.

Also, each resource page under the protection of a login session must check for out-of-nowhere access and redirect to the login page, which will forward the user to the target page after successful login.

8.10.3 Logout

Unlike logout under HTTP Basic Authentication (Section 8.6.3), which involves closing the browser or invalidating its authentication cache, logout from a PHP login session is simply a matter of removing the login status from the session state. In our example, that means

```php
unset($_SESSION['user']);
```

In practice, logout often means ending the login session altogether. Our example logout page (`logout.php`) resumes the session (line I) to terminate it (line II), removes the login marker, and switches back to HTTP (line III).

```php
<?php session_start();                                   // (I)
require_once("sessionfns.php"); session_end();           // (II)
require_once("loginfns.php"); logout_fn(); http();   // (III)
$title="Logout"; $css=array(""basic.css",form.css");
require("rfront.php");
?>
<h2>Logout Complete</h2>
<p>Logout is successful.</p>
<?php require("rback.php"); ?>
```

8.10.4 User Registration

User registration must be done before a user can log in. One way to conveniently register new users from anywhere is to do it on the Web. For example, the `register.php` page (Figure 8.13) registers new user accounts. It wants

FIGURE 8.13: Registration Page

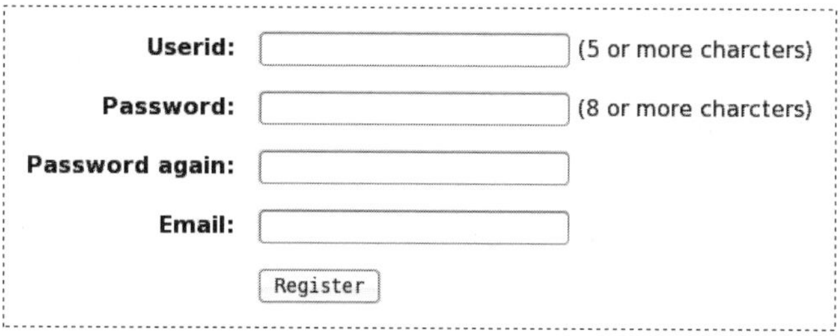

to do business over HTTPS (line A) and displays a form if not called via a POST request (line B). If so, the formdata are checked, valid data are recorded (C), and a confirmation page is displayed (line D). If something is wrong, a registration failed page with appropriate error messages is displayed (line E).

```php
<?php require_once("loginfns.php");  https();         // (A)
      require_once("regfns.php");

$title="New Member Account";
$css=array("basic.css","form.css");
$js=array("register.js"); require("rfront.php");

if(empty($_POST['submit'])) require("regform.php"); // (B)
elseif( check_reg_data() && record_pass() )           // (C)
{ ?>                                                   // (D)
```

```
<h2>Login Registration Confirmation</h2>
<p>Thanks. we your login account has been created.</p>
<?php } else { ?>                                      // (E)
<h2>Login Registration Failed</h2>
<p class="error">Sorry we failed to register your
login account.</p>
<p class="error"><?php echo $errmsg; ?></p>
<p><a href="">try again</a></p>
<?php } ?>
<?php require("rback.php"); ?>
```

The file `regfns.php` contains functions for user registration. The function `check_reg_data` validates all registration data by calling checking functions that return true/false and adds to the global error `$errmsg`.

```
<?php require_once("loginfns.php");
define("ID_LEN",5); define("PASS_LEN",8);
$errmsg="";

function check_reg_data()
{ global $errmsg;
  id_check($_POST['uid']);
  pass_check($_POST['pass']);
  pass_match($_POST['pass'],$_POST['pass2']);
  email_check($_POST['email']);
  return (empty($errmsg)) ;
}
```

A typical such checking function is `id_check` for valid userid that is not already used.

```
function id_check()
{ global $errmsg,$PASSWORD_FILE;
  if( empty($_POST['uid']) )
  { $errmsg .= "The userid is empty.";  return false; }
  if( strlen($_POST['uid']) < ID_LEN )
  { $errmsg .= "The userid ". $_POST['uid']
             . " is too short.";
    return false;
  }
  if( check_user($_POST['uid'],$PASSWORD_FILE) )
  { $errmsg = "The userid " . $_POST['uid'] .
           " exists. Please use a different userid.";
    return false;
  }
  return true;
}
```

The function `record_pass` calls `add_passwd` to save an entry for the registered user.

```php
function record_pass()
{ global $errmsg, $PASSWORD_FILE;
  if(! add_passwd($_POST['uid'], $_POST['pass'],
          $PASSWORD_FILE, $_POST['email']))
  { $errmsg .= "Failed to store password.";
    return false;
  }
  return true;
}
```

First, `add_passwd` makes sure the new user is not already in the password file (line F). It then uses the recommended *Blowfish encryption*, supplied by the `crypt()` function, for the password (line G) and prepends *email:userid:* to the encrypted password (line H). The password file is a PHP file where each entry starts with `//` to prevent any of it from ever delivered to the Web.

```php
function add_passwd(&$id,&$pass,&$file,$email="")
{ if ( check_user($id,$file) ) return false;          // (F)
  $pass_hash=crypt($pass,'$2a$09$dynamicwebdesign$'); // (G)
  $f=@fopen($file, "a");
  if ($f)
  { fwrite($f,"// $email:" ."$id:" .$pass_hash."\n"); // (H)
    fclose($f); return true;
  }
  return false;
}
```

In addition, we also need a page for changing password after login and a page to help users who have forgotten their passwords. Usually, the password is reset to a generated one and the new password sent to the user via the email address on record.

Here is a password generation function:

```php
function gen_pass($length=8, $strength=2)
{   $vowels = 'aeuyj9753';
    $consonants = 'bdghmnpqrstvz-';
    if ($strength & 1)
    { $consonants .= 'BDGHLMNPQRSTVWXZ_'; }
    if ($strength & 2)
    { $vowels .= "YUEAJ2468"; }
    if ($strength & 4)
    { $consonants .= ',.#'; $vowels .= '@$%'; }
    $password = '';
    $c_max=strlen($consonants)-1;
```

```
$v_max=strlen($vowels)-1;
$alt = time() % 2;
for ($i = 0; $i < $length; $i++)
{ $password .=
    $alt ? $consonants[rand(0,$c_max)]
        : $vowels[rand(0,$v_max)];
  $alt = !$alt;
}
return $password;
}
```

Although we do not do it in this example, it can be convenient for users in some cases if a successful registration will start a login session automatically without asking the user to log in. In such a case, successful registration can start a session and redirect the user to an appropriate target page such as the user's profile page.

8.10.5   User Profiles

Websites with user registration and login usually record essential data about each user such as addresses, phone numbers, and sometimes also order history and credit card information. A page such as myprofile.php would then be made available for users to access and update such information.

One critical function related to login in myprofile.php is changing the password. Figure 8.14 shows the change password form.

FIGURE 8.14: Changing Password

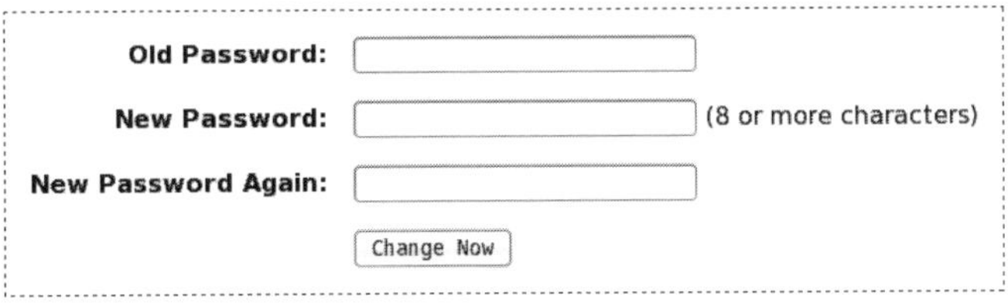

After checking for correct input, our change password processing calls the following PHP function to change the recorded password:

```
function change_passwd(&$id,&$oldpass,&$pass,&$file,&$err)
{ if (! check_pass($id,$oldpass,$file))
  { $err .= "Old password wrong."; return false; }
  $done=false;
  $lines=file($file);
  if ( $f=@fopen($file, "w") )
  { foreach($lines as $line)
    { if ( !$done && strstr($line, "$id:") )
```

```
   { $pass_hash=crypt($pass,
             '$2a$09$dynamicwebdesign$');
     $line="// " .$id .":" .$pass_hash . "\n";
     $done=true;
   }
   fwrite($f, $line);
 }
 fclose($f);
}
else { $err .= "Cannot open password file."; }
return $done;
}
```

Note that we must always collect and check the old password again to make sure that the person changing the password actually knows the current password before allowing the password change to proceed. Also, we often also want the new password to be sufficiently different from the old one.

Fully functional code for this example (**Ex:** `LoginSession`) can be found at the DWP website.

8.10.6   Session Timeout

Of course we want login sessions to end with logout. But it happens that users will be distracted during a session or sometimes forget to continue. The danger is in someone else working with a left-over session.

The PHP init parameter `session.gc_maxlifetime` (usually 1,440 seconds) controls the max time interval since last session save. The saved session file may be deleted (garbage collected) after this max lifetime setting. The probability of session garbage collection depends on `session.gc_probability` and `session.gc_divisor`. To expire a session after 30 minutes of user inactivity, you can use

```
ini_set('session.gc_maxlifetime',30*60);
ini_set('session.gc_probability',1);
ini_set('session.gc_divisor',1);
```

Because session garbage collection is not exactly predictable, you may want to control your own *max inactivity interval* by saving the time (via `time()`) in the session state and checking it against the next access time. If the delay exceeds the allowable interval, you can terminate the current session.

8.11   Image Manipulation

In addition to being a powerful tool for creating textual content in webpages and for processing forms, PHP also enables you to create and manipulate images easily.

The function

```
getimagesize($imgFile)
```

returns an array containing the width, height, and type of the given image, which can be a local or remote file.

The GD library, when compiled into PHP, provides a rich set of functions for common raster image processing tasks. The GD library supports JPEG, PNG, GIF, and XPM image formats and provides functions for image creation, drawing of text and graphics, as well as image translation, rotation, resizing, cropping, and color manipulations.

You can create a new image "object" from a file with

```
$img = imagecreatefromtype($filename);
```

where *type* can be jpeg, png, or gif. Or you can create an empty image (all black) with

```
$img = imagecreatetruecolor($width, $height);
```

If you happen to have the base64 encoded string for an image, you can obtain an image object from it too:

```
$data = base64_decode($str);
$img = imagecreatefromstring($data);
```

The output function

```
imagetype($img, [$filename]);
```

outputs an image of the *type* to standard output or to a file.

To resize an image object, you can use this function

```
function resize($width,$height,$img)
{ $new_img = imagecreatetruecolor($width, $height);
  list($w,$h)=
  imagecopyresampled($new_img, $img, 0,0,0,0,
     $width, $height, imagesx($img), imagesy($img));
  return $new_img;
}
```

to scale $img to the given width and height. The GD library offers many other useful image processing functions.

Let's look at some example applications.

8.11.1 Security Images

On the Web, when registering for an account, uploading a forum entry, or posting a book/product review, the site sometimes needs to make sure that a person rather than a program (such as a web bot) is at the other end. This is often done with a challenge-response test known as a *CAPTCHA*. Basically a CAPTCHA is a way to automatically generate challenges that are easy for people but hard for programs. The purpose is to prevent automated programs from uploading or submitting data via Web forms.

Figure 8.15 shows a 3-D CAPTCHA image used in a form. A GD-based PHP program 3DCaptcha.php can generate the 3-D display of a randomly generated character code. The code is stored in the form session so it can be checked later against the user input. Failing to match will cause form processing to fail (**Ex: Captcha**).

FIGURE 8.15: A CAPTCHA Image

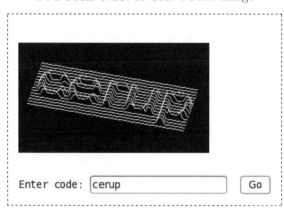

The code behind the form in Figure 8.15 is

```php
<?php session_start(); $title="Captcha Test";
$css=array("basic.css","form.css");
$js=array("captcha.js"); require("rfront.php");

if ( !empty($_POST['code']) )                          // (A)
{ if( $_SESSION['3DCaptchaText']==$_POST['code'] )
  { echo "<h2>Code Matched.</h2>";  }
  else
  { echo "<h2>Code Not Matched.</h2>";  }
  unset($_SESSION['3DCaptchaText']);
}
else { ?>                                              // (B)
```

It either processes the form (line **A**) or sends out the form (line **B**).

```
<h2>Please Enter the Code Displayed</h2>
<pre>
<form method="post" action="">
<img src="3dcaptcha/3DCaptcha.php"                    (C)
 alt="security image" onclick="changeimg(this)" />   (D)

<label>Enter code: <input name="code" value=""
   onfocus="this.value=''" /></label>  <input
type="submit" value=" Go " /></form>
</pre>
<?php } require("rback.php"); ?>
```

The security image src is a PHP URL (line C). The 3DCaptcha.php script must send an HTTP response that is an image. The HTTP header Content-Type is used of course to indicate the image type. Thus, a PHP script can dynamically produce a response in any content type as long as it first sends the correct HTTP content-type header.

The 3DCaptcha.php script also records the code string in

```
$_SESSION['3DCaptchaText']
```

for later verification. If the image is hard to read, a user can click on it and get a different one (line D). The JavaScript definition for changeimg is

```
var imgct=0;
function changeimg(img)
{  imgct++;
   src='3dcaptcha/3DCaptcha.php?'+imgct;
   img.src=src;
}
```

The working program (**Ex: Captcha**), together with the 3-D CAPTCHA code based on work by Marc S. Ressl, can be found at the DWP website.

8.11.2   Watermarking Photographs

With the GD library we can also apply watermarks to photos (**Ex: Watermark**) on-the-fly (Figure 8.16).

The function applyWatermark takes a jpeg image file ($image), a png watermark file ($watermark), an error message return parameter ($err), and an optional output file name ($targetFile) as arguments. The function superimposes the watermark (line 1) at the center of the image and outputs the new image either to the given output file (line 2) or as a valid HTTP response (line 3) with the correct image type (line 4).

```
<?php
function applyWatermark(&$image, &$watermark,
```

FIGURE 8.16: A Watermarked Image

```
                          &$err,$targetFile=null)
{ $im = imagecreatefrompng($watermark);
  $im2 = imagecreatefromjpeg($image);
  if (!$im2) { $err .= "Failed to open $image!";
              return false; }
  if (!imagecopy($im2, $im,
      (imagesx($im2)/2)-(imagesx($im)/2),
      (imagesy($im2)/2)-(imagesy($im)/2), 0, 0,
      imagesx($im), imagesy($im)) )                   // (1)
  { $err .= "Watermarking failed."; return false; }

  if ($targetFile != NULL)                            // (2)
  { $done = imagejpeg($im2, $targetFile, 65); }
  else                                                // (3)
  { $last_modified = gmdate('D, d M Y H:i:s T',
                            filemtime ($image));
    header("Last-Modified: $last_modified");
    header("Content-Type: image/jpeg");               // (4)
    $done = imagejpeg($im2, NULL, 65);
  }
  imagedestroy($im); imagedestroy($im2);
  return $done;
} ?>
```

8.12 Providing Content Management

Taking advantage of rich-text editing in HTML5, we introduce here a way
to enable site administrators to revise, modify, and update selected webpages
simply and easily.

In HTML5, setting the `contenteditable` attribute to `true` for an element makes the element and all elements enclosed in it editable. All major browsers support natural and intuitive editing operations for `contenteditable` elements such as inserting, deleting, and replacing text, changing font styles, sizes, weight, and color, inserting images, and so on. In addition, operations listed under the browser's `Edit` menu are supported for `contenteditable` elements. These can include copy, cut, paste, undo and redo operations.

An element remains editable until its `contenteditable` becomes `false`. To turn the entire document, or the whole document in an `iframe`, editable, use the javascript code `document.designMode=true;`. Our approach here does not need to use this whole-document feature.

8.12.1   AMP: Admin Modifiable Page

With PHP, we can give any particular page two modes: *normal mode* and *content-management mode*. In normal mode, the page is displayed for end-user consumption. In content-management mode, the editable parts of the page are subject to revisions and updates. Let's call such a page an *Admin Modifiable Page* (AMP). Accessing an AMP `xyz.html` displays it in normal mode whereas accessing it with a query string `edit=true`, say, displays it in content-management mode where editing buttons and dialogs allow the user to modify each individual editable part embedded in the AMP. The edited result is saved back to the server side making it available immediately for all users and for further editing.

Let's illustrate this content management scheme by applying it to a restaurant *specials* page (Figure 8.17) where two areas, lunch specials and dinner

FIGURE 8.17: Specials Page in Normal Mode

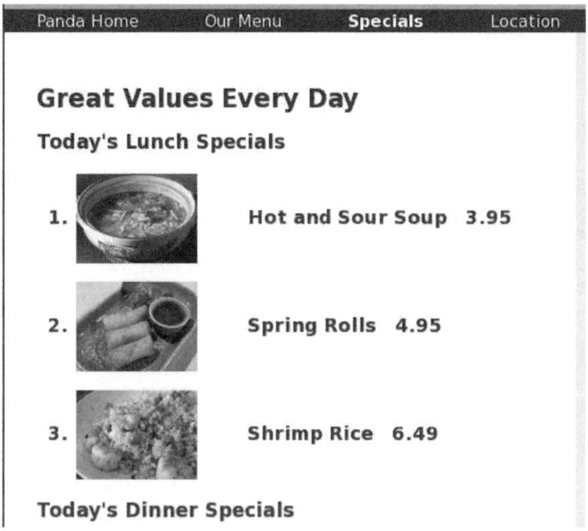

specials, are made editable (**Ex:** AMP). Figure 8.17 shows the lunch part. The dinner part is entirely similar.

An admin user finds the hidden `Edit` link at the bottom of the AMP (Figure 8.18). The link leads to the same page in content-management mode with buttons showing up for the editable parts in the page (Figure 8.19).

FIGURE 8.18: The Edit Button

The admin user can easily use browser-supported editing under HTML5 to make changes. Clicking the `Start` button begins editing operations on the particular part of the page (Section 8.13).

FIGURE 8.19: Content-Management Mode

The user can edit one editable section at a time. During editing, the `save` button can record the current state in JavaScript for a possible later `Restore`. Clicking the `Done` button ends editing and automatically sends the updated HTML code to the server side, updating the page. The admin user is brought to the live page in normal mode to double-check the changes made. To make no changes, the admin user may click a button at the end of the page (Figure 8.20) to go back to the AMP in normal mode.

The easily reusable implementation is a combination of HTML5, CSS, JavaScript, and PHP, as we will explain in the following sections. Let's first look at the AMP `specials.html`.

FIGURE 8.20: Go Back Button

---

footer   [Do Nothing and Go Back]

8.12.2   Making an AMP

The file editfns.php supplies a number of useful PHP functions for turning any page into an AMP.

Depending on the setting of $edit, the function

```
function editableSection($id)
{ global $target, $edit, $tid, $page, $btn_arr;
  $tid=$id; $target="$tid.inc";
  if ($edit)
  { echo '<section id="'.$tid.'" class="cedit">'; } // (A)
  require($target);
  if ($edit) { echo '</section>';
           require_once("controls.php"); }        // (B)
}
```

produces a section of HTML code that is stored in a separate file $target, with or without an edit wrapper (lines A and B).

Follow these simple steps to make any page an AMP:

1. Add PHP code at the beginning of the page to initialize editing variables and load edit functions (line C), and to modify the page title and to include style and JavaScript files for editing mode or normal mode (lines D and E):

```
<?php  $edit=isset($_GET['edit']);
$page=$_SERVER['PHP_SELF']; $target=""; $tid="";    // (C)
require_once("editfns.php");
?>
<title><?php if($edit) echo "Manage ";            // (D)
?>Panda Specials</title><?php if ( $edit ) {
?><link rel="stylesheet"
        type="text/css" href="editcontent.css" />
<script type="text/javascript" src="editcontent.js">
</script><?php } else { ?><link rel="stylesheet"
type="text/css" href="hide.css" /> <?php } ?>       // (E)
```

2. The function editableSection (line B) is called to produce a section parent element (line C) for any editable part. Specifically, place each part in the page to be edited into a separate file (lunch.inc, for example) and insert the following PHP code in its place:

```php
<?php $btn_arr[]="undoBtn.php";                    // (F)
      $btn_arr[]="htmlBtn.php";
      editableSection("lunch");                    // (G)
?>
```

The call on line G wraps code from lunch.inc in a section element with a given id (for JavaScript control) in the cedit class (for CSS effects). Basic edit buttons from the file controls.php, plus additional buttons given by files listed on the global $btn_arr array (line F), are added just below the section. By appending to $btn_arr before calling editableSection, we can add appropriate edit buttons for each individual editable section.

The dinner part of the page is done in the same way.

3. Place code for the Edit link somewhere at the end of the page:

```php
<?php if (! $edit) { ?>
<a class="hidden" href="?edit=true"> Edit </a><?php }
else { ?><button onclick="window.history.back()">
Do Nothing and Go Back</button> <?php } ?>
```

This hidden link Edit is only there in normal mode and reveals itself on mouseover. Clicking this link leads to the same page in content management mode. Admin user login ought to be required to manage the page unless we want everyone to be able to update the page. For simplicity, we focus on content management here. See Section 8.10 for PHP-based user login.

In content-management mode, a Go Back button takes the place of the Edit link (Figure 8.20).

8.13   HTML5 contenteditable API

The file controls.php supplies basic editing buttons connected to JavaScript event handlers (in editcontent.js) that make use of the HTML5 editing API to achieve browser-supported editing.

For example, the Start/Cancel button

```html
<input title="start/cancel editing" class="ec"
  onclick="ssEdit('<?php echo $tid; ?>',this);"
  style="color: green" type="button" value="Start" />
```

calls ssEdit with the id of the section (line A) to be edited and a reference to the button itself.

The ssEdit function (line I), calls editInit (line H) if theid (the recorded editable section id) is different from the id argument.

```
var el=null, bt=null, theid=null, oldInnerHTML, htmlcode;

function editInit(id, btn)                            // (H)
{ el=document.getElementById(id);
  oldInnerHTML=el.innerHTML;
  htmlcode=document.getElementById(id +'_code');
  theid=id; bt=btn;
}


function ssEdit(id,btn)                               // (I)
{ if ( btn.value=="Start" )
  { if ( id != theid )
    { if(el!=null) stopEdit(bt); editInit(id,btn); }
    el.setAttribute("contenteditable","true");
    btn.value="Cancel";  btn.style.color="red";
    return;
  }
  if ( id == theid )  stopEdit(btn);
}


function stopEdit(btn)                                // (J)
{  restore(theid);
   el.setAttribute("contenteditable","false");
   btn.value="Start";  btn.style.color="green";
   el=bt=theid=null;
}
```

Then, browser editing is turned on by setting the `contenteditable` attribute to `true`. The same button, when labeled `Cancel`, stops editing of the same section. Visual indications of editable contents are defined in `editcontent.css`.

Figure 8.21 shows editing of the **Today's Lunch Specials** header in progress.

FIGURE 8.21: Editing in Progress

A simple edit button, such as `undo` or `redo`, can tap directly into the HTML5 editing API:

```
<input onclick="document.execCommand('undo',false,null)"
       class="ec" type="button" value="Undo" />
<input onclick="document.execCommand('redo',false,null)"
       class="ec" type="button" value="Redo" />
```

The API `document.execCommand` method is used to perform various editing tasks, such as `Undo` and `Redo`, supported by the browser in editing mode. There are many (over 35) other commands including `bold`, `italic`, `fontSize`, `createLink`, `insertHTML`, and `insertImage`. The method has the general form:

```
execCommand(command, false, strValue)
```

The *strValue* provides a string for certain commands such as `fontSize` and `insertImage`. Use `null` if a command requires no value. See, for example, the Mozilla documentation on *Rich-text Editing* for a list of commands and the values they require.

The `undo`, `redo`, `cut`, `copy`, `paste`, and `delete` commands are usually also available on the browser's `Edit` menu. Hence, the `undo` and `redo` buttons (Figure 8.21) are nice but can usually be omitted without loss of editing capability. In our example (Figure 8.19), we provided only the `undo` button.

The following changes the font style of the current selection:

```
document.execCommand('bold',false,null);
document.execCommand('italic',false,null);
```

8.13.1   Editable Style

It is important to have clear visual indication when a section becomes editable. For our example, we used the following styles:

```
.cedit[contenteditable^="t"] { border: thin dashed blue; }

.cedit[contenteditable^="t"] *:hover { color: blue; }

.cedit[contenteditable^="f"] { border: 0; }

.cedit[contenteditable^="t"] img
{ border: thin blue dashed;  }

.cedit[contenteditable^="t"] img:hover
{ border: thin blue solid; }

div.rtedit
{ display: inline-block;  margin-top: 0px;
   color: black; font-weight: bold;
}

div.rtedit form { display: inline; }
```

8.14   Updating the AMP

Clicking the Done button sends the innerHTML code of the editable section back to the server side to overwrite its target .inc file (lunch.inc or dinner.inc in our example, line F).

The PHP generated code for the done button is

```
<form method="post" action="saveedit.php">
<input id="specials_code" name="htmlcode"              (K)
       value="" type="hidden" /><input
name="target" value="specials.inc" type="hidden" />
<input name="page" type="hidden"
       value="/dwp/exc8/manage/specials.html" />
<input title="finish and update page" type="submit"
       class="ec" value="Done" style="color: blue;"
       onclick="return(doneEdit('lunch'))" />          (L)
</form>
```

The PHP script saveedit.php can save editing results for any editable section.

The onclick handler for the Done button (line L) sets the value for htmlcode in the form (line M).

```
function doneEdit(id)
{ if(id==theid && window.confirm(
     "Are you sure you want to update the page?"))
  { el.setAttribute("contenteditable","false");
    htmlcode.value=el.innerHTML;                        // (M)
    el=btn=theid=null; return true;
  }
  return false;
}
```

The form action script, saveedit.php takes the received htmlcode and replaces the contents of the target file with it and then redirects the user to the live page (line N).

```
<?php /////    saveedit.php    /////
function saveFile()
{ if(!empty($_POST['htmlcode']) &&
     !empty($_POST['target']) && !empty($_POST['page']))
  { $f=$_POST['target'];
    if ( file_exists($f) )
    { rename($f, "$f.bak");     // back-up copy
      if ( $sp=fopen($f,"w") )
      { fwrite($sp, trim($_POST['htmlcode']));
        fclose($sp); chmod($f, 0666); return true;
      }
```

```
           else {  return false;  }
      }
  }
  return false;
}
if ( saveFile() )
{ header("Location: " . $_POST['page']); exit(); }     // (N)
else {  // send error page
?>
```

If `saveFile()` fails, an appropriate error page is sent.

8.14.1   Editing Images

The images in the specials page can be scaled and deleted easily. But we still need to supply a way for the admin user to change an image or add a new image. We can provide an `Image` button (Figure 8.22) to support image editing in an integrated way (**Ex:** `AMP_Img`).

FIGURE 8.22: Image Edit Button

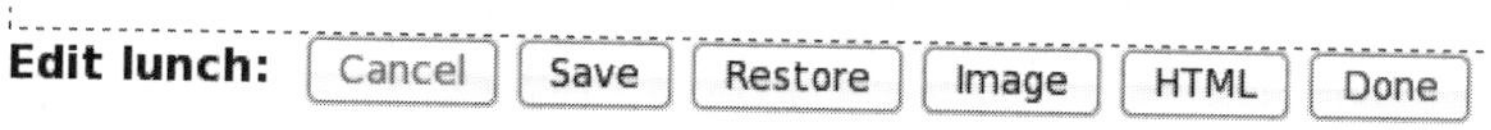

Edit lunch: Cancel   Save   Restore   Image   HTML   Done

After selecting a picture in the editable section, or getting ready to add a picture at the insertion point, the user clicks the `Image` button to upload an image. After the image is uploaded, the correct image URL is automatically inserted at the edit point.

The `imgBtn.php` is included via `$btn_arr[]="imgBtn.php";` (line F) to add the `Image` button:

```
<input title="Inserts a new image" value="Image"
  onclick="getImg('<?php echo $tid; ?>','form_ifr',          (O)
           'imgform.php?imgdir=specials_img');"
  class="ec" type="button" />
<iframe class="imgform" id="<?php echo $tid; ?>form_ifr"    (P)
 name="<?php echo $tid; ?>form_ifr" src=""></iframe>
```

Clicking `Image` (line O) reveals the hidden `iframe` (line P). The function

```
function getImg(id, fr_id, form)
{  if ( id == theid)
   {  ifr=document.getElementById(id+fr_id);
      ifr.src=form; ifr.style.display="block";
   }
}
```

places in the `iframe` this image upload page

```
imgform.php?imgdir=specials_img
```

which uploads an image to the designated folder specials_img (Figure 8.23, left). After succesful uploading, clicking the continue button in the result

FIGURE 8.23: Image Upload Iframe

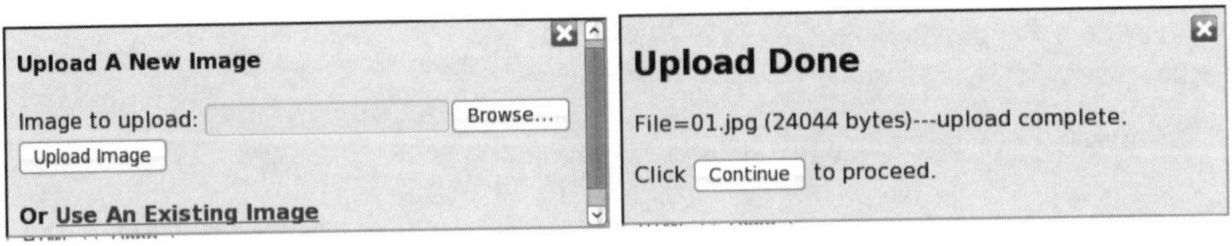

page (Figure 8.23, right) places the uploaded image at the edit insertion point and closes the iframe.

The onclick JavaScript code within the iframe for the continue button is parent.insertImage(*imgURL*), which calls the function in its parent page:

```
function insertImage(imgUrl)
{ if ( window.confirm("Use the upload image?") )
  { document.execCommand("insertImage",false,imgUrl); }
  ifr.style.display="none";
}
```

By the way, to access a JavaScript function from the parent page, you can use

```
ifr=document.getElementById('targetFrame');
ifr.contentWindow.targetFunction();
```

To use an image already uploaded before, the admin user would click on the use an existing image link (Figure 8.23, left), which leads to a display of existing images to pick.

8.14.2   Edit-Control Buttons

To make this content management approach general, it is important to easily incorporate more or less buttons to form the desired editing user interface for different editable content.

We use a single customizable controls.php as the edit control, which provides basic buttons Start/Cancel, Save, Restore, Done and any number of additional buttons given in an array (line Q).

```
<div class="rtedit">Edit <?php echo $tid; ?>: 
<input title="start/cancel editing" value="Start"
       onclick="ssEdit('<?php echo $tid; ?>',this);"
       style="color: green" class="ec" type="button" />
<input title="save work" class="ec" type="button" onclick=
```

```
            "saveEdit('<?php echo $tid; ?>');" value="Save" />
<input title="restore saved work"
       onclick="restore('<?php echo $tid; ?>');"
       class="ec" type="button" value="Restore" />
<?php if($btn_arr)
        { reset($btn_arr); addButtons($btn_arr); }?>          (Q)
<form method="post" action="saveedit.php">
<input id="<?php echo $tid; ?>_code" type="hidden"
       name="htmlcode" value="" />
<input type="hidden" name="target"
       value="<?php echo $target;?>" />
<input type="hidden" name="page" value="<?php echo $page;?>" />
<input title="finish and update page" style="color: blue"
       onclick="return(doneEdit('<?php echo $tid; ?>'))"
       class="ec" type="submit" value="Done" />
</form></div>
```

For example, to add an Image button and an HTML button, we can use

```
$btn_arr[]="imgcontrol.php"; $btn_arr[]="htmlcontrol.php";
```

Thus, to develop a new edit control *xyz*, we simply define it in *xyz*control.php and pass it to controls.php via the array $btn_arr.

8.14.3 Content Management Summary

By combining PHP, HTML5 editing API, JavaScript, and CSS we have achieved a practical, convenient, and user-friendly way to manage and update webpages referred to as *admin modifiable pages* (AMPs). Figure 8.24 gives a summary view of the methodology that puts HTML5 editing to good use.

FIGURE 8.24: Content Management Overview

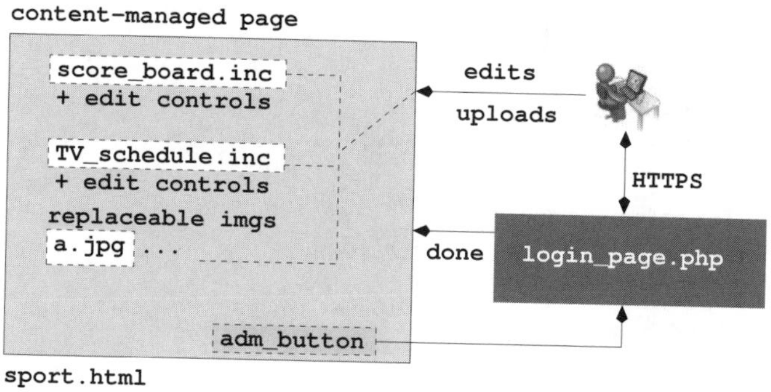

The edit controls for each editable part (.inc) in a page can provide just

enough edit capabilities for the desired tasks at hand. The updated results
are immediately available. Please visit the DWP website for the fully working
example (**Ex:** AMP_Img).

Code discussed here has been constructed for instructional use. In practice,
content management must be integrated with login and session control so only
admin users can make content changes.

8.15   PHP Classes and Objects

Among programming technologies, *object-oriented programming* (OOP) is one
of the most important. Unlike traditional procedure-oriented programming
where procedures or functions are the basic building blocks, OOP promotes
classes and *objects* as higher-level program modules.

At run-time, objects are self-contained computational units that interact
to perform desired tasks. Classes are blueprints for building objects. Each
class may define its own *properties* (data members) and *methods* (function
members) that are off-limits to code outside the class unless declared `public`.
A `private` member is accessible only to other members in the same class. A
`protected` member is accessible in its own class and in *derived* classes.

Originally a procedural language, PHP started to add simple class definition in version 4.0. Starting with version 5.0, PHP provides good support for
OOP. To define a class in PHP, use

```
class ClassName
{          members          }
```

A constructor `__construct(...)` is a special member of a class used to initialize new objects. For example, for a point on the screen we can define a
class `Point` (**Ex:** PointClass).

```
class Point
{  public function __construct($a,$b)
   {  $this->x=$a;  $this->y=$b;  }

   public function move($h,$v)
   {  $this->x+=$h;  $this->y+=$v;  }

   public function display()
   {  echo "($this->x, $this->y)";  }

   private $x, $y;
}
```

With the class `Point` defined, we can create `Point` objects using the `new`
operator:

```
$p1=new Point(4,5);
$p2=new Point(100, 200);
```

The supplied arguments are passed to the constructor. To access a member, use the notation *object->member*. Thus we can say -> is the member-of operator. For example,

```
$p1->display();
```

is a call to the method display() of point $p1, which produces the output (4, 5). Thus,

```
$p1->move(2,2);    // moves point $p1
$p1->display();    // displays (6, 7)
```

The class Point encapsulates data (x and y screen coordinates) and methods (move, display, and so on) to form a self-contained object. A method is different from an unattached function (one not in a class) in that a method can use the symbol $this to refer to the object itself and access the object state (values of all its properties).

We can add a method equal to the class Point for testing if two points are at the same location:

```
public function equal($p)
{ return ($this->x==$p->x && $this->y==$p->y ); }
```

Other methods for the class Point such as distance($p1,$p2) come to mind.

OOP offers great advantages for writing well-organized, flexible and reusable code. Principles and techniques of OOP are outside the scope of this text. Here we give some examples to show basic OOP support in PHP.

8.15.1   A Navbar Class

In Section 8.1 where we discussed page templates using PHP, we have seen code for a reusable navbar. Here, we will turn the code into a class Navbar that can be even more useful (**Ex: NavbarClass**).

```
<?php   /////    Navbar.php    /////
class Navbar
{ public function __construct($linkArray)
  { $this->links=$linkArray;
    $this->page=basename($_SERVER['PHP_SELF']);
  }

  public function addAttr($a) { $this->attr.= "$a "; }

  public function display()
  { echo "<nav " . $this->attr . " >\n";
```

```
    foreach($this->links as $label=>$href)
    { if($this->page==basename($href))
      { echo '<span class="self">' .$label. '</span>'; }
      else
      { echo '<a href="' .$href. '">' .$label. "</a>"; }
    }
    echo "\n</nav>\n";
  }


  private $links, $attr="", $page;
}
?>
```

Here is how we can use the class `Navbar`.

```
require_once('Navbar.php');                    // (a)
$left=array("Main Page"=>"index.php",          // (b)
            "Products"=>"products.php",
            "Services"=>"services.php",
            "News"=>"news.php",
            "Contact"=>"contact.php");
$left_nav=new Navbar($left);                   // (c)
$left_nav->addAttr('class="leftnavbar"');      // (d)
$left_nav.display();                           // (e)
```

The class definition is loaded (line a). We create a new `Navbar` object
`$left_nav` (line c) where the array `$left` supplies the links for the navbar
(line b). After adding a `class` attribute (line d), we can display the code for
the desired nabvar by calling the `display()` method (line e).

8.15.2   A `PageFront` Class

Let's look at another class `PageFront` that can be used to generate the front
part of pages in a website.

The constructor records the page title and a navbar object, belonging to
a class like `Navbar` defined in Section 8.15.1. The `display()` method (line 2)
is called to generate the desired HTML code (**Ex: PageFrontClass**).

```
class PageFront
{ public function __construct($t,$nav)          // (1)
  { $this->title=$t;
    $this->navbar=$nav;
  }

  public function display()                      // (2)
  { echo PageFront::$PP;                         // (3)
    echo '<title>' . $this->title . '</title>';
```

```
    $this->displayCss($this->css);
    $this->displayJs($this->js); $this->displayAny($this->any);
    $this->displayFile($this->srcfile); $this->displayBody();
  }


  // more members ...
}
```

The constant $PP (line 4) is the standard leading parts of any HTML5 page. A static member belongs to the class, not individual object instances, and is accessed (line 3) as

ClassName::*staticMemberName*

The displayCss() method (line 5) outputs code for each external CSS file recorded in the array $css by the method addCss (line 6).

```
private static $PP='<!DOCTYPE html>'.                         // (4)
'<html xmlns="http://www.w3.org/1999/xhtml"
     lang="en" xml:lang="en">
<head><meta charset="utf-8"/>';

private $css=array();
private function displayCss()                                 // (5)
{ reset($this->css);
  foreach($this->css as $file)
  { echo '<link rel="stylesheet" type="text/css" href="'
       . $file .  "\" />\n";
  }
}
public function addCss($css) { $this->css[] = $css; } // (6)
```

External JavaScript files are treated the same way. Other header code, such as meta description, keywords, and so on, in source files on the $this->srcFile array, as well as any strings on the $this->any array, can be included similarly.

```
public function addFile($f) { $this->srcfile=$f; }
private function displayFile()
{ reset($this->srcfile);
  foreach($this->srcfile as $file) { require($file); }
}

public function addAny($str) { $this->any[]=$str; }
private function displayAny()
{ reset($this->any);
  foreach($this->any as $str) { echo "$str\n"; }
}
```

Code generated by `displayBody()` closes the HTML head, begins the HTML body, and includes the navbar.

```
private function displayBody()
{ echo "</head>\n<body><div id=".
      '"centerpage"><section id="main">'."\n";
  $this->navbar->display();
  echo '<article id="content">' . "\n";
}
```

Putting `PageFront` to use, a page can instantiate a `PageFront` object (line 7), add desired CSS and JavaScript files, and call the `display()` method.

```
require_once("PageFront.php");
require_once("navbar.php");  // creates $left_nav
$pf=new PageFront("Join Our Club", $left_nav);          // (7)
$pf->addCss("basic.css"); $pf->addCss("form.css");
$pf->display();
```

The DWP website has the complete code for the examples **Ex: NavbarClass** and **Ex: PageFrontClass**.

8.16   System Calls and Script Security

8.16.1   System Calls in PHP

PHP scripts are easy and convenient, they are placed in regular document space accessible from the open Web, and PHP scripts can access local files (Section 8.4) as well as make system calls:

- `system($cmd)`—Executes the given shell command and sends its output (text lines) as part of the script's output back to the browser.

- `passthru($cmd)`—Executes the given shell command and sends its output (image or other binary data) directly as output.

- `exec($cmd, [$result])`—Executes the given shell command and returns the last line of its output. If the `$result` argument is supplied, then all output lines are also appended to it as an array. The `exec` function does not automatically add anything to the script's output. For example,

  ```
  $last_line=exec("ls -l", $all_lines_array);
  ```

- `shell_exec($cmd)`—Executes the given shell command and returns its output as a string. The function is the same as the *backtics operator* `'$cmd'`. For example, `$files='ls'` and `$files=shell_exec("ls")` are the same.

PHP also has a set of functions to run a command as a child process with I/O connected to the PHP script.

8.16.2 PHP Script Security

Security concerns include broken HTML, URL, or other code; access to files not intended for the public; running unexpected programs; loss of file or data; and unexpected run-time errors. Here are some measures to make your PHP scripts more secure:

- When constructing a URL consisting of variable values, call `urlencode` to make sure it is properly encoded. This is especially important when creating an HTTP `Location` header.

- Be careful when inserting variable values into HTML, JavaScript, or other code. Make sure the variable value does not break the syntax of the code you are generating. For example, if we have `$dim="6'3\" x 2'6\"";`, the code generated by

```
<input name="dimension" value="<?php echo $dim; ?>" />
```

would be

```
<input name="dimension" value="6'3" x 2'6"" />
```

which of course would be incorrect. Even this innocent-looking code

```
<img src='<?php echo $pic;?>' ... />
```

could be wrong if `$pic="Bob's Desk.jpg";`. The solution is to replace trouble-making characters with HTML entities or character references (Section 3.7). It is not hard to do. The PHP function call

```
$str=htmlspecialchars($str, ENT_QUOTES)
```

will replace the five characters < > ' " & by their HTML entities. Thus, `htmlspecialchars($dim, ENT_QUOTES)` returns

```
6&#039;3&quot; x 2&#039;6&quot;
```

- When making a system call where the command or part of the command involves values from the file system, database, or user input, we must also guard against special characters at the shell level such as |, &, >, <, comma, semi-colon, colon, *, and so on. Use `escapeshellarg($arg)` to ensure the integrity of each individual argument in a command and use `escapeshellcmd($cmd)` for a shell command string. These functions add a backslash (the shell escape character) to all potential problem-causing characters.

- The `escapeshellarg` is also useful somethimes when placing PHP-generated values in other code (such as JavaScript). Also, the PHP function `addslashes($str)` can be used to a BACKSLASH in front of quotes and backslashes.

- When using user input in database queries, you must guard against potential SQL injection (see Section 9.16).

- Be careful creating and using include files. Naming include or required files with the `.php` suffix ensures that the source code does not get out onto the Web. Do not include files by using names from user input or a database without fisrt checking that file name is OK. Another way is to to construct the filename string in your PHP code. For example

```
if ($page == "intro") include("./introduction.php);
```

PHP also offers a *safe mode*. When operating under safe mode, PHP applies file owner/group checks, system call restrictions, directory access restrictions and environment variable usage limitations. Other settings in the PHP static configuration file `php.ini` (Section 10.23) also control PHP security.

8.17   For More Information

Full language and function documentation for PHP can be found online at `php.net/docs.php`.

PEAR (PHP Extension and Application Repository) is an object library providing a large number of modules supporting many common Web application tasks. The PEAR base installation is part of PHP since version 5.0. See `pear.php.net` for more information.

For details on browser support of rich-text editing, see HTML5 contenteditable API and browser-specific documentations such as that found at `developer.mozilla.org/`.

Database access for websites via PHP will be discussed in Chapter 9.

8.18   Summary

PHP is a powerful tool for producing webpages on-the-fly and for supplying server-side functionalities such as page templates, form processing, file and folder management, file uploading, user registration and authentication, login, session control, email, image processing, security code verification, and CAPTCHA.

PHP as a programming language has excellent support for strings, arrays, regular expression patterns, local and remote file and directory access, password encryption, random number generation, raster graphics processing, networking, and OOP.

For debugging, running a PHP script from the command line can expose syntax errors and display page output quickly. Setting error reporting levels provides control over the display of error and diagnostic messages at runtime. For security reasons, take care when using PHP-generated or user input values when making system calls, accessing files/folders, or generating HTML or other code.

The chapter also demonstrated an innovative approach for content management through *admin modifiable pages* (AMPs) that integrates PHP, HTML5 rich-text editing API, JavaScript/DOM, and CSS to give owners an intuitive and user-friendly way to update key information on their websites.

Exercises

8.1. The PHP functions `include`, `require`, `include_once`, and `require_once` are used in page templating. Discuss their similarities and differences.

8.2. Is it possible to dynamically generate JavaScript or CSS files for webpages? If so, how would you code a `myjs.php` or a `mycss.php` file? What HTTP header is needed in each case?

8.3. If `$myvar` is a user-defined global variable, how do you access/set it from a PHP function.

8.4. In PHP, can you define `xyz` and `XYZ` as two different functions? Why or why not?

8.5. Outline the major steps for processing a form.

8.6. Describe the PHP error reporting system and how it can be used to debug PHP scripts and to avoid errors getting into webpages viewed by end users.

8.7. In order for a PHP script to create, modify, or delete files on the server side, what file permissions must be set? Explain in detail.

8.8. Using PHP, create a `reigsterTeam.php` page to display a form for registering a project or sports team, collecting team name, team leader, team members, emails, phone numbers, and other pertinent information for the team. The same `teamRegister.php` also checks and processes the submitted form and stores the information so that it is included in a `teamRoster.php` page.

8.9. Make the `teamRoster.php` page an AMP where each team's data can be edited and updated.

8.10. In session control, what is *out-of-nowhere* access? How should it be handled?

8.11. PHP session control can use both cookie and URL rewriting to handle the session ID. Find out how to make session control work if cookie support is denied by a browser.

8.12. A user must login before visiting a page for password changing. Why is it still necessary to collect the old password in the change-password form?

8.13. When a session consists of a linear sequence of steps, such as in online shopping, it is natural to handle going from one step to the next. But it can be convenient for users to be able to go back to the previous step and make changes. What complications do you see in such a session? How would you support going back to a previous step?

8.14. Write PHP code to allow uploading, adding, and deleting photos and captions that will automatically be shown using a horizontal thumbnail strip described by exercise **Scrolling Thumbnails** in Chapter 7.

8.15. Add login and session control to the specials AMP described in Section 8.12.

8.16. Write an object-oriented implementation for the AMP described in Section 8.12.

Chapter 9

Database-Driven Websites

Modern database systems make it easy to store, retrieve, and update data on the computer. The data are available where stored and usually also accessible online via computer networks. Databases and websites are often closely connected for two simple reasons:

1. Webpages can easily serve as user-friendly GUIs for databases. The Web can make any database accessible concurrently by many users anywhere on the Internet, 24/7.

2. A website can become much more functional and dynamic if it uses databases to support its operations such as user accounts, shopping carts, invoices, product inventories, calendars, schedules, geographical and map functions, and many other applications.

In fact, the collection of all websites worldwide can be thought of as a big loosely organized database where all kinds of information are just a few search queries away.

Sites such as Amazon® and Ebay® are totally database driven. In general you will find that most, if not all, e-commerce, e-business, e-government, and e-learning sites depend heavily on databases.

Figure 9.1 illustrates the Web to database interface.

FIGURE 9.1: Web to Database Interface

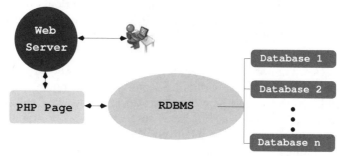

Connecting a website to a database involves these major steps:

1. Setting up the desired database to store and manage the data required in the target applications.

2. Writing programs, according to specific business logic, to retrieve, store, remove, and update data items in the database. The work usually involves database programming (SQL) and Web-to-database interface programming (PHP, for example).

3. Creating webpages that collect data going into the database, that display information extracted from the database, and that depend on the database for their operations.

We will introduce database systems, the *Structured Query Language* (SQL) for managing databases, and writing application programs that use databases. We will use the popular, and free, database system MySQL for our databases and PHP to interface databases to Web applications. The PHP-MySQL combination is popular and widely used by Web developers.

PHP access to MySQL will be presented in detail. Examples will show how to combine databases with the Web for practical applications such as selling and payment on the Web.

9.1 What Is a Database?

A database is a collection of data efficiently organized in digital form to support specific applications. A database is normally intended to be shared by multiple users.

A *Database Management System* (DBMS) is a software system that operates databases, providing storage, access, security, backup, and other facilities. A DBMS usually makes databases available to many users at the same time and handles user login, privileges (abilities to perform certain operations), and local or network access.

A *Relational Database* is one that uses *relations* or tables to organize data. Even though there are other ways to organize data, relational databases are by far the most popular and widely used in practice. Well-known relational DBMS (RDBMS) include IBM DB2, Oracle, MySQL, SQLite, Derby, and Microsoft Access.

Databases are important for organizations as well as websites. A database can help a site register users, maintain accounts, store product/service information, and perform countless other operations. A database can also store information about website content items, such as pictures, forum/blog posts, product reviews/ratings, and help deliver Web content dynamically.

Often a database becomes an integral part of a website. Such sites are the so-called *database-driven sites* and examples are everywhere: Amazon®, eBay®, Facebook®, and the list goes on. Thus, it is important for Web programmers to become proficient in dealing with databases and in interfacing webpages with databases.

9.2 Relational Databases

A relational database uses multiple tables, called *relations* in database theory, to efficiently organize data. A *relation* is a set of related attributes and their possible values. Figure 9.2 shows a typical database table.

FIGURE 9.2: A Simple Database Table

Last	First	Dept	Email
Wang	Paul	CS	pwang@kent.edu

Each table is defined by a *schema* that specifies

1. The names of the *attributes* (column headings)

2. The type of value allowed for each attribute

For example, the type of an attribute can be character string, date, integer, or decimal, and so on.

Each row in a table is a set of attribute values. A row is also called a *record* or a *tuple*, and no two records can be exactly the same in a table (the *no duplicate row* rule). The collection of all records in a table is called a *table instance* or *relation instance*. Immediately after a table is defined by its schema, it has no records. So the table instance is the empty set. As records are inserted into the table, the table instance grows. As records are removed from the table, the table instance shrinks. As data in table records are changed, the table instance changes. When dealing with database tables, it is important to keep in mind the difference between the table schema and the table instances.

Each database in an RDBMS is identified by a name. A relational database usually consists of multiple tables organized to efficiently represent and inter-relate the data. The RDBMS also stores, in its own management database, information about which users can access what databases and what database operations are allowed. Usually a user must first login to the RDBMS from designated hosts before access can be made. For example, if access is restricted to the *localhost*, then only access made by a program, running on the same host computer as the RDBMS, is allowed.

9.3 SQL: Structured Query Language

Of course, each RDBMS needs to provide a way for application programs to create, access, and manipulate databases, tables, records, and other database-related items. The *Structured Query Language* (SQL[1]) standardizes a programming language for this need. SQL, an ISO and ANSI standard, is a *declarative language* that uses sentences and clauses to form *queries* that specify database actions. SQL consists of a *Data Definition Language* (DDL) to specify schemas and a *Data Manipulation Language* (DML) for adding, removing, updating, retrieving, and manipulating data in databases.

Major RDBMSs are all SQL compliant to a great degree. Thus, programs coded in SQL can easily be made to work on different systems. However, there are still differences among different database systems. In this book, examples show SQL code for the MySQL system. The code can easily be repurposed for other RDBMSs.

A *database query*, or simply a query, is a command written in SQL that instructs the RDBMS to perform a desired task on a database. A data retrieval query usually returns a *resultset*, which is a table of records (Figure 9.3). A data update query does not return a resultset.

FIGURE 9.3: Database Query

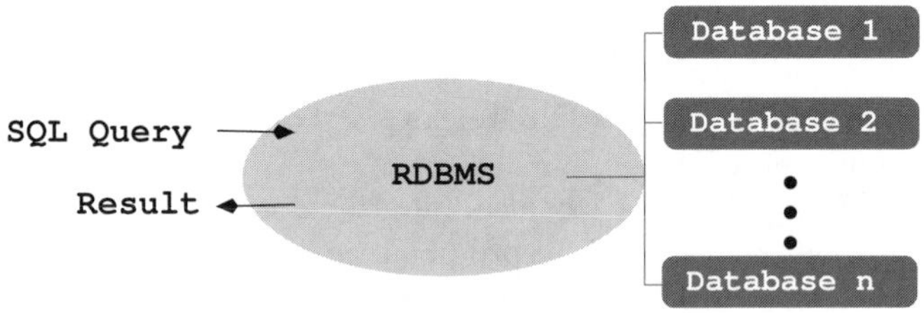

9.4   SQL Queries

Say we have a table named `member` to store members for a particular website (Figure 9.4). The SQL query

FIGURE 9.4: Sample Table: `member`

uid	last	first	email	passwd
jsmith	Smith	Joe	...	...
pwang	Wang	Paul	pwang@sofpower.com	...
jjones	Jones	Joel	...	...
...				

---

[1]SQL is pronounced by saying each of the three letters.

```
SELECT * FROM member WHERE last='Wang';
```

retrieves from table `member` all columns for all rows where `last` is `Wang`. The resultset of this query is a table of all records satisfying the `WHERE` clause. If there is no member by that last name, the resultset is empty. Because each table row is called a *tuple* in database theory, we can say that the resultset is a subset of tuples in `member` satisfying the condition given by the `WHERE` clause

```
WHERE last='Wang'
```

which requires the value of the `last` field to be equal to the string `Wang`. In an SQL query, a string literal can be given with either single quotes or double quotes.

Note the declarative style of a query. We give *keywords* in a query in all-caps for easy identification. But, keywords in SQL are actually case insensitive. Table and attribute names are usually case sensitive. String values are always case sensitive. A query is terminated by a *terminator*, which by default is the semicolon (;).

The symbol * of the `SELECT` query gets you a resultset with all columns (attributes) of the table from which you are selecting. Of course, you may choose a subset of columns, for example,

```
SELECT last, first, email FROM member; -- three columns
SELECT passwd FROM member
        where uid="pwang";  # password for pwang
```

Comments in SQL are given as follows:

- From a # character to the end of the line

- From a --SPACE sequence to the end of the line

- Within a /* and */ bracket (as in C code)

When selecting a proper subset of attributes, duplicate rows become possible in the resultset. To avoid that, add the `DISTINCT` keyword in front of the column names. For example,

```
SELECT DISTINCT zip FROM employee;
SELECT DISTINCT city, state, country FROM participant;
SELECT DISTINCT major, year FROM student WHERE year="Freshman";
```

The resultset of a query can also be sorted by the values in any specific column. Just put the `ORDER BY` clause at the end of a query. For example (**Ex:** SortQ),

```
SELECT * FROM client ORDER BY last_name;

SELECT id, grade FROM student_grade
    WHERE course_id="CS-101" AND semester="Fall"
```

```
                  AND year="2012" ORDER BY grade;
```

```
SELECT customer_name, amt FROM sale ORDER BY amt DESC;
```

Normally, the ordering is from low to high (ascending) unless the keyword DESC (descending) is given.

It is now obvious that when retrieving data from tables, we need to know their schemas (attribute names and their data types). If we have created the tables in the first place, we have that information. If we are working with tables others have created, we can obtain the schema of any table with these queries:

SHOW TABLES – lists all tables in the database
SHOW COLUMNS FROM *table* – lists attributes of table
DESCRIBE *table* – displays all aspects of table

9.4.1 MySQL Expressions

The WHERE clause is the keyword followed by an expression. If that expression evaluates to 0 or FALSE, then the clause is not satisfied. Otherwise, if the expression evaluates to nonzero or TRUE, then the clause is satisfied.

Table 9.1 lists MySQL relational operators that compare numbers or strings. SQL relational operators are not exactly Boolean. They do not return only TRUE or FALSE. They can also return NULL, a symbol used to indicate unknown values in database tables.

TABLE 9.1: MySQL Relational Operators

Operator	Meaning
=	Equal
<>,!=	Not equal
>	Greater than
<	Less than
>=	Greater than or equal
<=	Less than or equal

For a relational expression *e1 rop e2*, the *three-value logic* works as follows:

- If both *e1* and *e2* are not NULL, then return the result by testing their values.

- If at least one of *e1* and *e2* is NULL, then return NULL.

Thus, NULL = NULL, NULL != NULL, and 3 > NULL all return NULL. This makes sense. What other than 'unknown' can be the answer when operating with an unknown value? In SQL, 0, ' ' (the empty string), and NULL are treated as logical false, anything else is true.

But what if we need to test if a value is NULL? We can use the IS NULL or IS NOT NULL operator. For example,

```
NULL IS NULL    -- returns 1
3 IS NOT NULL   -- returns 1
```

There is also the so-called NULL-safe equal operator <=>. It is like the = operator, but returns 1 if both operands are NULL, and 0 if only one operand is NULL.

Logical operators are AND (&&), OR (||), NOT (!), and XOR (exclusive or). They operate on the three values TRUE, FALSE, and NULL, as you would expect. For example,

```
0 AND NULL /* is 0 */     1 AND NULL /* is NULL */
1 OR NULL  /* is 1 */     0 Or NULL  /* is NULL */
1 XOR 1    /* is 0 */     NOT NULL   /* is NULL */
```

The usual arithmetic operators + * - / are available. But / normally will return a floating-point result. Use the DIV operator for integer division (dropping the fractional part) and % (or MOD) for computing the remainder. Dividing by zero produces NULL.

The operator = is also used for assignment when inside a SET environment, as we will see.

9.4.2 More on SELECT Queries

You can give any SQL expression to SELECT for evaluation. For example,

```
SELECT 4*5;              -- is 20
SELECT "A" < "a";        -- is 0
```

It is possible to perform pattern matching in a WHERE clause. MySQL SQL pattern matching provides wildcard characters for use with the LIKE (NOT LIKE) operator. The character % matches any string of zero or more characters. The underscore (_) matches any single character. The default escape character is the backslash (\). For example,

```
SELECT email FROM member WHERE zip LIKE "44%";
SELECT * FROM student WHERE phone NOT LIKE '330-%';
```

In MySQL string comparisons and pattern matching are usually case insensitive unless one of the operands uses a case-sensitive *collating sequence*. You can force case-sensitive comparisons by specifying a collating sequence as in

```
SELECT "ABC" < "abc" COLLATE utf8_bin;  -- returns 1
SELECT filename LIKE '%.html' COLLATE utf8_bin;
```

MySQL also supports regular expression pattern matching much like that in JavaScript (Section 6.16.7) or PHP (actually Linux grep). The operators RLIKE and NOT RLIKE are used for regular expression pattern matching. For example,

```
filename RLIKE '\.html*'          -- foo.html or bar.htm
filename RLIKE '\.jpg$|\.JPG$'    -- foo.jpg or bar.JPG
zip RLIKE '^44224-[0-9]{4}$'      -- 44224 + 4
```

A SELECT may produce new columns in the resultset. For example,

```
SELECT name, vacation_taken,vacation_accrued,
(vacation_accrued-vacation_taken) AS       -- new column
 vacation_balance FROM employee;

SELECT CONCAT(last,', ',first) AS fullname
      from member ORDER BY fullname;       -- new column
```

CONCAT is a MySQL built-in function that joins strings together. Also available are the usual mathematical functions (log, trig, and so on) as well as string functions (substring, length, replacement, and so on). See MySQL documentation (dev.mysql.com/doc/refman/) for more information.

9.4.3   Aggregating Attribute Values

The SELECT query is not limited to getting a subset of rows and/or columns from tables. It also can *aggregate* data across records (rows) and give you a single combined result.

MySQL built-in aggregating functions include

- COUNT(*expr*)—Returns a count of the number of non-NULL values of *expr* in an aggregation; count(*) returns the number of rows in the resultset.

- AVG(*expr*)—Returns the average of *expr* values in an aggregation.

- MAX(*expr*)—Returns the max of *expr* values in an aggregation.

- MIN(*expr*)—Returns the min of *expr* values in an aggregation.

- SUM(*expr*)—Returns the total of *expr* values in an aggregation.

- GROUP_CONCAT(*expr*)—Returns the comma-separated string concatenation of *expr* values in an aggregation.

For example, if we want to know the total enrollment from a student table (Figure 9.5), we can use the following query:

```
SELECT COUNT(*) as enrollment FROM student;
```

The resultset contains a single record showing the number of records on the student table as the enrollment. Here are some more examples (**Ex: Aggregate**):

FIGURE 9.5: The student Table

ID	name	dept_name	tot_cred
00128	Zhang	Comp. Sci.	102
12345	Shankar	Comp. Sci.	44
19991	Brandt	History	80
23121	Chavez	Finance	110

```
SELECT COUNT(letter_grade) FROM  grade
  WHERE letter_grade='A';    /* how many As */

SELECT COUNT(DISTINCT major) FROM
  student;    /* how many different majors */

SELECT AVG(hw2) as hw2_avg,
       MAX(hw2) as hw2_max,
       MIN(hw2) as hw2_min FROM grade;

SELECT SUM(amt) as sales FROM
  order;    /* total sales from order table */

SELECT GROUP_CONCAT(first_name, '_', last_name)
  AS name_list FROM student;
```

The preceding examples aggregate over entire tables and each produces one result row. But it is also useful to aggregate over groups of rows in a table. For that, we use the GROUP BY clause:

```
GROUP BY attr    /* rows with same attr value */

/* rows with same (attr_1, attr_2) pair value */
GROUP BY attr_1, attr_2
```

For example, we can have the per-department enrollment from the student table (Figure 9.5) with the query

```
SELECT dept_name, COUNT(*) AS enrollment
  FROM student GROUP BY dept_name
  ORDER BY enrollment DESC
```

which produces a resultset like that shown on the left side of Figure 9.6. Here are some more GROUP BY examples:

FIGURE 9.6: Aggregation Results

dept_name	enrollment
Comp. Sci.	4
Physics	3
Elec. Eng.	2
Finance	1
History	1
Music	1
Biology	1

dept_name	enrollment
Biology	1
Comp. Sci.	4
Elec. Eng.	2
Finance	1
History	1
Music	1
Physics	3
NULL	13

```
SELECT state,      /* state by state zip code count */
  COUNT(DISTINCT zip_code) as zip_count
  FROM zip_table GROUP BY state;

SELECT year, AVG(midterm) as mid_avg,  /* stats by year */
  MAX(midterm) as mid_max,
  MIN(midterm) as mid_min
  FROM grade GROUP BY year;

SELECT customer_name SUM(amt) /* sales per customer */
  as sales FROM order GROUP BY customer_name;
```

If you wish to have both aggregating over distinct groups and over the entire table, just add the WITH ROLLUP modifier at the end of the GROUP BY clause. This causes a final summary row to be added to the resultset, which is computed as if without the GROUP BY clause. Thus,

```
SELECT dept_name, COUNT(*) AS enrollment
  FROM student GROUP BY dept_name
  WITH ROLLUP
```

produces a final row with the total enrollment as shown on the right side of Figure 9.6[2].

9.4.4   Selecting from Multiple Tables

So far we have been selecting from a single table. Selecting from multiple tables

---

[2]Note: WITH ROLLUP and ORDER BY cannot be user together.

```
SELECT * FROM t1, t2 WHERE ... ;
```

means selecting from the *cross project* t1*times*t2 which is a table with columns from t1 followed by columns from t2 and filled with all rows from t1 followed by each row in t2. If t1 has m rows and t2 n rows, then their cross product has $m * n$ rows. The JOIN operation in SQL forms the cross product as shown below.

```
A  B              1 2 3                     A  B 1 2 3
        JOIN                gives           C  D 1 2 3
C  D              4 5 6                     A  B 4 5 6
                                            C  D 4 5 6
```

Selecting from more than one table is important because it can pull information together from related tables. For example,

```
SELECT tb1.name, tb1.rank, tb2.salary FROM tb1, tb2
WHERE tb1.name = tb2.name;}
```

Selecting from more than two tables is defined recursively.

This not being a database book, it is not possible to cover all the fine details of the SELECT query.

Let's turn our attention to defining tables next.

9.5 Creating Tables

Before creating tables for a database, we must first analyze the specific application and identify the tables we need and how they are related. One way to do this is with *entity relationship* (ER) modeling. In this approach, we identify the actual and logical entities involved in an application. For instance, a book selling site would involve entities such as books, publishers, shippers, customers, sales, orders, returns, and so on. Each entity will be characterized by a set of attributes. And we will recognize and formalize the relationships among the entities. ER modeling produces a set of ER diagrams. Figure 9.7

FIGURE 9.7: Simple ER Diagram

shows a simple ER diagram indicating (1) a book is related to one publisher and (2) a publisher can be related to one or more books. It shows one of the several popular styles for ER diagrams. Often, each entity box also lists entity attributes.

With a complete set of ER diagrams, we are set to define table schemas based on the entities and to create our database.

Let's turn our attention to table creation queries.

9.6   CREATE TABLE Queries

A simple CREATE TABLE query has the general form

```
CREATE TABLE table
(    attr_1 data_type,
     attr_2 data_type,
     attr_3 data_type,


     . . . .


     PRIMARY KEY (attr_a, ...)
);
```

to specify a table schema with the comma-separated attribute names and allowed data types of their values. The PRIMARY KEY *constraint*, a list of one or more comma-separated attribute names given in parentheses, requires that

1. No key attribute may have NULL as value.

2. No two records (rows) in the table are allowed to have the save values for the primary key. In other words, a primary key value always determines at most one row in the table.

The constraint is enforced automatically by the RDBMS. It is recommended that you define a PRIMARY KEY for each table you create. It is used to establish a *search index* for the table records for fast and efficient access.

Here is a table creation example:

```
CREATE TABLE member
(  uid       varchar(10),      -- user id
   last      varchar(25),      -- last name
   first     varchar(15),      -- first name
   email     varchar(35),      -- email address
   passwd    varbinary(48),    -- encrypted password
   PRIMARY KEY (uid)
);
```

It is an error to create a new table with the same name as an existing table. You can use

```
CREATE TABLE IF NOT EXISTS table;
```

to avoid the error. Or you can drop an existing table

```
DROP TABLE IF EXISTS table;
```

before recreating it.

9.7 MySQL Data Types

When it comes to data types in databases, the declaration syntax and semantics are unfortunately system dependent. We will describe data types in MySQL.

Data types generally fall into three categories: numerical, string, and date/time. Here are some common types:

- Integers—TINYINT (1 byte), SMALLINT (2 bytes), MEDIUMINT (3 bytes), INT (4 bytes), BIGINT (8 bytes).

- Decimal numbers—DECIMAL(m, d), where m is the precision (total number of significant digits) and d is the number of digits after the decimal point; for example, price DECIMAL(5,2). The keyword NUMERICAL is the same as DECIMAL.

- Floating-point numbers—FLOAT (4 bytes) and DOUBLE (8 bytes). These represent approximate values such as 0.333333. If you can, avoid approximate values and choose DECIMAL instead.

- Fixed-length character strings—CHAR(n) stores a string of n (max 255) characters. When storing a shorter string, it is right-padded with spaces to the specified length. When it is retrieved, the trailing spaces are automatically removed.

- Variable-length character strings—VARCHAR(n) stores a string up to n (max 65535) characters. When storing a string, all its characters are stored (no padding).

- Byte strings—BINARY(n) and VARBINARY(n) types are the same as character string types except they deal with bytes rather than characters.

- Enumerated string—ENUM('s1', ...) specifies a type that stores one of the listed strings, for example, size ENUM('S','M','L','XL').

- Date and time—A DATE is given in the yyyy-mm-dd format ranging from 1000-01-01 to 9999-12-31. A TIME is given in the hh:mm:ss format ranging from $-838:59:59$ to $838:59:59$. A TIME value can be used to indicate the time of day or a time interval. A DATETIME is given as a DATE, a SPACE, and a TIME ranging from 1000-01-01 00:00:00 to 9999-12-31 23:59:59. A YEAR(2) or YEAR(4) type represents a year using 2 or 4 digits.

- Timestamps—A DATETIME usually to keep track of data changes in a database. For example,

```
stamp TIMESTAMP DEFAULT CURRENT_TIMESTAMP
               ON UPDATE CURRENT_TIMESTAMP
```

will enter the current time in the `stamp` column when a record is inserted into the table and will update the `stamp` when any attribute on the same row is changed.

9.7.1   Data Modifiers

A table record uses the special symbol `NULL` to indicate an unknown value for any particular attribute. Later when the value becomes known, the value can be updated. Unknown values cannot be avoided in practice. Consider the `major` attribute for records on the `student` table. We know a freshman is not required to declare a major. Therefore, that attribute must remain unknown until later.

Of course, not all values for a record can be unknown. For the must-have values, we can add the

```
NOT NULL
```

modifier after the attribute data type. For example, here is an improved `member` schema (**Ex: Member**).

```
CREATE TABLE member
(  uid      varchar(10),
   last     varchar(25) NOT NULL,
   first    varchar(15) NOT NULL,
   email    varchar(35) NOT NULL UNIQUE,    -- (A)
   passwd   varbinary(48) DEFAULT NULL,     -- (B)
   PRIMARY KEY (uid)
);
```

Thus, the names and email must be entered for every member. Because `uid` is a primary key, it is automatically `NOT NULL`. With the `UNIQUE` modifier (line `A`), we also insist that all email addresses must be different. If a user has not set up a password yet, it defaults to `NULL` (line `B`). The `DEFAULT` *val* modifier specifies a value when none is supplied for that attribute.

9.8   PHP Database Access

To connect databases to the Web, we can use a suitable active page language that offers support for the RDBMS being used. We will use PHP to access MySQL.

As a Web scripting language, PHP provides access to MySQL as well as other well-known RDBMS systems. In fact, PHP has a built-in database system called SQLite that can be a good choice for light-duty database applications. In this book, we will work with PHP and MySQL. Figure 9.8 shows the PHP-MySQL connection that is typical of such Web-to-database connections.

FIGURE 9.8: PHP Access to MySQL

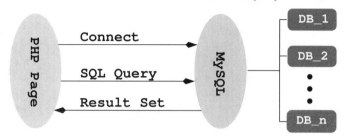

1. PHP sends a connection request to a target MySQL system (the database server), presenting a userid, password, and name of the target database to be used. The database server will authenticate the connection and either grant or deny the connection.

2. With a connect established, the PHP script can issue SQL queries and obtain results.

3. When the PHP page is done, the database connection is usually closed automatically.

There is a procedural interface from PHP to MySQL with a set of functions for making a connection, sending queries, receiving resultsets, and so on. In addition, there is also an object-based interface where you create a connection object that supplies methods for making queries and receiving resultset objects. Using one or the other for coding is a personal preference. The object-based support, provided by the `mysqli` package, is what we use in this book.

9.8.1 Connecting to a Database

Generally, you need four pieces of information to gain connection to a database: the database server host name, userid, password, and name of database. The first three items allow you to gain access to the RDBMS, and the last item indicates the database with which to work.

The PHP `mysqli` code for making a connection is

```
$db_obj = new mysqli($host,$user,$pass,$dbname);
```

```
if (mysqli_connect_errno())
{  printf("Can't connect to $host $dbname. Errorcode: %s\n",
   mysqli_connect_error());
   exit;
}
```

A successful connection produces a usable database connection object `$db_obj` that is used for making queries. Otherwise an error message is produced and the script exits. In this chapter, we always assume that `$db_obj` is a `mysqli` database connection object.

To close the database connection, simply call the `close()` method

```
$db_obj->close();
```

or it will be closed automatically when the PHP page is done.

The `$host` is the domain name or IP address of the host computer that runs the RDBMS. Here is an example of setting values in preparation for making the connection:

```
$host="localhost";  $user="pwang";
$pass="instructor"; $dbname="pwang";
```

Often, the database supporting a website runs on the same host as the Web server where the PHP page also resides. For better security, access to such a database is usually restricted to programs running on the same host. For such local access, we need to set `$host` to `localhost` instead of its domain name.

To set the host name correctly independent of local or remote access, we can use this code:

```
$host = "webdev.cs.kent.edu";
$phphost=trim('hostname');   // host of php script
if ( $phphost == $host )
{  $host="localhost"; }
```

9.8.2   Queries and Resultsets

After making a database connection, a PHP script can proceed to make queries, receive resultsets, and process them. We first create a desired query string in PHP, for example,

```
$query = "SELECT * FROM car WHERE make='Ford'
     && model='Focus' && year='2012'";
```

Note how we used double quotes for the PHP string and single quotes for strings in the query. It is of course possible to replace each single quote with a \".

To send the query to the RDBMS, we call the `query` method of the connection object

```
$result_obj = $db_obj->query($query);
```

which sends the given $query to the connected database system.

If the query fails, the query method returns FALSE. For a successful SELECT, SHOW, DESCRIBE, or EXPLAIN query, the query method returns a result object. For other SQL queries such as INSERT, UPDATE, DELETE, and DROP, the query method returns TRUE on success or FALSE on error. Any error message is available in $db_obj->error.

Having obtained a resultset, we can display it with a call to the htmlTable function (see **Ex:** HTMLCode), which constructs and returns the HTML table code for any given resultset or an empty string if the resultset has zero rows.

```
function htmlTable(&$result_obj, $class)
{  if ($result_obj->num_rows == 0) return "";    // (0)
   $tb = "<table border='1' class='" .
                             $class . "'>\n";     // (1)
   $result_obj->data_seek(0);                     // (2)
   $header_done = false;
   while( $data = $result_obj->fetch_assoc() )    // (3)
   { if (!$header_done)                           // (4)
     {  $tb .= '<tr>';
        foreach($data as $attr => $value)
        {   $tb .= "<th>$attr</th>"; }
        $tb .= "</tr>\n";
        $header_done = true;                       // (5)
     }
     $tb .= "<tr><td>";                            // (6)
     $tb .= implode("</td><td>", $data);
     $tb .= "</td></tr>\n";                        // (7)
   }
   return ($tb . "</table>\n");
}
```

It is always a good idea to check for an empty resultset (no rows) before processing its entries (line 0). The $tb variable starts with the table tag with a given CSS class setting (line 1). The data_seek method is called with 0 to begin accessing records in the resultset from row 1 (line 2). Of course, you can move the seek point to any desired starting position within the resultset. The value

```
$result_obj->num_rows
```

gives you the total number of records in a resultset.

For each iteration of the while loop, we call the fetch_assoc() method to obtain the next record as an attribute-value associative array (line 3). The table header cells are produced as part of the processing of the first record, and the attribute names of the resultset are used as HTML table headings (lines 4 and 5). Then, the attribute values are placed in table cells (lines 6 and 7). As the `while` loop proceeds, all rows are entered into the HTML table cells. Figure 9.9 shows a sample display of an HTML table generated by the PHP function `htmlTable`.

FIGURE 9.9: HTML Table from Database

first	last	address	position
Bob	Smith	128 Here St, Cityname	Marketing Manager
John	Roberts	45 There St, Townville	Secretary
Brad	Johnson	34 Nowhere Blvd. Snowtown	Doorman

In rare cases, a resultset can have different columns using the same attribute name. In such a situation, you can use the `$result_obj->fetch_row` method instead of `fetch_assoc`. The `fetch_row` method returns the next data record as an array of attribute values indexed from 0.

The call `$row=$result_obj->fetch_object()` returns the next result row in an object `$row`, and you can access values with `$row->attr_name`.

9.9   User Authentication Using Database

We have discussed PHP controlled login sessions in detail (Section 8.10). There we used flat files to store userids and passwords. We now show how to save the authentication information in a database and how to perform user authentication with it (**Ex: DbAuth**).

9.9.1   Adding a New User

Here is how to save a new user, together with the userid and password, into the `member` table (Section 9.6). Assuming that we have created the global database connection object `$db_obj` and that the user registration formdata passed validation, we can call `add_member` to save the information to the database table `member`.

```
function add_member()
{ global $db_obj;
  $uid=$db_obj->escape_string($_REQUEST['uid']);        // (A)
  $last=$db_obj->escape_string($_REQUEST['last']);
  $first=$db_obj->escape_string($_REQUEST['first']);
  $email=$db_obj->escape_string($_REQUEST['email']);
```

```
$pass=$db_obj->escape_string($_REQUEST['passwd']);   // (B)
$query="INSERT INTO member VALUES ('$uid','$last',   // (C)
          '$first', '$email', PASSWORD('$pass'))";   // (D)
return ($db_obj->query($query));                      // (E)
}
```

The method `$db_obj->escape_string` makes a string safe to use in a query. It prepends a BACKSLASH to any potential trouble-making character such as ", ', RETURN, NEWLNE, and BACKSLASH. It is highly recommended that you use this method to render safe all strings spliced into a query (lines A through B). This is especially important on strings coming from the end user to guard against *SQL injection* (Section 9.16).

The INSERT query adds new records (rows) to a table (Section 9.10). On line C, we add a new record by supplying a value for each attribute in table-definition order.

The PASSWORD (line D) is a one-way encryption function that turns any string into a 41-byte binary string on modern MySQL systems. It is not advisable to store unencrypted passwords. The password column datatype must be long enough for the encryption to be used. The PASSWORD function is the encryption MySQL uses for its own user authentication. MySQL also supports several AES (Advanced Encryption Standard) and SHA (Secure Hash Algorithm) encryption algorithms.

9.9.2 Authenticating a User

A website can display its own login form, collect the userid and password entered by the user, and then check them against those stored in the database. The function

```
function authenticate($uid, $pass)
{  global $db_obj;
   $userid=$db_obj->escape_string($uid);
   $pass=$db_obj->escape_string($pass);
   $query="SELECT * FROM member WHERE uid ='$uid'"
      . " AND passwd = PASSWORD('$pass')";            // (F)

   if ( ($result = $db_obj->query($query))            // (G)
           && $result->num_rows == 1 )
   { return $uid; }
   else
   { return ""; }
}
```

can be called to check the login information `$uid` and `$pass` against those in the `member` table. If the query on line F produces exactly one row (line G) then the `$uid` is returned. Otherwise, an empty string is returned.

9.9.3   Changing Passwords

Many login systems will support password changing. It is relatively easy for us to update the password entry in the `member` table.

After receiving and validating the userid, current password, and new password from user input, we can call the function

```
function change_pw($uid, $oldpw, $newpw)
{ global $db_obj;
  $uid=$db_obj->escape_string($uid);
  $oldpw=$db_obj->escape_string($oldpw);
  $newpw=$db_obj->escape_string($newpw);
  if ( authenticate($uid,$oldpw) )              // (I)
  {  $query="UPDATE member " .                  // (II)
          "SET passwd=PASSWORD('$newpw') " .
          "WHERE uid='$uid'";
     return ($db_obj->query($query));  // TRUE or FALSE
  }
  return FALSE;
}
```

to make the password change. For example,

```
if ( change_pw($uid, $oldpw, $newpw) )
{  echo "<p>Password change for $uid has been done</p>"; }
else
{  echo "<p>Password change failed.</p>"; }
```

Note: We must first authenticate the user's current password (line I) before installing the new password (line II). The `UPDATE` query used on line II has the general form

```
UPDATE table_names
      SET attr_1=val_1, SET attr_2=val_2, ...
      WHERE ... ;
```

You can specify one or more tables to update (comma-separated list). When specifying an attribute name, you can always qualify it with a table name (`member.uid` for example) to disambiguate it from attributes in other tables.

9.10   INSERT and DELETE Queries

Adding and removing table records are important updates in any database operation.

The `DELETE FROM` query has a form similar to that of `UPDATE`. The query

```
DELETE FROM table_names WHERE ... ;
```

removes all matching rows from the given tables. If the `WHERE` clause is omitted, all rows will be removed and the tables become empty. The decision to remove a record must not be taken lightly. Once removed, the data may be lost forever. Often, instead of removing a record, we can mark a record inactive. For example, if a person on the `member` table decides to quit or not pay dues, we can set the `active` column for that member's record to 0. When selecting from `member`, the clause

```
WHERE active=1
```

ensures that only active members are chosen. When an inactive member decides to return, the `active` value can be set back to 1.

The `INSERT INTO` query has the form

```
INSERT INTO table_name (attr_1, ...)
        VALUES (val_1, ...), ... ;
```

It adds one or more new rows with the values provided. The values for each row form a parenthesized list and one or more such lists can be supplied. The list of attributes specifies the attribute names corresponding to each value supplied. The attribute list may give, in any order, a subset of the table attributes. Unspecified attributes get set to their default values or `NULL`. If the attribute list is omitted, then each value list must supply all the attribute values for a row in the same order as in the table schema. In a value list, each value supplied must conform to the data type specified in the table schema. This means a string must be enclosed in single or double quotes, a date/time must be quoted and in the required format, a number is not quoted and in the right form. The values `DEFAULT` and `NULL` can be used.

If a new row violates a primary or unique key constraint (called a duplicate), then an error is produced and the `INSERT` query stops inserting the offending row and any more rows after it. Add the `IGNORE` keyword after `INSERT` to skip any duplicates and continue the insertion. Or you can use the `REPLACE` query, which is the same as `INSERT` except it will delete any existing duplicate record before inserting a new one.

9.11 Database Supporting Product Orders

Consider a website with online purchases. With each purchase, we need to create a shopping cart and a checkout receipt on the user interface side. Customers may also retrieve information on their previous orders. On the server side, we need to store all the relevant information on products, customers, and orders in a database. There are different ways to structure and organize the data, and decisions on what tables and attributes to have can impact the efficiency and usability of the database.

9.11.1   Linking Related Tables

It is tempting to throw all pieces of data into a big table that would look like a spreadsheet. But that is almost always wrong as a database design. A well-organized relational database usually consists of many different tables that are interrelated. These tables are usually produced by a process known as *database normalization*. It is beyond the scope of this text to cover normalization techniques or many other important aspects of database design. It suffices to say here that the purpose of normalization is to define table schemas in such a way that *anomalies*, operational difficulties and data inconsistencies, are eliminated and the efficiency and flexibility of the database are increased.

Figure 9.10 shows our database design for customer orders that calls for four related tables (**Ex:** `OrdersDatabase`):

FIGURE 9.10: Customer Order Tables

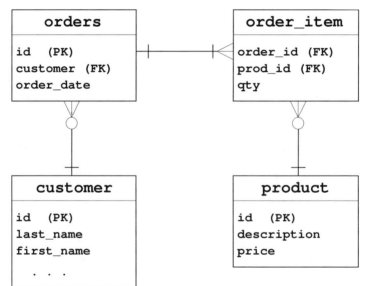

- `orders`—Each order entry has a unique `id` that serves as the primary key (PK), a `customer` to whom the order belongs, and an order date. The `customer` attribute refers to a specific customer, via the `id` key in the `customer` table. An attribute that refers to an entry, through a primary or unique key, in another table is known as a *foreign key*. Skillful use of foreign keys to interrelate tables can improve the organization of a database.

- `order_item`—Each entry in this table connects a quantity of a specific product to an order id. Multiple entries with the same `order_id` can associate several product items with any single order. The `order_id` attribute is a foreign key, so is the `prod_id`.

- `product`—Products for sale.

- `customer`—Known customers for the website.

The four tables in Figure 9.10 are further characterized by graphical *cardinality relationships*:

- Each `order` entry is associated with one customer entry.

- Each `customer` entry may be associated with zero or more `order` entries.

- Each `order` entry is associated with one or more `order_item` entries.

- Each `order_item` entry is associated with one `order` entry and one `product` entry.

- Each `product` entry may be associated with zero or more `order_item` entries.

The MySQL code for these tables is as follows (`orders.sql` in **Ex: OrdersDatabase**).

```
drop table IF EXISTS order_item, orders, customer, product;

CREATE TABLE customer
(   id VARCHAR(10), lastname VARCHAR(25),
    firstname VARCHAR(15), email VARCHAR(35),
    address VARCHAR(35), city VARCHAR(15),
    state CHAR(2), zip5 CHAR(5), zip4 CHAR(4),
    PRIMARY KEY (id))  ENGINE = InnoDB;

CREATE TABLE product
(   id VARCHAR(12), description  VARCHAR(256),
    price DECIMAL(7,2), PRIMARY KEY (id)
)  ENGINE = InnoDB;



CREATE TABLE orders
(   id VARCHAR(10), customer VARCHAR(10),
    order_date DATE NOT NULL, PRIMARY KEY (id),
    FOREIGN KEY (customer) REFERENCES customer (id)
)  ENGINE = InnoDB;

CREATE TABLE order_item
(   order_id VARCHAR(10), prod_id VARCHAR(12),
    qty SMALLINT DEFAULT 1,
```

```
FOREIGN KEY (order_id) REFERENCES orders (id)
   ON DELETE CASCADE ON UPDATE CASCADE,          -- (A)
FOREIGN KEY (prod_id)  REFERENCES product (id),
PRIMARY KEY (order_id, prod_id)
) ENGINE = InnoDB;
```

The MySQL InnoDB storage engine supports and enforces the FOREIGN KEY constraint, which is given in the general form

```
FOREIGN KEY (attr_1, ...) REFERENCES parent_table
(attr_a, ... )
```

The referencing table (the one with the FOREIGN KEY clause) is called the *child table*, and the table being referenced is called the *parent table*. Note the following conditions when using the FOREIGN KEY constraint:

1. Both parent and child must be InnoDB tables.

2. The parent table must already exist when defining the child table.

3. The corresponding attributes in both parent and child tables must have compatible or same data types.

4. The set of referenced attributes must correspond to a PRIMARY or UNIQUE key in the parent table.

The ON DELETE and ON UPDATE clauses (line A) will be explained in Section 9.11.4.

9.11.2   Displaying Orders in HTML

By extracting information from the four order-related tables we can display information for the Web user with relative ease. Say we want to display information about a particular order from our database.

Given the order id $order, we can extract the customer name as follows:

```
SELECT CONCAT(firstname, ' ', lastname) FROM customer
WHERE id=(SELECT customer from orders WHERE id='$order')
```

The MySQL built-in function CONCAT concatenates strings. The SELECT inside the WHERE clause is known as a *subquery*.

To obtain the product and pricing information, a query similar to

```
SELECT product.id AS Product_ID,
       product.description AS Description,
       product.price AS Price,
       qty AS Quantity,
       qty*product.price AS Amt
FROM order_item, product
WHERE order_id="ord_01009" && prod_id=product.id;   -- (B)
```

can be used. Here we display information for order ord_01009 from the join of the order_item and product tables by choosing entries with product id contained in the order (line B). Figure 9.11 shows the HTML display of the resultset.

FIGURE 9.11: Displayed Order Information

Product_ID	Description	Price	Quantity	Amt
prod_99004	Tennis Racquet	95.85	1	95.85
prod_99008	Tennis Balls	3.85	4	15.40
prod_99009	Tennis T-shirt	15.85	1	15.85

9.11.3 Managing Orders

The display shown in Figure 9.11 is needed when a customer inquires about a particular order made in the past. The HTML code for the order display can be produced by a PHP function orderTable (see **Ex: OrderUserInterface**), and here is a sample call to it:

```
$ot = orderTable("ord_01009");
if ( $ot ) echo $ot;
else {  echo "Order display failed."; }
```

The orderTable function retrieves information from the database (line C) and calls htmlOrder to obtain the HTML code (line D).

```
function orderTable($order)
{ global $db_obj;
  $query = "SELECT concat(firstname, ' ', lastname)
    FROM customer WHERE
    id=(SELECT customer from orders WHERE id='$order')";
  if ( !($customer=value_from_db($query)) )
      return "<p>No such order: $order</p>";
  $ans="<p>Customer: $customer,  Order: $order</p>\n";
  $query = "SELECT product.id as Product_ID,
    product.description AS Description,
    product.price AS Price, qty AS Quantity,
    qty*product.price AS Amt
    FROM order_item, product
    WHERE order_id='$order' && prod_id=product.id";
  if ( ($result=$db_obj->query($query)) )                // (C)
```

```
  {  $ans.=htmlOrder($result,"Amt","db");                  // (D)
     $result->close(); return $ans;
  }
  else return FALSE;
}
```

The `htmlOrder` function is similar to the `htmlTable` function discussed earlier (Section 9.8.2). But, `htmlOrder` needs to sum up the amounts (line E), format it correctly (line F), and display a total at the end of the table (line G).

```
function htmlOrder(&$result_obj, $amt_name, $class)
{ if ( $result_obj->num_rows == 0 ) return "";
  $tb = "<table border='1' class='" .$class. "'>\n";
  $result_obj->data_seek(0);
  $header_done = false;
  $total=0.0;   $cols=0;
  while( $data = $result_obj->fetch_assoc() )
  {  if (!$header_done) // table headers
     {  $tb .= '<tr>';
        $cols=count($data);
        foreach($data as $attr => $value)
        {  $tb .= "<th>$attr</th>";  }
        $tb .= "</tr>\n";
        $header_done = true;
     }
     $total += $data[$amt_name];                            // (E)
     $tb .=("<tr><td>". implode("</td><td>", $data));
     $tb .= "</td></tr>\n";
  }
  $total=number_format($total, 2, '.', '');                 // (F)
  $tb .="<tr class='total'><td colspan='" .($cols-1)        // (G)
        . "'>Total: </td><td>$total</td></tr>";
  return ($tb . "</table>\n");
}
```

Managing orders is of course more than just displaying them. We also need to change the quantity for products in an order (line H) and if the quantify becomes zero, we need to remove the item from the order (line I). By the way, we always want to make sure that `$qty` is an integer.

The PHP functions `intval($str)` and `floatval($str)` convert a given string into an integer and float value, respectively. For example,

```
$str = '232.98 The';
echo floatval($str);   // 232.98
echo intval($str);     // 232
```

When removing a product item from an order (line J), we will check to
see if the last item on the order has been removed (line K) and if so, remove
the empty order from the orders table (line L).

```
function changeQty($order, $product, $qty)        // (H)
{ global $db_obj;
  if ( ($qty=intval($qty)) == 0 )                 // (I)
  { removeItem($order, $product);
    return;
  }
  elseif ( $qty > 0 )
  { $query="INSERT INTO order_item VALUES
        ('$order', '$product', $qty)
        ON DUPLICATE KEY UPDATE qty=$qty";
    return ( $db_obj->query($query) );
  }
}

function removeItem($order, $product)             // (J)
{ global $db_obj;
  $query="DELETE FROM order_item WHERE
    order_id='$order' && prod_id='$product'";
  if ( $db_obj->query($query) &&
      $db_obj->affected_rows == 1 ) // one row
  { $query="SELECT * FROM order_item WHERE
            order_id='$order'";
    $result=$db_obj->query($query);
    if ($result && $result->num_rows == 0 )       // (K)
    { $result->free();
      return (removeOrder($order));               // (L)
    }
    if ($result) $result->free(); return true;
  }
  return false;
}
```

When adding a product item to an order (line M), we first need to make sure
the $qty is correct (line N) and then we need to either add a new entry in the
order_item table or update the quantity if the item is already on the table.
The query with the ON DUPLICATE KEY UPDATE clause is just right for this
situation (line O).

```
function addItem($order, $product, $qty)          // (M)
{ global $db_obj;
  $qty=intval($qty); if ($qty <=0) return FALSE;  // (N)
  $query="INSERT INTO order_item VALUES           // (O)
```

```
    ('$order', '$product', $qty)
    ON DUPLICATE KEY UPDATE qty=qty+$qty";
  return ( $db_obj->query($query) );
}
```

When a purchase is finalized (paid for), a website will want to update the orders tables and enter the new order in the orders table. (See Section 9.14 for online payments.)

When recording a new order, the function enterOrder is called with a unique order id, the customer id, and the items in $cart (line P) to perform:

1. Creating a new entry on the orders table (line R)

2. Adding each item in $cart to the order_item table (line S and T)

All queries involved in the tasks must succeed or fail as a unit. Such a group of queries is known as a *transaction*. MySQL InnoDB tables support transactions.

We begin a transaction by disabling autocommit (line Q) so that each executed query will not change the data in the database until an explicit commit is issued. Any changes made by an executed but yet uncommitted query in a transaction are only visible to subsequent queries in the same transaction. Also, the RDBMS allows no concurrent queries from others to interfere with the tables used in a transaction[3].

If all is well at the end of a transaction, we can commit the group of queries at once by calling $db_obj->commit() (line V) or, if something goes wrong, we can cancel all the queries by calling $db_obj->rollback() (lines U and W).

```
function enterOrder($order, $customer, &$cart)      // (P)
{ global $db_obj;
  if ( empty($cart) ) return false; // empty order
  $err=FALSE;
  $db_obj->autocommit(FALSE);                       // (Q)
  if ( newOrder($order, $customer) )                // (R)
  { foreach($cart as $product=>$qty)                // (S)
    { if (! addItem($order,$product,$qty))          // (T)
      { $err=TRUE; break; }
    }
    if ( $err ) $db_obj->rollback();                // (U)
    else $db_obj->commit();                         // (V)
  }
  else { $db_obj->rollback(); return FALSE; }        // (W)
  $db_obj->autocommit(TRUE);                        // (X)
  return !$err;
}
```

---

[3]See the ACID rule for database transactions for more information.

```
function newOrder($order, $customer)
{ global $db_obj;
  $query="INSERT INTO orders VALUES
          ('$order', '$customer', CURRENT_DATE)";
  return ( $db_obj->query($query) );
}
```

We turn `autocommit` back on before returning from the function (line X). To test `enterOrder`, you can use the simple code:

```
$cart=array("prod_99007"=>4, "prod_99008"=>4, "prod_99004"=>4);
if (enterOrder("ord_30994", "cus_12002", $cart))
      echo orderTable("ord_30994");
else echo "failed";
```

9.11.4 The `FOREIGN KEY` Constraint

The `FOREIGN KEY` constraint in a table schema induces a parent and child table relationship (Figure 9.12). The InnoDB storage engine enforces the `FOREIGN KEY` constraint and will not allow the insertion of a child row if it references a nonexistent parent row. For the same reason, neither will InnoDB allow the removal of a parent row if it still has one or more child rows referencing it.

FIGURE 9.12: `FOREIGN KEY` Induced Parent-Child Relationships

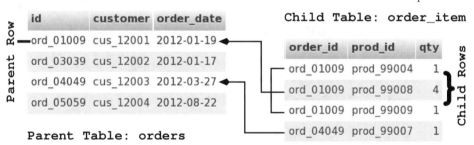

The `ON DELETE CASCADE` clause (Section 9.11.1, line A) in the `order_item` table schema tells InnoDB to allow the deletion of an order (a parent row) from the parent `orders` table (line Q) by automatically removing all its child rows from the `order_item` table.

```
function removeOrder($order)                      // (Q)
{ global $db_obj;
  $query="DELETE FROM orders WHERE id='$order'";
  return ( $db_obj->query($query) );
}
```

On the other hand, the ON DELETE SET NULL option means, instead of deleting child rows, InnoDB will set foreign key references in any child row to NULL when its parent row is deleted.

The ON DELETE clause takes care of deletion of parent rows. But what about updates made to parent rows? There is also the ON UPDATE CASCADE clause (Section 9.11.1, line A), which means that if the value of a referenced key attribute in a parent row, id of an orders table entry for example, is changed, the referencing key values in all child rows, the child row order_id values in our example, will be automatically updated accordingly. The other option is ON UPDATE SET NULL with the obvious semantics.

9.12   Displaying Shopping Carts

As a user adds items to an online shopping cart, a PHP associative array $cart, as part of the session state, can hold the product and quantity information.

```
$_SESSION['cart']=array();
$cart=&$_SESSION['cart'];    // reference to session state
$cart[$product]=$qty;        // adds product to cart
```

The $cart array and the product table can then be used to produce the HTML code for displaying the familiar cart to the end user as needed (**Ex: OrderUserInterface**). Figure 9.13 shows the display of a generated shopping cart with the array $cart=array("prod_99007"=>2, "prod_99008"=>3);

FIGURE 9.13: Shopping Cart Display

Product_ID	Description	Price	Quantity	Amt
prod_99007	Tennis Shoes	75.95	2	151.90
prod_99008	Tennis Balls	3.85	3	11.55
			Total:	**163.45**

Update   Checkout

The cartTable function takes the items in the $cart array and a form action URL (line 1) and constructs an HTML form containing a table of items. Each row of the table represents one item in the shopping cart where the quantity can be modfied. A UNION of SELECT queries is used to obtain, from the database, the desired resultset (lines 2 through 3), which is passed to the function htmlCart for the HTML code (4).

```
function cartTable(&$items,$action)              // (1)
{ global $db_obj;
  $query="";
  if (empty($items))
     return "<p>Shopping cart is empty.</p>";
  foreach($items as $prod=>$qty)                 // (2)
  { $query .= "UNION SELECT id as Product_ID,
           description AS Description,
           price AS Price, $qty AS Quantity,
           $qty*price AS Amt
           FROM product  WHERE id='$prod'";      // (3)
  }
  $query=substr($query,6);
  if ( ($result=$db_obj->query($query)) )
  {  $ans=htmlCart($result,"Product_ID", "Amt", // (4)
                   "Quantity", $action, "db");
     $result->close(); return $ans;
  }
  else return FALSE;
}
```

The `htmlCart` is similar to the `htmlOrder` function (Section 9.11.3) except that each Quantity column cell contains an html input element like this:

```
<input size='3' name='cart[product_id]' value='qty'
   type='number' step='1' min='0' required='' />
```

When the update button is clicked, the formdata (`$_REQUEST['cart']`) submitted to the designated `$action` page can easily be used to update the cart (`$cart`) kept in the session state.

```
if (!empty($_POST['update']) && !empty($_POST['cart']))
{ $newcart=&$_POST['cart'];
  foreach($cart as $prod=>$qty)
  { $nqty=0;
    if (!empty($newcart[$prod]))
       $nqty=intval($newcart[$prod]);
    if ( $nqty <= 0 ) { unset($cart[$prod]); }        // (5)
    else if ( $nqty!=$qty ) { $cart[$prod]=$nqty; }   // (6)

  }
}
```

It is important to be careful when accepting user input into the session state or a database query. Make sure to check the validity of the user input before using it (lines 5 and 6). The complete example (**Ex: Cart**) can be found at the DWP website.

Our simple example here does not take into account price changes or discounts, nor does it record payment types. In practice, these would be important. See the chapter-end exercises for improving the order tables in these respects.

9.13   Handling Query Results and Errors

We already know how a database connection is made with a new `mysqli` object.

```
$db_obj = new mysqli($host, $uid, $pass, $dbname);
if (mysqli_connect_errno())  // connection failed
{  printf("Can't connect to $host $dbname. %s:%s\n",
   mysqli_connect_errno(), mysqli_connect_error());
   exit();
}
```

In case of a connection error, `$db_obj` is `FALSE` and the functions `mysqli_connect_errno()` (error number) and `mysqli_connect_error()` (error message) provide diagnostic information. Of course, instead of `printf`, we may use `error_log` instead (Section 8.5).

Having established a connection, we can issue queries using the object `$db_obj`:

```
$result=$db_obj->query($some_query);
```

If `$result` is `FALSE`, then the query has failed and the error message is available as `$db_obj->error`.

Depending on the query type, a successful query may or may not produce a resultset. For a successful resultset query (`SELECT`, `SHOW`, `DESCRIBE`, or `EXPLAIN`), `$result` is a resultset object, and the number of records (rows) and attributes (columns) in the resultset are given by

```
$result->num_rows
$result->field_count
```

For a successful non-result-set query (`INSERT`, `UPDATE`, `REPLACE`, or `DELETE`), the `result` variable is set to `TRUE` and the quantity

```
$db_obj->affected_rows
```

tells you the number of rows affected.

A resultset object belongs to the class `mysqli_result` that provides properties and methods for retrieving and handling meta data and records in the resultset. These include

- `$result->fetch_assoc()`—Returns the next record as an associative array.

- `$result->fetch_row()`—Returns the next record as an enumerated array.

- `$result->fetch_object()`—Returns the next record as an object.

- `$meta=$result->fetch_field_direct($i)`—Returns the metadata such as type and length for the particular attribute `$i` as properties in the returned object `$meta`. For example, we can use `$meta->type` and `$meta->length`. See MySQL documentation for details.

Here is a useful function that obtains a single value from the database produced by a given query `$q`.

```
function value_from_db($q)
{   global $db_obj;
    $r = $db_obj->query($q);
    if (  $r && $r->num_rows == 1 )
    {   $row = $r->fetch_row();
        return($row[0]);
    }
    else return FALSE;
}
```

It is important to note that a successful query producing a nonempty resultset is no proof that you have the correct query. The query may simply work correctly for the database instance you used to test it, or the resultset is missing records that should be included, or the resultset includes records that should be excluded, or the query may break if a table becomes empty or certain attributes are NULL. A developer must examine the logic of the query carefully, avoid any assumptions of values in the database, and test each query extensively.

9.13.1 Paginated Display of Larger Result Sets

It is not unusual for a resultset to contain many records. Displaying such results in a webpage becomes a problem. Often, the solution is to break the result into pages and allow the user to easily browse through the pages.

We will implement a `Paginate` class in PHP that can be used as follows. First we set the pagination parameters:

```
$tb="orders";            // (1) table name
$str="order_date DESC";  // (2) sort string
$rpp=2;                  // (3) records per page
$page_total=0;           // (4) total page count, if known
```

Here we want to display all records on the `orders` table (line 1) sorted by order_date, most recent first (line 2), with two records per page (line 3). The number of pages needed, set to 0 for now, will be determined later.

We then create a `Paginate` object (**Ex:** `PaginateDemo`) with these parameter values:

```
$paging_obj=new Paginate($rpp,$tb,$str,$page_total);
```

Then we can call the `page_table` method to produce the table display for the page:

```
$paging_obj->page_table($page);
```

and call the `page_nav` method to produce the page-to-page navigation:

```
$paging_obj->page_nav($page);
```

Figure 9.14 shows two typical pages of our pagination scheme.

FIGURE 9.14: Paginated Display

Page 1 of orders

id	customer	order_date
ord_05059	cus_12004	2012-08-22
ord_04049	cus_12003	2012-03-27

Go to page: next, 1, 2, 3

Page 2 of orders

id	customer	order_date
ord_12345	cus_12003	2012-01-24
ord_01009	cus_12001	2012-01-19

Go to page: prev,next, 1, 2, 3

The `Paginate` constructor saves the incoming argument values in private variables as the object state. It also sets `$this->PAGE_TOTAL`, the total number of pages needed for the given table `$tb`, to an incoming nonzero `$tt` (line c), or to a computed value (lines a and b).

```
class Paginate
{ public function __construct($rpp,$tb,$str,$tt=0)
   {  $this->RECORDS_PER_PAGE=$rpp;
      $this->TABLE_NAME=$tb;
      $this->SORT_STR=$str;
      if ( $tt == 0 )
      { $query = "SELECT COUNT(*) FROM $tb";              // (a)
        $this->PAGE_TOTAL= ceil(value_from_db($query)    // (b)
                     /$this->RECORDS_PER_PAGE);
      }
      else   $this->PAGE_TOTAL=$tt;                       // (c)
   }

   private $RECORDS_PER_PAGE, $TABLE_NAME,
           $SORT_STR, $PAGE_TOTAL;
```

After instantiation, a `Paginate` object can create the HTML table code for any page directly from the database with the `page_table` method:

```
public function page_table($page)
{ global $db_obj;
  $start = ($page-1) * $this->RECORDS_PER_PAGE;         // (d)
  $query = "SELECT * FROM $this->TABLE_NAME
             ORDER BY $this->SORT_STR
             LIMIT $start, $this->RECORDS_PER_PAGE";   // (e)
  $result=$db_obj->query($query);
  if ( $result ) echo (htmlTable($result,"db"));
  else return FALSE;
}
```

The SQL LIMIT clause (lines d and e) makes it simple to select the records for any given page ($page). In general, the LIMIT clause limits the resultset to no more than *count* records, beginning with the given *start_row*:

```
LIMIT start_row, count
LIMIT 10,20 -- records 10 up to 29
```

The `htmlTable` function was described in Section 9.8.2.

The page navigation (see Figure 9.14) is produced by the `page_nav` method, which links back to the same page with query string values indicating the page total (`tt`) and target page (`page`).

```
public function page_nav($page)
{ $total=$this->PAGE_TOTAL;
  echo "<p class='pagenav'>Go to page: ";
  if ($page > 1) echo "<a href='?tt=$total&page=".
                        ($page-1)."'>prev</a>,";
  if ($page < $total) echo "<a href='?tt=$total&"
                "page=".($page+1)."'>next</a>,";
  for ($i=1; $i<$total; $i++)
  { if ($i==$page)
    echo "<span class='current'> ".$i."</span>,";
    else
    echo "<a href='?tt=$total&page=".$i."'> "
       . $i . "</a>,";                                // (f)
  }
  echo "<a href='?tt=$total&page=".$total."'> "
       .$total."</a></p>\n";
}

}
```

Here is a typical link generated by the code on line **f**:

```
<a href='?tt=12&page=6'> 6</a>
```

Clicking on this link goes to page 6 of 12 pages. The complete example (**Ex:** PaginateDemo) can be found on the DWP website.

9.14   Receiving Payments

The orders, products, customer, and order_item tables form the core of any database required for selling on the Web. To receive payments online, a website must first establish an account with a *payment gateway*.

A payment gateway interfaces the customer order information provided by the selling website to the online payment processing system established by banks and credit card associations (for example, Visa^{TM} or MasterCard^{TM}). After obtaining success or failure of payment authorization from the payment processing system, the gateway redirects back to an appropriate order processing program at the selling website to complete or cancel the sale. Thus, you may think of a payment gateway as the online equivalent of a checkout station.

To receive payments online, a Web store needs

- A business bank account.

- A merchant account to receive credit card payments—A merchant account is a bank account established with a payment processor for the settlement of credit card transactions. Any merchant who wants to take credit card payments from a certain type of credit card must establish a merchant account dealing with that brand of credit card. Online merchants need a *Card Not Present Merchant Account.*

- A payment gateway account to process card transactions automatically online.

Many payment gateways compete for a share of the online payment processing pie. Most charge a one-time account set-up fee and then a monthly fee. Others, such as PayPal^{TM}, charges no set up or monthly fee but opt for a percentage of every sale made through their service. It is easy to understand why PayPal-type services are popular with small online stores. We will illustrate how a selling website interfaces order processing with payment processing by using PayPal as a concrete example. Other payment gateways follow similar approaches:

1. Secure transaction—A customer completes an order on a Web store. This must be done on a secure webpage, under HTTPS, to protect personal information including credit card numbers. HTTPS and TLS should always be used for online payment processing.

2. Order information to payment gateway—When a customer clicks on the checkout button, order information such as order id and total payment amount required is securely transmitted to the payment gateway, following interface protocols specified by that particular gateway.

3. Credit request—The payment gateway checks the information received from the Web store, collects credit card information from the customer if necessary, determines what company manages that credit card, and transmits an authorization request for the card to be charged.

4. Approval/Denial—The customer's credit card company, usually a credit card issuing bank, validates the card and the account. The payment gateway gets back from the payment processing system either an *authorized* response or some other code indicating the cause of failure.

5. Payment result processing—The payment gateway redirects back to a specified program at the store website with authorization approval or failure and perhaps other data related to the sale. A payment success program and a payment failure program are usually needed on the store website to receive the redirection.

6. Settlement or credit return—Web store managers can login to the payment gateway account to issue refunds or to collect payments after the delivery of goods or services.

Figure 9.15 illustrates the way a merchant website interfaces to a payment gateway through three webpages:

FIGURE 9.15: Interfacing to a Payment Gateway

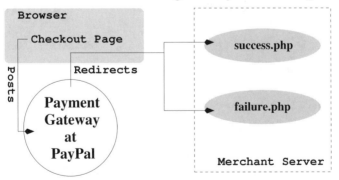

1. A checkout page—A customer clicking a checkout or enter order button initiates the posting, over HTTPS, of relevant order data to the payment gateway. The customer often also interacts with the payment gateway to provide billing, shipping and other information. To make the payment

gateway completely transparent to the customer, a site may decide to collect all the necessary information before contacting the payment gateway. All the shopping cart and other information related to the order are kept in the session state.

2. A payment success page—The payment gateway will redirect back to the success page if payment authorization is completed successfully. We will show a php page for payment success that will also enter the sale in the orders database.

3. A payment failure page—The payment gateway will redirect back to the failure page if payment authorization failed for some reason. This page will inform the customer of the payment failure and suggest what to do next. Any order information relating to the failed sale will not be entered into the database.

When the customer is ready, clicking a "Proceed to Checkout" button will lead to a checkout confirmation page similar to the one in Figure 9.16. This checkout page contains a form that interfaces to the payment gateway. Clicking the "Make Payment" button posts the required data to the gateway server for processing.

FIGURE 9.16: Checkout

Product_ID	Description	Price	Quantity	Amt
prod_99007	Tennis Shoes	75.95	2	151.90
prod_99008	Tennis Balls	3.85	3	11.55
			Total:	**163.45**

Make Payment   Edit Cart

To interface the cart shown in Figure 9.16 to a PayPal gateway,

```
https://www.paypal.com/cgi-bin/webscr}
```

we need to specify interface commands for sending cart details (lines A and B), the business ID (line C), a payment success callback URL (known as return, line D), and a failure URL (known as cancel_return, line E):

```
<input type="hidden" name="cmd" value="_cart" />      (A)
<input type="hidden" name="upload" value="1">         (B)
<input type="hidden" name="business"
       value="business-ID>" />                         (C)
```

```
<input type="hidden" name="return"
      value="https://success_url.php" />              (D)
<input type="hidden" name="cancel_return"
      value="https://failure_url.php" />              (E)
<input type="hidden" name="image_url"
      value="https://business_logo_url" />            (F)
<input type='hidden' name="item_name_1"              (G)
      value="Tennis Shoes">
<input type='hidden' name="amount_1" value="151.90">
<input type='hidden' name="quantity_1" value="2">
<input type='hidden' name="item_name_2"
      value="Tennis Balls">
<input type='hidden' name="amount_2" value="11.55">
<input type='hidden' name="quantity_2" value="3">     (H)
```

A business logo (line F) can be supplied to make the PayPal payment pages better connected visually to the website of the business. The information on lines A through F are fixed. The shopping cart details given (lines G through H) must be computed based on the customer's cart.

The `form` code for the checkout page can be produced by a PHP function `htmlCheckout` (**Ex: ManageOrder**) similar to `htmlCart` (Section 9.12) that generates a form with the above hidden input fields to be posted to a given gateway `$gateway` with `$gwValues` the fixed part of the hidden input (line 1).

```
function htmlCheckout(&$result_obj, $item_name,
    $amt_name, $qty_name, $gateway, $gwValues, $class)  // (1)
{ if ( $result_obj->num_rows == 0 ) return "";
  $payform = "<form method='post' action='$gateway'>     // (2)
          <table border='1' class='" . $class . "'>\n";
  $result_obj->data_seek(0);
  $header_done = false; $total=0.0; $cols=0;
  $item_count=0; $gwValues .= "\n";
  while( $data = $result_obj->fetch_assoc() )
  { $item_count++;
    if (!$header_done)  // cart table headers
    { $payform .= '<tr>'; $cols=count($data);
      foreach($data as $attr => $value)
      { $payform .= "<th>$attr</th>";  }
      $payform .= "</tr>\n"; $header_done = true;
      reset($data);
    }
    $n=$data[$item_name];   $q=$data[$qty_name];      // (3)
    $amt=$data[$amt_name];  $total += $amt;
    $gwValues .= "<input type='hidden' value=\"$n\"
          name=\"item_name_$item_count\">
      <input type='hidden' value=\"$amt\"
```

```
                name=\"amount_$item_count\">
        <input type='hidden' value=\"$q\"
                name=\"quantity_$item_count\">\n";              // (4)
  // table rows
    $payform .= '<tr>';
    foreach($data as $attr => $value)
    { $payform .= "<td>$value</td>"; }
    $payform .= "</tr>\n";
  }
  $total=number_format($total, 2, '.', '');
  $payform .="<tr class='total'><td colspan='" .($cols - 1)
        . "'>Total: </td><td>$total</td></tr>";
  return ($payform . "</table>\n$gwValues
    <input type='submit' value='Make Payment' /></form>\n");
}
```

The form action is set to the given gateway URL (line 2). The cart details are placed in hidden input fields (lines 3 through 4).

As with most gateways, PayPal will interact with the customer to collect payment and then redirects back to the `return` callback for payment success or to the `cancel_return` for failure.

The `cancel_return` script can simply acknowledge failure and offer the customer to try again, perhaps through a link to the cart display. Depending on the type of goods or services involved, the `return` script usually needs to perform a number of tasks, such as

1. Obtaining payment transaction data from the gateway

2. Verifying correctness of payment amount and currency

3. Entering order and payment data into the website database

4. Sending order details by email to the customer

5. Displaying order confirmation page

6. Initiating order fulfillment

Often the `return` script can securely request the gateway, via HTTPS, to transmit payment data for the successful transaction and then to use the data in the above steps. PayPal, for example, provides a *Payment Data Transfer* (PDT) service for exactly this purpose and it can provide a complete set of payment data, including total amount, currency, type, date, payer ID, as well as data for each order item.

Figure 9.17 depicts the PayPal PDT flow:

1. Customer makes payment on PayPal.

FIGURE 9.17: PayPal PDT Flow

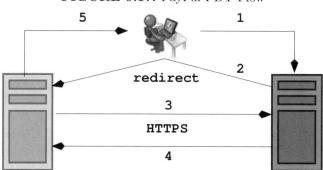

2. When payment is successfully completed, the PayPal server redirects the customer's browser to the `return` script on the merchant server and sends a per-transaction token.

3. The `return` script accesses the PayPal server using HTTPS POST following the PDT protocol.

4. The PayPal server replies to the HTTP POST with URL-encoded payment data.

5. After checking and processing the payment data and the order, a receipt is displayed to the customer.

9.14.1 HTTP Requests from PHP

In order to access and receive data from other Web servers, a server-side script often needs to initiate HTTP requests. PHP has good support for this.

The following code uses the PEAR HTTP_Request2 package (line I) to make an HTTP POST request for the PayPal PDT. We set the method of the `HTTP_Request2` object to `POST` (line II), formulate the request body (lines III-IV), send the request to PayPal (line V), and then place the response string into a PHP array (line VI).

```
require_once('HTTP/Request2.php');                     // (I)
$requrl='https://www.paypal.com/cgi-bin/webscr';
$req = &new HTTP_Request2($requrl);
$req->setConfig('ssl_verify_peer', FALSE);
$req->setMethod(HTTP_Request2::METHOD_POST);           // (II)
$auth_token = "<use string provided by PayPal>";       // (III)
$req->addPostParameter("at" , $auth_token);
$req->addPostParameter('cmd', '_notify-synch');
$req->addPostParameter('tx', $_GET['tx']);             // (IV)
```

```
$response = $req->send();
if (200 != $response->getStatus()) // HTTP ERROR    // (V)
{  echo "Sorry: Order Processing failed!\n";
   exit(1);
}
else
{   $msg = $response->getBody();
    $dataArray=pdtArray($msg);                       // (VI)
    echo "<pre>\n"; print_r($dataArray);
    echo "</pre>\n";
}
```

If you use a PHP version with the *HTTP extension* installed, then you do not need to use PEAR. The function `pdtArray` and the full `return` script (**Ex:** PDT) `processOrder.php` as well as the cancel return script `failedOrder.php` can be found at the DWP website.

Interfacing to PayPal is just an example. But other payment gateways follow the same principles and work in very similar ways. After verifying correct payment, an order processing script can proceed to enter the finalized order into the database. Order management functions, discussed in Section 9.11.3, can be used.

9.15   Prepared Statements

When a database application program, one written in PHP for your website for example, needs to issue the same query repeatedly with different values, *a prepared statement* can be more efficient and convenient. A prepared statement is simply a query template that is ready to be executed when its *parameters* (blanks) have been bound (filled) with values.

You first create a prepared statement, bind parameters to it, and then run it repeatedly with different parameter values. Because the prepared statement query has been parsed and is ready to run, you gain speed and efficiency.

For example, we can write a version of the `addItem` function (Section 9.11.3) for order management that uses a prepared statement so it will work more efficiently when adding multiple items. The increase in efficiency is, of course, directly proportional to the number of items added.

The `addItem_p` function creates a prepared statement (lines B to C) if one has not been done already (line A), using question marks to mark the blanks where parameter values are needed. If the `prepare` method of the `mysqli` object `$db_obj` is successful, it returns a prepared statement object or `FALSE` otherwise.

```
function addItem_p($order, $product, $qty)
{ global $db_obj, $stmt, $p_o, $p_p, $p_q;
```

```
    $qty=intval($qty);
    if ( !$stmt )                                   // (A)
    { $stmt = $db_obj->prepare(                     // (B)
        "INSERT INTO order_item VALUES (?, ?, ?)
        ON DUPLICATE KEY UPDATE qty=qty+?");         // (C)
      if ( $stmt )                                  // (D)
      { if (! $stmt->bind_param('ssii',
                    $p_o, $p_p, $p_q, $p_q)
            ) $stmt=FALSE;                          // (E)
      }
    }
    if ( $stmt && $qty > 0 )
    { $p_o=$order; $p_p=$product; $p_q=$qty;        // (F)
      return $stmt->execute();                      // (G)
    }
    return FALSE;
}
```

Variables are bound to a prepared statement with the `bind_param` method
(lines D through E) called with a a *type string* followed by variables, one for
each of the parameters (blanks). Note that the parameters must be bound
with variables and not with literal values. In fact, the variables do not need to
have values at the time of the `bind_param` call. The type string is a sequence
of type characters indicating the types of the values of the PHP variables, one
character for each parameter variable: `s` (string), `i` (integer), `d` (double), and
`b` (binary blob).

After the parameters are bound to variables, a prepared statement is a
query ready to run as soon as the parameters are assigned actual values
(line F). To run it, we call its `execute` method (line G), which returns `TRUE`
(success) or `FALSE` (failure).

For a prepared statement with a value producing query, the method
`$stmt->get_result()`[4] can be called to obtain the resultset, and the property `$stmt->$num_rows` gives the number of rows in the resultset. If the query
is non-value-producing, the property `$stmt->$num_rows` gives the number of
rows in the database affected.

With a prepared statement, you have an efficient way to collect query
results. After executing a prepared statement, you can call

`$stmt->bind_result(`*$var1, $var2, ...*`)`

to supply a variable for each resultset attribute. The call

`$stmt->fetch()`

puts attribute values from the next resultset record (row) into those variables.
In summary, the sequence of steps to use prepared statements are

---

[4]This method may not work in older versions of PHP.

1. Call `$stmt=$db_obj->prepare($query)` for a query with parameters.

2. Call `$stmt->bind_param("types", $v1, $v2, ...)` to bind variables to the parameters.

3. Assign values to each of the bound variables.

4. Call `stmt->execute()` to run the query.

5. Call `stmt->bind_result($rv1, $rv2, ...)`, then `stmt->fetch()` to obtain each row in the resultset. Or, alternatively, call `stmt->get_result()` to obtain a resultset object.

The complete example (**Ex:** `OrdersDatabase`) can be found at the DWP website. See the `mysqli_stmt` class documentation for more information.

9.16 Guarding against SQL Injection

Whenever you compose an SQL query from user input, there is the danger of *SQL injection attacks*. Consider the following php code for updating the password of a user:

```
$query = "UPDATE usertable SET pwd='$pwd' WHERE uid='$uid';";
```

where `$uid` and `$pwd` are from user-submitted formdata.

A malicious user might submit the following `$uid` string:

```
' or uid like '%admin%'; --
```

resulting in the query

```
UPDATE usertable SET pwd='...'
    WHERE uid='' or uid like '%admin%'; --;
```

which sets the password for any `uid` with `admin` in it to the password the user sent.

Or a malicious user might submit this `$pwd` string:

```
"hahaha', admin='yes', trusted=100 "
```

resulting in the query

```
UPDATE usertable SET pwd='hahaha', admin='yes',
    trusted=100 WHERE ...;
```

Even if there is no attack, we need to worry about user input containing semicolons, single quotes, or double quotes.

It is not hard to guard against SQL injection. We simply need to call

```
$str=$db_obj->escape_string($any_user_supplied_str);
```

on any user-supplied string to obtain a safe version of it to use in queries.

If a prepared statement is used, then its parameter values cannot be confused with its query syntax. Thus, using user-supplied values in a prepared statement is another way to defend against SQL injection. For example,

```
$stmt=$db_obj->prepare(
   "SELECT * FROM usertable WHERE username=?
      AND password=PASSWORD(?) AND active=1");

$stmt->bind_param("ss", $uid, $pass);
$uid=$_POST['uid']; $pass=$_POST['password'];
$stmt->execute();
if ( $stmt->$num_rows != 1 ) {  /* login failure * }
else {  /* login success */  }
```

9.17   The phpMyAdmin Tool

PhpMyAdmin uses the Web to provide a GUI tool for managing MySQL databases. This tool usually runs on the same server as the MySQL database server. After login to phpMyAdmin, a database administrator can access all databases under his/her control and perform database maintenance and update functions interactively. The tool also makes it easy to experiment with queries, consult MySQL documentation, import and export tables and databases, and generate PHP code for tested queries.

Many find the phpMyAdmin tool to be more powerful, easier to use, and much more user friendly than the command-line tool **mysql**. See Section 10.26 for more on phpMyAdmin.

9.18   PHP Built-In Database System

Recent versions of PHP include a built-in light-weight RDBMS called SQLite. According to its website,

> SQLite is a software library that implements a self-contained, serverless, zero-configuration, transactional SQL database engine. SQLite is the most widely deployed SQL database engine in the world. The source code for SQLite is in the public domain.

Use of SQLite within PHP requires no connection to an external database system, such as MySQL or login. It is suitable for working with small databases for simple applications. Starting with PHP 5.1, the *PHP Data Objects* (PDO) interface is used to access and manipulate SQLite3 databases.

The **sqlite3** is an application providing a command-line interface to the SQLite library that can evaluate queries interactively and display the results in multiple formats. On Linux, for example, the command

sqlite3 `myproject.db`

creates a new sqlite3 database. To access this database from PHP, use

```
try
{ $db_obj = new PDO('sqlite:/path/to/myproject.db');
} catch (Exception $e) { die($e); }
```

Then use the PDO object `$db_obj` and its methods to access and manipulate the database. See the PHP PDO documentation for details.

9.19   For More Information

The DWP website resource page contains links to information and documentation for SQL, MySQL, PHP interface to MySQL, SQLite, as well as other useful pages.

Chapter 11 presents LAMP, the Linux, Apache, MySQL, and PHP Web hosting environment.

9.20   Summary

A *Relational Database Management System* (RDBMS) uses tables (relations) to store data. It manages multiple databases for authorized users and provides concurrent network access to and efficient operations on the databases.

The *Structured Query Language* (SQL) is a standard API for RDBMSs and uses sentence-like declarations to specify queries for database operations. *Entity Relation Diagrams* (ERDs) are used to characterize attributes and relationships among physical and logical entities in a database application. ERDs lead to schemas, attribute names, and their possible values, which define tables.

The SELECT query can retrieve results from one or more tables and form a resultset that can have values from these tables and values computed from such values listed under specific column names. The resultset's size (number of records) can be limited and the records can be sorted. The clauses in a SELECT, or any other query, are given in a specific order. A general form for SELECT is

```
SELECT expr_1 as name_1, expr_2 as name_2, ...
   FROM table_1, table_2, ...
   WHERE conditions
   GROUP BY name_a, name_b, ...
```

```
ORDER BY name_A, name_B, ...
LIMIT offset, count
```

The PHP `mysqli` class provides excellent support for interfacing to MySQL databases. You can make database connections, issue queries, obtain query results, and detect errors. You can also start, commit, or rollback transactions and use prepared statements to efficiently repeat the same query with different parameters. Thus, PHP can form the ideal bridge between the Web and databases. The phpMyAdmin tool for database administrators is a good example. The PHP MySQL combination makes it easy to create database-driven websites, as shown by practical examples such as shopping carts, a database for customer orders, and an interface to the PayPal payment gateway.

PHP provides the ability to make HTTP requests from the server-side and also offers built-in support for SQLite, a light-weight RDBMS for simple tasks.

Exercises

Exercises for this chapter assume that you have access to a Web host *webhost* that supports PHP and MySQL and that you have a database *yourdb*.

9.1. What is a relation? A relation schema? A relation instance? Explain clearly.

9.2. What is SQL? DDL? and DML?

9.3. What queries produce resultsets? What queries do not?

9.4. What are the pieces of information needed to login and connect to a database?

9.5. What is a database *transaction*? Why is it needed? How is a transaction started? Ended?

9.6. Define the SQL to create a `member` table with at least these attributes: `uid`, `last_name`, `first_name`, `email`, `password`, `address`, `state`, `zip`, `status`. The `uid` is a unique login name. Everyone must have a different `email`. Passwords are stored in encrypted form, and `status` can be either active or inactive.

9.7. Write a PHP page with which you can enter records into the `member` table in Exercise 9.6. Use it to enter a dozen or so members into the table.

9.8. Revise the login system and its PHP and HTML files discussed in Section 8.10 to use the database table `member` in Exercise 9.6 instead of a flat password file. Make sure user registration, login, and forgot password features are supported. Remember: Only active members can login. Hint: See Section 9.9.

9.9. Implement a PHP webpage, protected by login of course, for updating user profiles that stores the changes in the member table.

9.10. (**Order Tables**) Take the online orders tables in Section 9.11 and the orders.sqlcode from **Ex: OrdersDatabase** and create the tables in *yourdb*. Enter a handful of records into each table.

9.11. Consider adding a payment_type attribute and an amt_received attribute to the orders table.

9.12. Refer to the orders tables. Specify the SQL query to find the "most popular products" sold within a given date range and the quantities sold for each.

9.13. Refer to the orders tables. Specify the SQL query to find the "most valuable customer" within a given date range and the amount of sales for each such customer.

9.14. (**Shopping Cart**) Implement webpages with products and a shopping cart to allow users to add products to the shopping cart, to view/modify the shopping cart, and to checkout. Your PHP script will update tables from the **Order Tables** exercise. Remember to use session control and login.

9.15. Add an "order history" feature to the **Shopping Cart** exercise.

9.16. Take the orderTable function (Section 9.11.3) and re-implement it using a prepared statement (Section 9.15).

9.17. (**Slide Show**) Set up a slidePicture table to hold records, each containing an image URL, a caption, and a short story for each picture. The URL locates an image file. Use this table to support a page that displays all active pictures in a scrolling thumbnail strip. Clicking on a thumbnail will display a larger image, the caption, and a short story.

9.18. (**Slide Show Admin**) Add to the **Slide Show** exercise an admin page (login required) to add new pictures (with caption and story); to delete old pictures (or make them inactive).

9.19. Products may have availability considerations in situations such as airline seats, one-of-a-kind goods, number of items in stock, and so on. How would you revise the design of the product table and how orders, carts, etc., are handled in relation to product availability?

9.20. The PHP access discussed in this chapter is particular to MySQL. Find out about the *PHP Data Object* (PDO), which allows you to use the same PDO methods to access different database systems.

Chapter 10

Web Hosting: Apache, PHP, and MySQL

The Web has become central to modern life. A key factor for this great success is the low cost of putting information on the Web. You simply find a Web hosting service to upload files for your website. Any Internet host can provide Web hosting if it has a good Internet connection and runs a *Web server* and other related programs. According to netcraft.com's March 2012 survey, among all Web servers, a full 65.24% are Apache^{TM}. Microsoft's server is a distant second with 13.81%.

A majority of Apache servers run on Linux systems. A Linux-Apache Web hosting environment usually also supports PHP and MySQL. The Linux, Apache, MySQL, and PHP combination (known as LAMP) works well to support Web hosting. A well-rounded Web developer needs to be introduced to these programs, their configurations, and operations.

In addition to understanding the big picture and the underlying principles, a practical hands-on approach guides you through the installation, configuration, testing, and administration of Apache, PHP, and MySQL so you can learn Web hosting and configuration through doing. Root access on your Linux is convenient, but not necessary.

For Web developers, materials in this chapter can also illuminate the context and operating environment for websites and their development.

10.1   What Is a Web Server?

A Web server is a program that listens to a specific networking port on the server host and follows the HTTP (HTTPS) to receive requests and send responses. The standard HTTP port is 80, but can be some other designated port such as 8080. The standard HTTPS port is 443.

In response to an incoming request, a server may return a static document from files stored on the server host, or it may return a document dynamically generated by a program, such as a PHP script, indicated by the request (Figure 10.1).

A single-thread server handles one incoming request at a time, while a multi-thread server can handle multiple concurrent requests. A server host

FIGURE 10.1: Web Server Overview

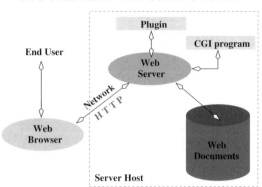

may have multiple copies of a Web server running to improve the handling of requests.

Many different brands of Web servers are available from companies and from open-source organizations. *GlassFish* is a free Web server that comes with the Java EE distribution from `glassfish.java.net`. The *Apache* Web server, available free from the *Apache Software Foundation* (`apache.org`), is the most popular Web server. Apache usually comes pre-installed on Linux distributions.

10.2   URL and URI

An important cornerstone of the Web is the *Universal Resource Locator* (URL, Section 1.4) that allows Web clients to access diverse resources located anywhere on the Web. For example, the HTTP URL

`http://dwp.sofpower.com`

leads to the DWP website. An HTTP URL (Figure 10.2) identifies a Web server running on a particular host computer and provides the following information:

- A *Universal Resource Identifier* (URI) that corresponds to a local pathname leading to a target resource (a file or program) stored on the server host

- An optional *pathinfo* indicating a target file/folder location as input data to the target resource

- An optional *query string* providing *key=value* pairs as input data to the target resource

FIGURE 10.2: HTTP URL Structure

`http://host:port`**`/folder/.../file`**`/path-info?query-str`

URI

The part of the URL immediately after the host:port segment (Figure 10.2) is referred to as the URI. The Web server uses the URI to locate the target resource, which can be a static page, an active page, or an executable program. A static page is returned directly in an HTTP response. Any pathinfo and query string are made available, as input, to an active page or an executable program. The resulting output is then returned in an HTTP response.

The set of files and directories made available on the Web through a Web server is known as its *document space*. The *document root* is the root directory for the document space, and it corresponds to the URI "/". In addition to the document root hierarchy, there can be other files and directories in the document space, for example, the `/cgi-bin` and the ~*userid* usually map to directories outside the document root hierarchy.

A Web server also works with other special directories (outside of its document space) for server configuration, passwords, tools, and logs. A URI is interpreted relative to the document root, cgi-bin, or another directory, as appropriate. The Web server can enforce access restrictions, specified in the Web server configuration files, on any file/folder in the document space.

10.3   Request Processing

For each incoming HTTP request, a Web server executes the following request processing cycle:

1. Accepts client connection, via TCP/IP (Section 1.2) for HTTP or via TLS/SSL for HTTPS (Section 10.20)

2. Receives the incoming request and processes the request by fetching the URI-indicated file or by invoking the URL-indicated program, feeding it any pathinfo and query string

3. Sends back an HTTP response

4. Closes the connection (or keeps it alive under HTTP1.1)

While processing a request, a busy website often will receive many new requests. It is normal to use multiple servers (multiprocessing) and/or multiple threads within the same server (multithreading) to handle concurrent requests.

10.4 Response and Content Types

For each incoming request, the Web server sends back a response containing the requested resource or an indication of error or some other condition.

An HTTP response has two parts: the headers and the body. The server specifies the `Content-Type` header to indicate the media type of the response body. Standard MIME (Multipurpose Internet Mail Extensions) content types (Chapter 1, Table 1.1) are used. The most common content type is `text/html`, but there are many other types. For a static file, the Web server uses the filename extension to infer its media type using a list often found on Linux systems in the file `/etc/mime.types`. The location of this content type list is configurable.

In case of dynamic content, those generated by server-side programs, the Web server relies on that program to set content type. We already know that PHP automatically sends the `text/html` content type header if a PHP script sends no content type header before outputting the response body.

10.5 Apache Web Server

Apache is the most popular Web server. You can download and install the Apache HTTP server from the Apache Software Foundation (`apache.org`) free of charge (Section 10.18). A Linux distribution will most likely have Apache already installed.

Apache is derived from the NCSA[1] httpd project and evolved through a series of code *patches* (thus, *a patchy* server). Apache, written in the C language, is open source and runs on almost all platforms. Apache is fast, reliable, multithreaded, full-featured, and HTTP/1.1 compliant. Although Apache 1.3 is still available, the most recent stable Apache 2 version is the one to use.

Apache has many components, including

- *Server executable*—The runnable program **httpd**

- *Utilities*—For server control, passwords, and administration

- *Files*—Including server configuration files, log files, password files, and source code files

- *Dynamic loadable modules*—Pre-compiled library modules that can be loaded into the **httpd** at run-time

- *Documentation*—About compilation, configuration, installation, and usage.

---

[1]National Center for Supercomputing Applications at the University of Illinois, Urbana-Champaign.

10.6   Accessing Linux Server Host

After files for a website have been developed and tested, they are placed on a Web host for access under a certain domain name. The easiest way is to use FTP/SFTP to transfer these files back and forth between the Web host and your own computer where the website files are developed and maintained.

Many Web hosts also allow SSH access where you are able to securely login to your hosting account and gain Linux Shell access, allowing you to perform a variety of tasks difficult or impossible to do with just FTP/SFTP. See the Appendix *Secure Communication with SSH and SFTP* at the DWP website for installing and using these applications.

Successful remote login via SSH to a Linux Web host results in your SSH window being connected to a *login Shell* running on the remote Linux (Figure 10.3).

FIGURE 10.3: Login via SSH

When you see the Shell prompt, you are ready to issue Shell commands. After you are done, you will need to logout from the remote Linux. To logout, first close any programs that you have been running and then issue the Shell-level command **exit** or **logout**. It is a good practice to first close all running programs manually instead of relying on the logout process to close them for you.

Here is a typical Shell prompt:

```
-bash-3.2$
```

After connecting to your Web host either by FTP/SFTP or SSH, you are positioned at your *account home folder*, which contains a subfolder, your Web folder, where your files for the Web should go. Commercial Web hosts provide

Web folders for individual domains, usually named `www` or `html`. Multi-user Linux systems in colleges and other organizations may provide individual users personal Web spaces under a `public_html` folder in the user's home folder.

10.6.1   Understanding the Linux *Shell*

The Linux Shell displays a prompt to signal that it is ready for your next command, which it then interprets and executes. On completion, the Shell re-signals readiness by displaying another prompt. The standard Shell for Linux is the *Bash*.

10.6.2   Entering Commands

You can give commands to the Shell to start application programs or manage files and folders. Virtually anything you want done in Linux can be accomplished by issuing a command to the Shell.

A command consists of one or more words separated by blanks. A *blank* consists of one or more spaces and/or tabs. The first word is the *command name*, and the remaining words of a command line are *arguments* to the command. A command line is terminated by pressing the RETURN (or ENTER) key. This key generates a NEWLINE character, the actual character that terminates a command line. Multiple commands can be typed on the same line if they are separated by a semicolon (;). For example, the command

ls *folder*

lists the names of files in a folder (directory) specified by the argument *folder*. If a directory is not given, **ls** lists the *current working directory* (Section 10.6.3).

Sometimes one or more *options* are given between the command name and the arguments. For example,

ls -F *folder*

adds the -F (*file type*) option to **ls**, telling **ls** to display the name of each file, or each *filename*, with an extra character at the end to indicate its file type: / for a folder, * for an executable, and so on.

At the Shell level, the general form for a command looks like

command-name [*options*] . . . [arg] . . .

The brackets are used to indicate *optional* parts of a command that can be given or omitted. The ellipses (. . .) are used to indicate possible repetition. These conventions are followed throughout the text. The brackets or ellipses themselves are not to be entered when you give the command.

Command options are usually given as a single letter after a single hyphen (-). For example, the *long listing option* for the **ls** command is -1. Such single-letter options can sometimes be hard to remember and recognize. Many Linux commands also offer full-word options given with two hyphens. For example, the `--help` option given after most commands will display a concise description of how to use that particular command. Try

```
ls --help
```

to see a sample display.

After receiving a command line, the Shell processes the command line as a character string, transforming it in various ways. Then, the transformed command line is executed. After execution is finished, the Shell will display a prompt to let you know that it is ready to receive the next command. Figure 10.4 illustrates the Shell command interpretation loop. *Type ahead* is allowed, which means you can type your next command without waiting for the prompt, and that command will be there when the Shell is ready to receive it.

FIGURE 10.4: Command Interpretation Loop

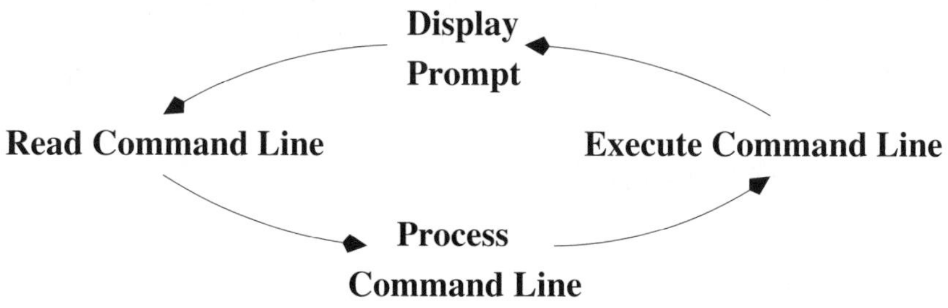

10.6.3   Current Working Directory and Filenames

To access a file or directory in the file system from the command line, you must call it up by its name, and there are several methods to do this. The most general, and also the most cumbersome, way to specify a *filename* is to list all the nodes in the path from the root to the node of the file or directory you want. This path, which is specified as a character string, is known as the *absolute pathname*, or *full pathname*, of the file. After the initial /, all components in a pathname are separated by the character /. For example, the file `index.html`, the entry page to the `sofpower.com` website, may have the full pathname

```
/home/sofpower/www/index.html
```

The full pathname is the complete name of a file. As you can imagine, however, this name often can be lengthy. Fortunately, a filename also can be specified relative to the *current working directory* (also known as the *working directory* or *current directory*). Thus, for the file /home/pwang/note.txt, if the current working directory is /home, then the name pwang/note.txt suffices. A *relative pathname* gives the path on the file tree leading from the working directory to the desired file. The third and simplest way to access a file can be used when the working directory is the same as the directory in which the file is stored. In this case, you simply use the filename. Thus, a Linux file has three names:

- A full pathname (for example, /home/sofpower/www/index.html)

- A relative pathname (for example, www/index.html)

- A (simple) name (for example, index.html)

The ability to use relative pathnames and simple filenames depends on the ability to change your current working directory. If, for example, you wish to access the file index.html, you may specify the full pathname

/home/sofpower/www/index.html

or you could change your working directory to /home/sofpower/www and simply refer to the file by name, index.html. When you log in, your working directory is automatically set to your home directory or your account root folder. The command

pwd   (print working directory)

displays the absolute pathname of your current working directory. The command

cd *directory*   (change working directory)

changes your working directory to the specified directory (given by a simple name, an absolute pathname, or a relative pathname).

Two *irregular files* are kept in every directory, and they serve as pointers:

File .   is a pointer to the directory in which this file resides.
File ..   is a pointer to the *parent* directory of the directory in which this
          file resides.

These pointers provide a standard abbreviation for the current directory and its parent directory, no matter where you are in the file tree. You also can use these pointers as a shorthand when you want to refer to a directory without having to use, or even know, its name. For example, the command

cd .

has no effect, and the command

cd ..

changes to the parent directory of the current directory. For example, if your working directory is `/home/sofpower/www`, and you issue the command **cd** ... Then your working directory becomes `/home/sofpower`.

You name your files and subdirectories when you create them. Linux is lenient when it comes to restrictions on filenames. In Linux you may name your file with any string of characters except the character /. But, it is advisable to avoid white-space characters and any leading hyphen (-).

10.6.4   Handling Files and Directories

Generally, there are two kinds of regular files: text and binary. A Linux text file stores characters in ASCII or UNICODE and marks the end of a line with the NEWLINE character[2]. HTML, JavaScript, and PHP files are all text files.

A binary file stores a sequence of bytes. Raster images and audio and video clips are usually binary files. Files may be copied, renamed, moved, and destroyed; similar operations are provided for directories. The command **cp** will copy a file and has the form

cp *source destination*

The file *source* is copied to a file named *destination*. If the destination file does not exist, it will be created; if it already exists, its contents will be overwritten. The **mv** (move) command

mv *oldname newname*

is used to change the file *oldname* to *newname*. No copying of the file content is involved. The new name may be in a different directory—hence the name "move." If *newname* already exists, its original content is lost.

Once a file or subdirectory has outlived its usefulness, you will want to remove it from your files. Linux provides the **rm** command for deleting files and **rmdir** for removing directories:

rm *filename1 filename2* ...
rmdir *directoryname1 directoryname2* ...

The argument of **rm** is a list of one or more filenames to be removed. **rmdir** takes as its argument a list of one or more directory names; but note that **rmdir** only will delete an empty directory. Generally, to remove a directory, you must first clean it out using **rm**.

To create a new directory, use the **mkdir** command, which takes as its argument the name of the directory to be created:

mkdir *name*

---

[2]On Windows or DOS systems, end of line is indicated by RETURN followed by NEWLINE.

When specifying a file or directory name as an argument for a command, you may use any of the forms outlined. That is, you may use either the full pathname, the relative pathname, or the simple name of a file, whichever you prefer.

On Linux, there are many text editors for creating and editing text files, including **gedit**, **nano**, **vim/gvim/vi**, and **emacs**. The editor **nano** is simple and easy to use but not very powerful. See the appendices on the DWP website for how to use **vim** and **emacs**.

10.7   Linux Files Access Control

Every file has an owner and a group designation. Linux uses a 9-bit code to control access to each file. These bits, called *protection bits*, specify access permission to a file for three classes of users. A user may be a *super user*, the owner of a file, a member in the file's group, or none of the above. There is no restriction on super user access to files.

u   (The owner or creator of the file)
g   (Members in the file's group)
o   (Others)

The first three protection bits pertain to u access, the next three pertain to g access, and the final three pertain to o access.

Each of the three bits specifying access for a user class has a different meaning. Possible access permissions for a file are

r   (Read permission, first bit set)
w   (Write permission, second bit set)
x   (Execute permission, third bit set)

10.7.1   Super User

Root refers to a class of super users to whom no file access restrictions apply. The *root* status is gained by logging in under the userid `root` (or some other designated root userid) or through the **su** command. A *super user* has read and write permission on all files in the system regardless of the protection bits. In addition, the super user has execute permission on all files for which anybody has execute permission. Typically, only system administrators and a few other selected users ("gurus" as they are sometimes called) have access to the super user password, which, for obvious reasons, is considered top secret.

10.7.2   Examining the Permission Settings

The nine protection bits can be represented by a three-digit octal number, which is referred to as the *protection mode* of a file. Only the owner of a file or a super user can set or change a file's protection mode; however, anyone can see it. The **ls -l** listing of a file displays the file type and access permissions. For example,

```
-rw-rw-rw- 1 smith 127 Jan 20 1:24 primer
-rw-r--r-- 1 smith  58 Jan 24 3:04 update
```

is output from **ls -l** for the two files **primer** and **update**. The owner of **primer** is **smith**, followed by the date (January 20) and time (1:24 A.M.) of the last change to the file. The number 127 is the number of characters contained in the file. The *file type*, *access permissions*, and *number of links* precede the file owner's userid (Figure 10.5). The protection setting of the file **primer** gives read and write permission to u, g, and o. The file **update** allows read and write to u, but only read to g and o. Neither file gives execution permissions. There are ten positions in the preceding mode display (of **ls**). The first position specifies the file type; the next three positions specify the r, w, and x permissions of u; and so on (Figure 10.5). Try viewing the access permissions for some real files on your system. Issue the command

ls -l /bin

to see listings for files in the directory **/bin**.

FIGURE 10.5: File Attributes

file type	user access	group access	other access	links	userid	size	date	time	file name
↓	↓	↓	↓	↓	↓	↓	↓	↓	↓
-	rw-	r--	r--	1	smith	127	Jan 24	2:04	update

10.7.3 Setting Permissions

A user can specify different kinds of access not just to files, but also to directories. A user needs the x permission to enter a directory, the r permission to list filenames in the directory, and the w permission to create/delete files in the directory.

Usually, a file is created with the default protection

```
-rw-------
```

so only the file owner can read/write the file. To change the protection mode on a file, use the command

chmod *mode filename*

where *mode* can be an octal (base 8) number (for example, 644 for **rw-r--r--**) to set all 9 bits specifically or can specify modifications to the file's existing permissions, in which case *mode* is given in the form *who op permission op2 permission2 ...*

Who represents the user class(es) affected by the change; it may be a combination of the letters u, g, and o, or it may be the letter a for all three. *Op* (operation) represents the change to be made; it can be + to add permission, − to take away permission, and = to reset permission. *Permission* represents the type(s) of permission being assigned or removed; it can be any combination of the letters r, w, and x. For example,

chmod o-w *filename*
chmod a+x *filename*
chmod u-w+x *filename*
chmod a=rw *filename*

The first example denies write permission to others. The second example makes the file executable by all. The third example takes away write and grants execute permission for the owner. The fourth example gives read and write permission (but no execute permission) for all classes of user (regardless of what permissions had been assigned before).

10.7.4 File Access Permissions for the Web

Folders and files for the Web, including .html, .php, .css, .js, .jpg, and other files, must be placed in the Web server's *document space* (Section 10.2). These files and folders are accessed by the Web server (Apache, for example), running on the same server host, for delivery to the Web.

Such Web-bound files and folders must grant the right permissions for access by the Web server. This usually means:

1. Each file must have the o+r permission.

2. Any folder holding files for the Web, directly or indirectly, must have the o+x permission.

3. Any immediate folder to receive uploaded files must have the o+wx permission.

4. Any file that can be overwriten from the Web must have the o+w permission.

On a SELinux (Security Enhanced Linux), files/folders in the server document space not only need the right access permissions but also must have the correct SELinux security context, which is usually the context type httpd_sys_content_t. The following Linux **chcon** command is used to set a target file/folder in this context:

chcon -t httpd_sys_content_t *target*

For a folder `$dir` to receive uploaded files, you need to make sure that `$dir` has the `httpd_sys_script_rw_t` context type. To make sure, you can run the Linux command

`chcon -t httpd_sys_script_rw_t` *target*

or something similar, to set the security context type for a *target* file or directory. To check the security context of a file or folder, use the Linux command

`ls -Zl` *file*
`ls -Zld` *dir*

Instead of setting the security context, you can choose to change the ownership of target files/folders to the userid (usually `www` or `apache`) assigned to the Web server on a particular server host. For example,

chown `apache` *target*

You do not need Linux root privilege to perform this particular **chown**. Making the Web server the owner of files and folders in the document space also means that they need just proper access permissions by **u** and no access by **g** or **o**.

10.8   Running Apache on Linux

It is often convenient to run a Web server on your desktop or laptop. Having full control over your own Web server can make Web development easier. Networking servers, the Web server included, are automatically started as the server operating system boots and stopped as it shuts down. We will focus on Apache running under Linux, knowing that Apache works in largely similar ways under other platforms.

On Linux, the boot-time *init scripts* for networking services are normally kept in the folder `/etc/init.d/`. For Apache, the script is `/etc/init.d/httpd`. The **system-config-services** is a tool (Figure 10.6) to turn network services on/off as well as to start/stop/restart any particular service. You will find **httpd** among the service entries listed. The same operations can also be done with the **service** command from the Linux shell. For example,

service `httpd start`
service `httpd graceful`   (restarts without service interruption)
service `httpd stop`

Both **system-config-services** and **service** invoke the proper `/etc/init.d/` scripts to do the job.

On newer Linux distributions, the **systemctl** command is preferred to enable/disable, start/stop, and restart services:

FIGURE 10.6: Service Control/Configuration Tool

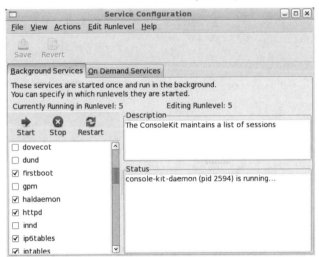

```
systemctl start httpd.service
systemctl stop httpd.service
systemctl restart httpd.service
```

On Ubuntu/Debian, you may have to first install this package (Section 10.28) to use the **service** command.

```
sudo apt-get install sysvconfig
```

The command **apt-get** is the automatic software package installer/updater on Ubuntu/Debian. Normally, it can only be run by `root`. The **sudo** command is a way of accessing commands normally off limits (see Section 10.18.1 for more information).

On CentOS/Fedora, you can add/delete any program to this services list with

```
chkconfig --add serviceName
chkconfig --del serviceName
```

On Debian/Ubuntu, use **sysv-rc-conf** instead.

You will usually find the document root at `/var/www/html/` and the Apache main configuration file at

```
/etc/httpd/conf/httpd.conf          (CentOS/Fedora)
/etc/apache2/apache2.conf           (Ubuntu/Debian)
```

Often, the main configuration file will include other component configuration files such as `php.conf` and `ssl.conf`.

To check if **httpd**, or any other process, is running, you can use

pidof httpd
pidof *process_Name*

and see if one or more process ids are found.

10.8.1   Controlling the Apache Server

The command **apachectl** (CentOS/Fedora) or **apache2ctl** (Ubuntu/Debian), usually found in `/usr/sbin`, can be used to control the **httpd**

apachectl *action*
apache2ctl *action*

Possible *actions* are listed in Table 10.1. The init script **/etc/init.d/httpd**

TABLE 10.1: Actions of **apachectl**

Action	Meaning
start	Starts **httpd** if not already running
stop	Stops **httpd** if running
restart	Starts/restarts **httpd**
graceful	Restarts **httpd**, respecting ongoing HTTP requests
configtest or -t	Checks the syntax of configuration files

checks a few things before it actually calls **apachectl** to take care of business.

10.9   Apache Run-Time Configuration

Features and behaviors of the Apache **httpd** can be controlled by *directives* kept in configuration files. The main configuration file is usually `httpd.conf` (or `apache2.conf`). When **httpd** starts, it reads the configuration files first. After making changes to the configuration, the **httpd** needs to be restarted before the new configuration takes effect. Unless you have installed your own Apache as an ordinary user (Section 10.19), you'll need `root` privilege to modify the Apache configuration or to restart it.

10.9.1   Apache Configuration File Basics

An Apache configuration file (`httpd.conf`, for example) is a text file that contains *configuration directives*. Each directive is given on a separate line that can be continued to the next line by a \ character at end of the line.

Lines that begin with the char # are comments and are ignored. A comment must occupy the entire line. No end-of-line comments are allowed. There are many different directives. Directive names are not case sensitive, but their arguments often are. A directive applies globally unless it is placed in a *container* that limits its scope. When in conflict, a local directive overrides a global directive.

The main configuration file is `httpd.conf`, and other *component configuration files* may exist and are included by the main file with the `Include` directive. For example, on many Linux systems, the configuration directory `/etc/httpd/conf.d/` stores component configuration files such as `ssl.conf` for SSL and `php.conf` for PHP (Section 10.21). The directive

```
Include conf.d/*.conf
```

is used to include all such component configuration files.

To test your Apache configuration for syntax errors, use either one of the following commands:

service `httpd configtest`
apachectl `configtest`
httpd `-t`

In addition to the central (main and component) configuration files, there are also *in-directory configuration files* known as *access files*. An access file, often named `.htaccess`, is placed in any Web-bound folder (your `public_html`, for example) to provide configuration settings applicable for the file hierarchy rooted at that particular folder. Directives in an access file *override* settings in the central configuration files. The possibility of an access file and what directives it may contain are controlled by the `AllowOverride` directive in the main configuration file. The `.htaccess` files are especially useful for individual users to configure their own Web spaces, usually the `public_html` under their home directories.

10.9.2 About Configuration Directives

Configuration directives control many aspects of the Apache Web server. The `httpd.conf` file has three main parts: *Global Environment*, *main server configurations*, and *virtual hosts configurations*. Comments are provided for each configuration directive to guide its usage. Apache has reasonable and practical default settings for all the directives, making it easy to configure a *typical server*. Additional directives specify how loaded components work. Commonly used directives include

- Server properties: host identification (`ServerName` *name*), file locations (`ServerRoot`, `DocumentRoot`, `ScriptAlias`), network parameters (`Listen` [*IP*:]*port*), and resource management (`StartServers`, `KeepAlive`)

- Enabling optional server features (`Options`) and in-directory configuration overrides (`AllowOverride`)

- Access restrictions and user authentication (`Allow`, `Deny`, `Require`, `Satisfy`, `AuthName`, `AuthType`, `AuthFile`)

- Content handling (`AddHandler`, `AddType`, `AddOutputFilter`)

- HTTP caching and content deflation (`ExpiresActive`, `ExpiresByType`, `DeflateCompressionLevel`, `AddOutputFilterByType DEFLATE`)

- Virtual hosts (`NameVirtualHost`)

For example, the directive

```
DirectoryIndex index.html  index.php
```

says `index.html` (or `index.php`) is the *directory index file* that is displayed if the folder containing it is the target resource of an incoming URI. Without an index file, a listing of filenames in that folder is generated (*index generation*) for display only if the `Indexes` option has been enabled. Otherwise, an error is returned.

10.9.3 Loading Modules

Apache is a modular server. Only the most basic functionalities are included in the core **httpd**. Many extended features are implemented as dynamically loadable modules (`.so`) that can be selectively loaded into the core server when it starts. This organization is very efficient and flexible.

The loadable modules are placed in the `modules` folder under the *server root* directory, that is defined in the main configuration file with the `ServerRoot` directive. To load a certain module, use the directive

```
LoadModule  name_module modules/moduleFileName.so
```

For example,

```
LoadModule dir_module modules/mod_dir.so        (loads module dir)
LoadModule php5_module modules/libphp5.so       (loads module php5)
```

The `dir` module enables Apache to generate a directory listing. The `php5` module provides PHP version 5 for dynamic webpages.

Configuration directives may be included conditionally, depending on the presence of a particular module, by enclosing them in an `<IfModule>` container. For example,

```
<IfModule mod_userdir.c>
UserDir public_html
</IfModule>
```

says if we are using the `userdir` module, then the Web folder for each Linux user is `public_html`.

10.9.4 Global Directives

Table 10.2 shows some more directives relating to how the Apache server works globally (see **Ex: apacheGlobal.conf**). The Alias and ScriptAlias directives map an incoming URI to a designated local folder.

TABLE 10.2: Apache Global Directives

Directive	Effect
ServerRoot "/etc/httpd"	Server root folder
KeepAlive On	Keeps connection for next request
MaxKeepAliveRequests 100	
KeepAliveTimeout 15	
User apache	Server userid is apache
Group apache	Server groupid is apache
ServerName monkey.cs.kent.edu	Domain name of server host
ServerAdmin pwang@cs.kent.edu	Email of administrator
DocumentRoot "/var/www/html"	Server document space root
UserDir public_html	Folder name of per-user Web space
AccessFileName .htaccess	In-directory configuration filename
TypesConfig /etc/mime.types	MIME types file
ScriptAlias /cgi-bin/	
"/var/www/cgi-bin/"	CGI program folder
Alias /special/ "/var/www/sp/"	Special URI-to-folder mapping

10.9.5 Container Directives

Configuration directives can be placed inside a container directive to subject them to certain conditions or to limit their scope of applicability to particular directories, files, locations (URIs), or hosts. Without being limited, a directive applies globally.

For example,

```
<IfModule mod_userdir.c>
    UserDir public_html
</IfModule>
```

enables the per-user Web space (**Ex: peruser.conf**) and designates the user folder to be public_html only if the userdir module is loaded.

Also, consider these typical settings (**Ex: docroot.conf**) for the document root /var/www/html:

```
<Directory "/var/www/html">
    Options Indexes FollowSymLinks          (1)
    Order allow,deny                        (2)
```

```
    Allow from all                      (3)
    AllowOverride None                  (4)
</Directory>
```

Within the directory `/var/www/html`, we allow index generation and the following of symbolic links (line 1). The order to apply the access control directives is `allow` followed by `deny` (line 2), and access is allowed for all incoming requests (line 3) unless it is denied later.

The `AllowOverride` (line 4) permits certain directives in `.htaccess` files. Its arguments can be `None`, `All`, or a combination of the keywords `Options`, `FileInfo`, `AuthConfig`, and `Limit`. We return to this topic when we discuss access control in detail (Section 10.10).

You will also find the following typical setting (**Ex: htprotect.conf**) in your `httpd.conf`:

```
<Files ~ "^\.ht">
    Order allow,deny
    Deny from all
</Files>
```

It denies Web access to any file whose name begins with `.ht`. This is good for security because files such as `.htaccess` are readable by the Apache Web server, but we do not want their contents exposed to visitors from the Web.

As `<Directory>` and `<Files>` work on the file pathnames on your computer, the `<Location>` container works on URIs. We also have `<DirectoryMatch>`, `<FileMatch>`, and `<LocationMatch>` that use regular expressions very similar to those in PHP.

10.10   Apache Web Access Control

10.10.1   What Is Web Access Control?

Running a Web server on your Linux system means that you can make certain files and folders accessible from the Web. However, you also want to control how such files can be accessed and by whom.

To make a file/folder accessible from the Web, you must place it somewhere in the document space configured for your Web server. This usually means placing a file/folder under the document root or inside a `public_html` folder and also making the file readable (the folder readable and executable) by the Web server via **chmod a+r** *file* (**chmod a+rx** *folder*). Files on your system not placed under the server document space or not having the right access modes (Section 10.7.3) will not be accessible from the Web.

The Web server can be configured to further limit access. Web access control, or simply access control, specifies who can access which part of a website with what HTTP request methods. Web access control can be specified based on IP numbers, domains, and hosts, as well as passwords. Access restrictions can be applied to the entire site, to specific directories, or to individual files.

Apache access control directives include `Allow`, `Deny`, `Order`, `AuthName`, `AuthType`, `AuthUserFile`, `AuthGroupFile`, `Require`, `Satisfy`, `<Limit>`, and `<LimitExcept>`.

10.10.2 Access Control by Host

If a file in the server document space has no access control, access is granted. The `order` directive specifies the order in which `allow` and `deny` controls are applied. For example,

```
order allow,deny
```

only access allowed but not denied are permitted. In the following, if access is first denied then allowed, it is allowed.

```
order deny,allow
deny from all
allow from host1 host2 . . .
```

On `monkey.cs.kent.edu`, we have a set of pages reserved for use inside our departmental local area network (LAN). They are placed under the folder `/var/www/html/internal`. Their access has the following restriction (**Ex:** folderprotect.conf):

```
<Location /internal>
  order deny,allow
  deny from all
  allow from .cs.kent.edu
</Location>
```

Thus, only hosts in the `.cs.kent.edu` domain are allowed to access the location `/internal`. The IP address of a host can be used. For example,

```
allow from 131.123
```

grants access to requests made from any IP with the prefix `131.123`.

To enable users to control access to files and folders under their per-user Web space (`public_html`), you can use something such as (**Ex:** htaccess.conf)

```
<Directory /home/*/public_html>
    AllowOverride All
    Order allow,deny
    Allow from all
</Directory>
```

in `httpd.conf`. This means users can place their own access control and other directives in the file `~/public_html/.htaccess`.

10.11   Requiring Passwords

Allowing access only from certain domains or hosts is fine, but we still need a way to restrict access to registered users either for the whole site or for parts of it. Each part of a site under its own password control is known as a *security realm*. A user needs the correct userid and password to log into any realm before accessing the contents thereof. Thus, when accessing a resource inside a realm, a user must first be *authenticated* or verified as to who that user is. The Apache Web server supports two distinct HTTP authentication schemes: the *Basic Authentication* and the *Digest Authentication*. Some browsers lack support for Digest Authentication, which is only somewhat more secure than Basic Authentication.

Let's look at how to set user login.

10.11.1   Setting Up User Login under Apache

To illustrate how to password protect Web files and folders, let's look at a specific example where the location `/WEB/csnotes/` is a folder we will protect.

We first add the following authentication directives to the `httpd.conf` file (**Ex:** `validuser.conf`):

```
<Location "/dwp/protected/">
    AuthName   "DWP Protected"
    AuthType   Basic
    AuthUserFile /var/www/etc/dwph5
    require valid-user
</Location>
```

The `AuthName` gives a name to the realm. The realm name is displayed when requesting the user to log in. Thus, it is important to make the realm name very specific so that users will know where they are logging into. The `AuthType` can be either `Basic` or `Digest`. The `AuthUserFile` specifies the full pathname of a file containing registered users. The optional `AuthGroupFile` specifies the full pathname of a file containing group names and users in those groups. The `Require` directive defines which registered users may access this realm.

valid-user	(all users in the `AuthUserFile`)
user id1 id2 id3 ...	(the given users)
group grp1 grp2 ...	(all users in the given groups)

The `AuthUserFile` lists the userid and password for each registered user, with one user per line. Here is a sample entry in `/var/www/etc/dwph5`.

`PWang:RkYf8U6S6nBqE`

The Apache utility **htpasswd** (**htdigest**) helps create password files and add registered users for the Basic (Digest) authentication scheme. (See the Linux man page for these utilities for usage.) For example,

```
htpasswd -c /var/www/etc/wdp1 PWang
```

creates the file and adds an entry for user `PWang`, interactively asking for `PWang`'s password. If you wish to set up a group file, you can follow the format for `/etc/group`, namely, each line looks like

group-name: *userid1 userid2 ...*

It is also possible to set up login from an `.htaccess` file. For example, put in `.htaccess` under user `pwang`'s `public_html`

```
AuthUserFile /home/pwang/public_html/.htpassword
AuthName "Faculty Club"
AuthType Basic
Require valid-user
```

Then place in `.htpassword` any registered users.

If more than one `Require` and/or `allow from` conditions is specified for a particular protected resource, then the `satisfy any` (if any condition is met) or `satisfy all` (all conditions must be met) directive is also given. For example **Ex:** `flexibleprotect.conf`,

```
<Location /internal>
    order deny,allow
    deny from all
    allow from .cs.kent.edu
    AuthName  "CS Internal"
    AuthType  Basic
    AuthUserFile /var/www/etc/cs
    require valid-user

    satisfy any
</Location>
```

means resources under the `/internal` can be accessed by any request originating from the `cs.kent.edu` domain (no login required) or a user must log in.

10.12   How HTTP Basic Authentication Works

Upon receiving an unauthorized resource request to a realm protected by *Basic Authentication*, the Web server issues a *challenge*:

```
HTTP/1.0 401 Unauthorized
WWW-Authenticate: Basic realm="CS Internal"
```

Upon receiving the challenge, the browser displays a login dialog box requesting the userid and password for the given realm. Seeing the login dialog, the user enters the userid and password. The browser then sends the same resource request again with the added authorization HTTP header

```
Authorization: Basic QWxhZGRpbjpvcGVuIHNlc2FtZQ==
```

where the base64 encoded *basic cookie* decodes to *userid:password*. From this point on, the browser automatically includes the basic cookie with every subsequent request to the given realm. This behavior persists until the browser instance is closed.

10.13   How HTTP Digest Authentication Works

Unless conducted over a secure connection, such as SSL (Secure Socket Layer) used by HTTPS, the Basic Authentication is not very secure. The userid and password are subject to easy eavesdropping over HTTP. The *Digest Authentication* is an emerging HTTP standard to provide a somewhat more secure method than Basic Authentication.

With Digest Authentication, the server sends a challenge (on a single line):

```
HTTP/1.1 401 Unauthorized WWW-Authenticate:
        Digest realm="Gold Club" nonce="3493u4987"
```

where the *nonce* is an arbitrary string generated by the server. The recommended form of the nonce is an *MD5 hash*, which includes the client's IP address, a timestamp, and a private key known only to the server.

Upon receiving the challenge, the browser computes

```
str1 = MD5(userid + password)
str2 = MD5(str1 + nonce + Resource_URI)
```

The browser then sends the authorization HTTP header (on one line)

```
Authorization: Digest realm="Gold Club", nonce="...",
 username="pwang", uri="/www/gold/index.html",
 response="str2"
```

The server verifies the response by computing it using the stored password.

From this point on, the browser includes the Digest Authentication header with every request to the same realm. The server may elect to rechallenge with a different nonce at any time.

10.13.1   Basic versus Digest Authentication

Basic Authentication is simple and works with all major browsers. Digest Authentication is somewhat more secure, but browser support is less complete.

Web servers, including Apache, tend to support both authentication schemes. When security is a concern, the best practice is to move from Basic Authentication over HTTP directly to Basic Authentication over HTTPS (Secure HTTP over SSL).

10.14   Password Encryption

The Apache-supplied **htpasswd** tool uses the same Linux/Unix password/data encryption method as implemented by the C library function crypt. In this encryption scheme, a *key* is formed by taking the lower 7 bits of each character from the password to form a 56-bit quantity. Hence, only the first 8 characters of a password are significant. Also, a randomly selected 2-character *salt* from the 64-character set [a-zA-Z0-9./] is used to perturb the standard Data Encryption Algorithm (DEA) in 4096 possible different ways. The key and salt are used to repeatedly encrypt a constant string, known only to the algorithm, resulting in an 11-character code The salt is prepended to the code to form a 13-character encrypted password that is saved in the password file for registered users. The original password is never stored.

When verifying a password, the salt is extracted from the encrypted password and used in the preceding algorithm to see if the encrypted password is regenerated. If so, the password is correct.

10.15   Automatic File Deflation

Apache takes advantage of many HTTP 1.1 features to make webpages faster to download. One such feature is automatic compression of a page before network transfer, resulting in significantly reduced file size and delivery time. This is especially true for textual pages whose compression ratio can reach 85% or more. A compressed page is uncompressed automatically by your browser.

The mod_deflate module for Apache 2.0 supports automatic (dynamic) file compression via the HTTP 1.1 Content-Encoding and Accept-Encoding headers. These two configuration directives (**Ex:** deflate.conf)

```
DeflateCompressionLevel 6
AddOutputFilterByType DEFLATE text/html text/plain \
    text/xml text/css application/x-javascript \
    application/xhtml+xml application/xslt+xml \
    application/xml application/xml-dtd image/svg+xml
```

indicate a list of content types for dynamic compression (using zlib) at the indicated compression level. Deflation adds a bit of processing load on the server side and the higher the compression level, the heavier the processing load.

Compression will only take place when the incoming HTTP request indicates an acceptable compression encoding. The detection of browser compression preferences and the sending of compressed or uncompressed data are automatic. Of course, any compressed outgoing page will carry an appropriate `Content-Encoding` response header.

The `AddOutputFilterByType` directive needs `AllowOverride FileInfo` to work in `.htaccess`.

10.16 URI to Document Space Mapping

Perhaps the single most basic function of a Web server is associating an incoming URI with a specific file, folder, or program on the server host. For example, it maps the URI / to the document root.

The `alias` module supplies a number of directives to associate designated URI paths to specific directories or files.

With URI aliasing, files can be stored outside the document root. The general form of the `Alias` directive is

```
Alias uri-path dir-or-file
```

For example,

```
Alias /projx /local/projects/project_x
```

enables any URI /projx/... to lead to the file system path

```
/local/projects/project_x/...
```

Access control to `projx` can be applied with either `<Directory>` or `<Location>`, the latter is preferred.

The `ScriptAlias` directive maps the location for CGI programs. For example,

```
ScriptAlias /cgi-bin  /var/www/cgi-bin
```

The `AliasMatch` and `ScriptAliasMatch` allows the prefix to be a pattern and the matching pattern can be used in the *dir-or-file* part.

When webpages are moved, replaced, or removed, the server can automatically make redirections: `Redirect` [*status*] *URI-path new-URI-path* For example,

```
Redirect permanent /symbolicnet.mcs.kent.edu
        http://symblicnet.org
```

The status codes are: `permanent`, `temp`, `seeother`, `gone`, and HTTP status numbers. The default is `temp`.

With `Redirect` we can also insist on HTTPS access on the whole or a part of a site:

```
Redirect permanent / https://superstore.com/
Redirect permanent /secure  https://bigbusiness.com/secure
```

The Apache `rewrite` module supplies powerful ways to transform URIs based on rewrite rules using regular expression matching. For example, you may put these lines in a .htaccess file in the folder dyn

```
options +FollowSymlinks
RewriteEngine on
RewriteRule ^(.+)\.html?$ http://big.com/dyn/$1.php [NC]
```

and any incoming *xyz*.htm or *xyz*.html URI will get rewritten to *xyz*.php. See the Apache `mod_rewrite` documentation for more information.

10.17   Virtual Hosting

Typically, the domain name or host IP part of an URL leads to a physical host computer located at the explicit IP or the IP associated to the domain name through the DNS (Section 1.2.1). Thus, a Web server running on *host_domain_Name* with *host_IP* serves webpages only for

http://*host_domain_name*
http://*host_IP*

Such a Web host is a real host (Figure 10.7). To run Apache for a real host, simply configure it to listen to port 80 on that host.

FIGURE 10.7: Real Host

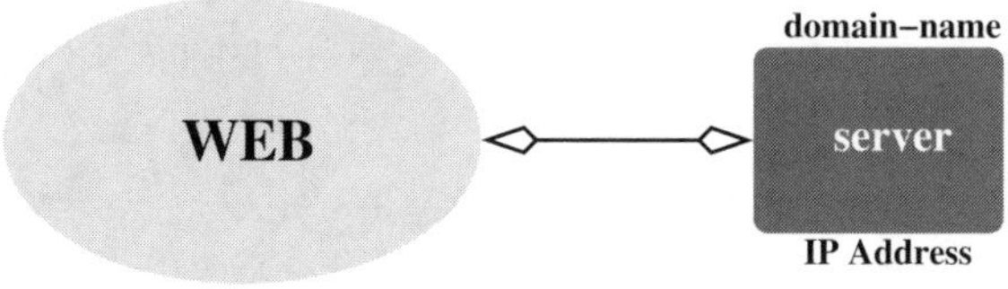

Often, a Web hosting service needs to deliver websites for more than one company. Virtual hosting is a technique to achieve that goal. One way is to associate multiple IPs to a single computer so the computer can serve each domain mapped to one of the IP numbers with a separate Apache server daemon. This approach is known as *IP-based virtual hosting*.

Name-based virtual hosting (Figure 10.8) can serve different domain names that are mapped, via DNS, to the same IP for the host computer. Because the host part of the URL is delivered to the server under HTTP 1.1, the Web server program can check the host name and deliver different websites to different domains.

For example, the host `eagle.cs.kent.edu` serves the domain

FIGURE 10.8: Name-Based Virtual Hosting

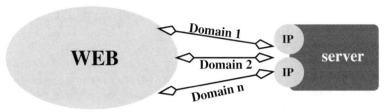

```
SymbolicNet.org
```

a service for the Symbolic Mathematical Computation research community, with a name-based virtual server under Apache with the configuration directives

```
NameVirtualHost 131.123.41.85
```

```
<VirtualHost 131.123.41.85>
   ServerAdmin pwang@cs.kent.edu
   ServerName eagle.cs.kent.edu
   DocumentRoot /var/www/html
   DirectoryIndex index.html index.html.var


   ...


</VirtualHost>
```

The `VirtualHost` container encloses all configuration settings applicable for a particular website. Each virtual host is defined by its own `VirtualHost` container.

10.18 Installing Apache with Package Management

We have mentioned that most Linux distributions come with Apache installed. With root access, you can use the Linux package management (Section 10.28) commands

CentOS/Fedora:
> **yum** install httpd
> **yum** update httpd

Ubuntu/Debian:
> **sudo apt-get** install apache2
> **sudo apt-get** update apache2

to install/update your Apache server.

If you wish to have the very latest Apache release, or if you do not have root access, you can install Apache manually as described in Section 10.19.

10.18.1   Sudo

Linux administration tasks such as setting up new user accounts, installing and updating system-wide software, and managing network services must usually be performed by privileged users such as **root**. This is secure but not very flexible.

Sudo is a method to allow regular users to perform certain tasks temporarily as **root** or as some other more privileged user. The command name **sudo** comes from the command **su** (substitute user), which allows a user to become another user. Putting **sudo** in front of a complete command you wish to execute says: *"allow me enough privilege to execute the following command."* If the given command is allowed, **sudo** lets the given command execute and enters it into a log for security.

The file `/etc/sudoers` contains data governing who can gain what privileges to execute which commands on exactly what host computers and whether or not your password is required. Thus, the same **sudoers** file can be shared by many hosts within the same organization. The file can only be modified via the privileged command **sudoedit**. You can read about **sudoers** and its syntax rules by

man 5 sudoers

The general form of a user entry in **sudoers** is

r_user hosts=(*s_user*) *commands*

meaning *r_user* can execute the given *commands* as *s_user* on the listed *hosts*. The (*s_user*) part can be omitted if *s_user* is **root**.

Here are some example **sudoers** entries:

```
pwang   localhost=/sbin/shutdown -h now

pwang   localhost=/sbin/service httpd start, \
                  /sbin/service httpd start, \
                  /sbin/service httpd restart, \
                  /sbin/service httpd graceful

pwang   localhost=/sbin/apt-get update apache, \
                  /sbin/apt-get install apache, \
                  /sbin/apt-get remove apache

root    ALL=(ALL)        ALL
```

Each entry must be given on one line. The `root` entry is always there to give `root` the ability to **sudo** all commands on hosts as any user.

Even if you log in (or **su**) as `root`, you may prefer to use **sudo** so as to leave log entries for the tasks performed.

10.19   Manual Installation of Apache

If you prefer not to install Apache with package management, you may install Apache manually. The installation procedure follows the standard Linux configure, make, install sequence.

If you have `root` access, you will be able to install Apache in a system directory such as `/usr/local` and assign port 80 to it. If not, you still can install Apache for yourself (for experimentation) in your own home directory and use a nonprivileged port, such as 8080. Let `$DOWNLOAD` be the download folder, for example, either `/usr/local/apache_src` or `$HOME/apache_src`, and let `$APACHE` be the installation folder, for example, `/usr/local/apache` or `$HOME/apache`.

To download and unpack the Apache HTTP server distribution, follow these steps.

1. Download—Go to `httpd.apache.org/download.cgi` and download the

   `httpd-`*version*`.tar.gz`

   or the `.tar.bz2` file, as well as its MD5 fingerprint file, into your `$DOWNLOAD` folder.

2. Integrity check—Use **md5sum** on the fingerprint file to check the downloaded file.

3. Unpack—From the `$DOWNLOAD` folder, unpack with one of these commands:

   tar zxvpf `httpd-`*version*`.tar.gz`
   tar jxvpf `httpd-`*version*`.tar.bz2`

   You will find a new Apache source folder, `httpd-`*version*, containing the unpacked files.

10.19.1   Configure and Compile

Now you are ready to build and install the Apache Web server. Follow the `INSTALL` file and the *Compiling and Installing* section of the Apache documentation `httpd.apache.org/docs/`*version-number*. You will need an ANSI C compiler (**gcc** preferred) to compile, Perl 5 to make tools work, and DSO

(Dynamic Shared Object) support. These should already be in place on newer Linux distributions.

From the Apache source folder, issue the command

./configure *options*

to automatically generate the compilation and installation details for your computer. The INSTALL file has good information about configuration. To see all the possible options, give the command

./configure `--help`.

For example, the `--prefix=`*serverRoot* option specifies the pathname of the server root folder, and the option `--enable-mods-shared=all` elects to compile all Apache modules into dynamically loadable shared libraries.

The recommended method (**Ex:** `makeapache.bash`) to configure and compile is

./configure `--prefix=$APACHE --enable-mods-shared=all \`
            otherOptions
make
make `install`

Here, the Apache server root folder has been set to your installation folder **$APACHE** as the destination for the results of the installation. The recommended *otherOptions* are

```
--enable-cache --enable-disk-cache \
--enable-mem-cache --enable-proxy \
--enable-proxy-http --enable-proxy-ftp \
--enable-proxy-connect --enable-so \
--enable-cgi --enable-info \
--enable-rewrite --enable-spelling \
--enable-usertrack --enable-ssl \
--enable-deflate --enable-mime-magic
```

Each of the preceding three commands will take a while to run to completion.

After successful installation, it is time to customize the Apache configuration file **$APACHE/conf/httpd.conf**. Follow these steps:

1. Check the `ServerRoot` and `DocumentRoot` settings. These should be the full pathnames as given by **$APACHE** and **$APACHE/htdocs**, respectively.

2. Set the listening port:

   `Listen 80`   (requires **root** privilege)
   `Listen 8080`   (no need for **root** privilege)

3. Make any other configuration adjustments as needed.

Now you can start the Apache server with

$APACHE/bin/apachectl `start`

If the start is successful, you can then use a Web browser on the same host computer to visit

`http://localhost.localdomain:`*port*

and see the Apache welcome page, which is the file

`$APACHE/htdocs/index.html`

Then, test the server from another host on the same LAN, with

`http://`*host:port*

where *host* is the domain name of your server. Make sure the firewall on the server allows both HTTP and HTTPS access. Otherwise, the Apache Web server will not be accessible from other hosts. On CentOS/Fedora, firewall configuration is an option on the `system->admin` menu. For Ubuntu, the **gufw** tool is handy for the same purpose.

It is recommended that you install PHP together with Apache. See Section 10.21 for details.

10.20   SSL/TLS

Web servers support HTTPS for secure communication between the client and the server. HTTPS is HTTP over the *Secure Socket Layer* or *Transport Layer Security* protocol (Figure 10.9). SSL/TLS may be placed between a reliable

FIGURE 10.9: HTTP and HTTPS

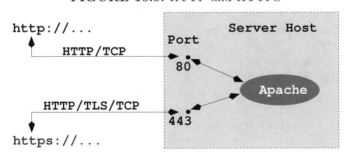

connection-oriented network layer protocol (TCP/IP, for example) and the application protocol layer, HTTP for example (Figure 10.10). SSL/TLS provides for secure communication between client and server by allowing mutual authentication, the use of digital signatures for integrity, and data encryption for privacy. SSL/TLS provides the following security functions

FIGURE 10.10: HTTPS Protocol Layers

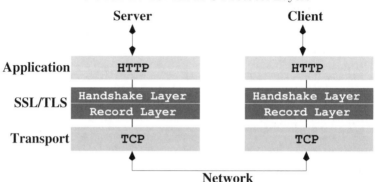

- Communication privacy—by *symmetric cryptography* with key exchange via *public-key cryptography.*

- Communication reliability and integrity—by *signed message digest*

- Mutual authentication of peers—by *digital certificates.*

- Message reliability and integrity—by signed message digest.

- The handshake layer allows peers to agree upon security parameters for the record layer, authenticate themselves, instantiate negotiated security parameters, and report error conditions to each other.

- Record Protocol takes the cleartext, fragments the data into manageable blocks, optionally compresses the data, applies a Message Authenticity Check (MAC), encrypts, and transmits the result.

In secure communication, the very first concern is that the parties are actually who they say they are. A *digital certificate* is a document (computer file) signed by a *certificate authority* (CA), such as VeriSign, that can vouch for the identity of the certificate holder. A CA issues a digital certificate for a client after carefully verifying the client's identity and legitimacy. Each certificate is digitally signed by a CA and contains the identity of the client, the client's public key, the expiration date of the certificate, and details of the issuing CA. Certificates can be issued for different purposes and different domains: Web server, email, code signing, payment systems, and so on. Digital certificates follow standardized formats, for example X509v3, and are used by security programs, such as Web servers and Web browsers, to authenticate a user or a Web server. Certificates not issued by widely recognized CAs or that have expired can cause a warning about the certificate's status requiring the user to decide whether or not to accept the certificate.

Web servers supporting HTTPS will need to obtain a valid digital certificate and enable SSL/TLS.

10.21   PHP Module for Apache

An Apache server is generally expected to support PHP, and it is not hard to add the PHP module for Apache. With the PHP module, the Apache Web server will be able to interpret PHP codes embedded in textual documents of any type as they are being delivered to the Web (Figure 5.16). Most Linux distributions will already have Apache installed with PHP. For example, you may find the PHP module `libphp5.so` already in the Apache modules folder (usually `/etc/httpd/modules`).

You can also use the Linux package management facility to install/update Apache+PHP:

yum `install httpd php`	(CentOS/Fedora)
yum `update httpd php`	(CentOS/Fedora)
sudo apt-get `install apache2 php5 \`	
`libapache2-mod-php5`	(Ubuntu/Debian)
sudo apt-get `update apache2 php5 \`	
`libapache2-mod-php5`	(Ubuntu/Debian)

10.21.1   Installing the PHP Module

This section describes how to install the PHP module manually and add it to your Apache server. If you already have Apache+PHP installed, please skip this section.

First, download the current PHP release (**php-***version*`.tar.gz` or `.tar.bz2`) from `www.php.net/downloads.php`, check the MD5 fingerprint, and unpack into your `$DOWNLOAD` folder as before (Section 10.19).

Next, go to the PHP source code folder `$DOWNLOAD/php-`*version* to configure the PHP module. For example (**Ex: confphp**),

```
cd $DOWNLOAD/php-version
./configure --with-apxs2=$APACHE/bin/apxs \
  --prefix=$APACHE/php  --enable-shared=all \
  --with-gd --with-config-file-path=$APACHE/php \
  --enable-force-cgi-redirect --disable-cgi \
  --with-zlib --with-gettext --with-gdbm \
  > /tmp/conf.output 2>&1
```

Then check the `conf.output` to see if you get these lines:

```
checking if libtool supports shared libraries... yes
checking whether to build shared libraries... yes
checking whether to build static libraries... no
```

If you need to redo the configuration step, please first clean up things with

```
make distclean
```

After successful configuration, you are ready to create the PHP module. Enter the command

make

It will take a while. After it is done, you should check the .libs/ folder to see if the PHP module libphp5.so has been created. If so, then issue the command

make install

The install directory is $APACHE/php as specified by the --prefix option. The install process also moves libphp5.so to the folder $APACHE/modules/ and modifies $APACHE/conf/httpd.conf for the **httpd** to load the PHP module when it starts by adding the Apache configuration directive

```
LoadModule php5_module modules/libphp5.so
```

In addition, you need to add a few other directives to tell Apache what files need PHP processing:

```
AddHandler php5-script .php
AddType text/html .php
DirectoryIndex index.php index.html
```

As stated, any time a change is made to the configuration, you need to restart Apache (Section 10.8.1) in order to get the new configuration to take effect.

10.22   Testing PHP

To test your Apache+PHP installation, you can create the page info.php (**Ex:** info.php)

```
<html><head><title>php info</title></head>
<body> <?php phpinfo(); ?>
</body></html>
```

and place it under the document root folder. Then, visit

```
http://localhost.localdomain/info.php
```

from your Web browser. The phpinfo() function generates a page of detailed information about your PHP installation, including version number, modules loaded, configuration settings, and so on.

As Apache starts, it loads the PHP module and also any PHP-specific configuration in a file usually named php.ini. The location of this file (usually /etc/php.ini) is given as the *Loaded Configuration File* in the phpinfo() generated display.

10.23 PHP Configuration

The file (php.ini)indexPHP!php.ini@php.ini containing *configuration direc-tives* is read when the PHP module is loaded as the Web server (**httpd**) starts. Any changes made to php.ini will only take effect after Apache is restarted (Section 10.8.1).

PHP has toggle (on/off) and value configuration options. You edit the php.ini, which contains a set of reasonable defaults, to make any adjustments.

For example, if you are running a Web development site where seeing error messages will help debugging PHP scripts, then you would set (**Ex: PHPErr.ini**)

```
;;;; Enables error display output from PHP
display_errors = On
display_startup_errors = On
```

For a production Web server, you would definitely want to change these to

```
display_errors = Off
display_startup_errors = Off
log_errors = On
;;;; Enables all error, warning, and info msg reporting
error_reporting = E_ALL
;;;; Sends msgs to log file
error_log = <pathname of a designated error.txt file>
```

PHP also allows you to open any local or remote URL for generating page content. However, if your site has no need for opening remote URLs from PHP, you may increase security by setting

```
allow_url_fopen = Off
```

PHP also has very good support for HTTP file uploading. If you wish to allow that, then use

```
file_uploads = On
```

```
;;;; Use some reasonable size limit
upload_max_filesize = 2M
```

PHP *extensions* provide optional features for many different purposes. For example, the gd extension supports manipulation of fonts and graph-ics from PHP, and the mysql extension provides a PHP interface to MySQL databases. Dynamically loadable extensions are collected in a PHP modules folder (usually /usr/lib/php/modules), but are set in the php.ini by the extension_dir directive. On many Linux systems, the extensions are loaded by default through extension-specific .ini files in the folder /etc/php.d/. By editing these files you control which extensions are loaded when Apache+PHP starts.

See PHP documentation for a list of all configuration directives. To examine the setting of all PHP configurations directives, you can simply look at the `phpinfo()` display (Section 10.22).

10.23.1   Changing PHP Configurations

While the `php.ini` file provides static values for configuration options, it is possible to override certain settings in PHP scripts and in `.htaccess` files.

From a PHP script, you may call the function

```
ini_set($option, $value)
```

to set an option. For example,

```
ini_set("upload_max_filesize", "3M");
```

We have seen the use of `ini_set` for error reporting (Section 8.5) and for session control (Section 8.10).

Each PHP configuration option has a *change mode* designation that specifies where the option can be set. The modes are

- `PHP_INI_ALL`—Option can be set anywhere.

- `PHP_INI_USER`—Option can be set in user scripts.

- `PHP_INI_PERDIR`—Option can be set in `php.ini`, `.htaccess` or `httpd.conf`.

- `PHP_INI_SYSTEM`—Option can only be set in `php.ini` or `httpd.conf`.

To set a PHP option in the Apache configuration file `.htaccess` or `httpd.conf`, use

```
php_flag toggle_option on (or off)
php_value value_option value
```

For example, to automatically compress PHP output, you may place

```
php_flag zlib.output_compression On
php_value zlib.output_compression_level 5
```

in a `.htaccess` or the `httpd.conf` file. This is useful because Apache only deflates static files (Section 10.15) automatically. PHP-generated pages/contents can only be automatically compressed by the PHP handler.

The **Ex:** `HostConf` example lists many Apache and PHP configuration code we have discussed.

10.24 HTTP Caching

Sending deflated pages can reduce network traffic and increase page load speed. We know it can be done for static pages automatically by the Web server and for PHP pages by zlib output compression.

But, for any URL that a browser has previously visited, there is something even faster—not requesting or sending the page at all if it has not changed. With HTTP caching, standardized in HTTP 1.1, Web servers can attach expiration times to URLs so browsers and proxies can use their cached (saved) versions that have not expired[3]. A *Web proxy* is an agent computer requesting or serving Web resources on behalf of actual clients or *origin servers*. Proxies are widely used to make the Web more flexible, secure, efficient, and responsive. A proxy can often deliver cached Web resources directly to requesting clients without contacting the origin servers.

The Apache server provides configuration directives for default expiration settings on files, files in folders, and files of particular media types. For example,

```
ExpiresActive On
ExpiresByType application/pdf "modification plus 5 days"

<Location "/images">
ExpiresDefault "access plus 2 months"
Header append Cache-Control "public"
</Location>
```

where the `Cache-Control "public"` tells browsers and proxies to cache the resource that can be used, before its expiration time, without contacting the *origin server*. For an expired cached page, a client can request the URL with the added header

```
If-Modified-Since "<Last-Modified time>"
```

using the stored `Last-Modified` time from the origin server when the resource was cached. Upon receipt of such a request, the origin server will determine if the requested URL has changed since the `Last-Modified` time. If it has not changed, the server sends a very short response with an `HTTP 304 Not Modified` status, allowing the client to use the cached version, and perhaps also with a new `Expires` time. If the resource has changed, the new version, perhaps with deflation, is sent with a new `Expires` time or `max-age`.

Additionally, HTTP 1.1 offers a caching method where an origin server assigns an *entity tag* (ETag) for a particular URL to be cached by clients. When a client, with a cached copy, needs to retrieve the same URL again, it will send a request with the added HTTP header

```
If-None-Match: "<the cached ETag>"
```

---

[3]Cached pages no longer current are known as *stale*.

Upon receipt of such a request, the origin server will compare the incoming ETag with one it computes using the same hashing algorithm. If the two ETags match, the server sends a very short response with an HTTP 304 Not Modified status allowing the client to use the cached version. If the ETags do not match, the updated target resource (perhaps deflated) and its new ETag will be sent back in a response. The client will then replace its cached copy with the new version. Figure 10.11 summarizes cached page validation with ETag.

FIGURE 10.11: Cached Page Validation with ETag

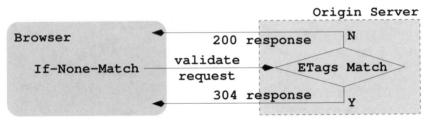

Thus, the ETag functions as a message digest or fingerprint of the resource at a given URL. We can apply entity tag based caching to avoid unnecessary loading of webpages that change infrequently, but at unpredictable times. For example, a PHP page is often produced by combining several files that do not depend on dynamic data, such as values from the file systems, databases, or Web services (Section 11.12). Such files often do not change frequently and are good candidates for optimization with ETags generated by PHP.

As an actual implementation example, let's apply ETags to the sliding menu example (**Ex: EtagPage**) described in Section 6.14.

The sliding menu page is composed by loading files several files. They are now listed on the $thefiles array at the very beginning of the page:

```
<?php
$thefiles=array(basename($_SERVER['PHP_SELF']),
  "etag.inc", "label.php", "cooling.inc", "heating.inc");
require_once("etag.php"); ?>
```

The file etag.php supplies functions and code needed to handle the ETage for this or any other such page (**Ex: PHPEtag**).

```
<?php  /////    etag.php    /////
$hash="";

function makeHash(&$files)                        // (1)
{ $str="";
  foreach($files as $f)
  { $str .= filemtime($f); }
```

```
      return md5($str);
}


// checks ETag request header
function check_etag(&$files)
{ global $hash;
    $hash = makeHash($files);                          // (2)
    if ( !empty($_SERVER['HTTP_IF_NONE_MATCH']) )      // (3)
    { $browser_hash=$_SERVER['HTTP_IF_NONE_MATCH'];
        if (  strstr($browser_hash, $hash)  )          // (4)
        { header("HTTP/1.1 304 Not Modified");         // (5)
            cache_control_headers(); exit;             // (6)
        }
    }
}


function cache_control_headers()
{ global $hash;
    header("Cache-Control: max-age=0, must-revalidate");
    header("ETag: \"{$hash}\"");
}

check_etag($thefiles);
cache_control_headers();                               // (7)
header("Content-Type: text/html; charset=UTF-8"); // (8)
?>
```

The makeHash function (line 1) computes an MD5 message digest of a string consisting of all the last-modified times of the involved files. This becomes the ETag value that changes if and only if one or more of the files are modified.

The check_etag function computes a new ETag value (line 2) and compares it with the ETag in the incoming request (lines 3 and 4). If the ETags match, we send just the HTTP 304 status line and the desired cache headers, then exit (lines 5 and 6).

If the ETags do not match, we send the desired cache headers (line 7) and the content type header (line 8) before page output. The Cache-Control HTTP header may take one or more of the following values:

- public—tells the browser and proxies along the network path that the page may be cached.

- private—asks the proxies along the network path not to cache the page but allows browsers to cache the page. Thus, the page is private between the origin server and any particular user.

- must-revalidate—tells browsers to revalidate with the origin server before serving the cached page.

- `max-age=S`—tells browsers to revalidate if the age of the cached page exceeds S seconds.

- `No-cache`—allows browsers to cache the page but not use the page without revalidating. Note: Some browsers, IE in particular, treat `No-cache` the same as `No-store`.

- `No-store`—asks browsers not to save the page in its cache store. Use this for pages with sensitive information.

See and test the complete example (**Ex:** `PHPEtag`) on the DWP website.

10.25   MySQL

MySQL is often the right choice for Web applications, especially in combination with Linux and PHP because PHP also has excellent built-in support for connecting and querying MySQL databases.

MySQL is a freely available open-source relational database management system that supports SQL. It runs on Linux, MS Windows®, Mac OS X®, and other systems and can be used from many programming languages, including C/C++, Eiffel, Java, Perl, PHP, Python, and Tcl. The MySQL database server supports both local and network access. It supports a *privilege and password system* to specify who can access/modify what in the database system.

Most Linux distributions come with MySQL installed. If you can locate the command **mysql** on your system, then, most likely, you have MySQL already. To be sure look for **mysqld** by starting **system-config-services**. If not, or if you wish to install the latest version of MySQL, please refer to Section 10.27.

10.25.1   Initializing, Starting, and Stopping MySQL

MySQL uses a **default database** named `mysql` for its own purposes, such as recording registered users (userid and password), managing databases, and controlling access privileges. The command **mysql_install_db** (in `usr/bin/`) is run once to initialize the MySQL default database (usually located in `/var/lib/mysql/mysql/`) and is done automatically when the MySQL server **mysqld** is started for the very first time. The **mysql_install_db** script contains many initialization settings for MySQL, and adjusting these settings allows you to customize various aspects of MySQL.

Starting **mysqld** can be done with the **system-config-services** GUI tool or the command

```
service mysqld start
```

The same GUI and command-line tools can be used to stop/restart the **mysqld**.

With **mysqld** started, MySQL client programs can communicate with it to access/manipulate databases served by it (Figure 10.12).

FIGURE 10.12: MySQL Server and Clients

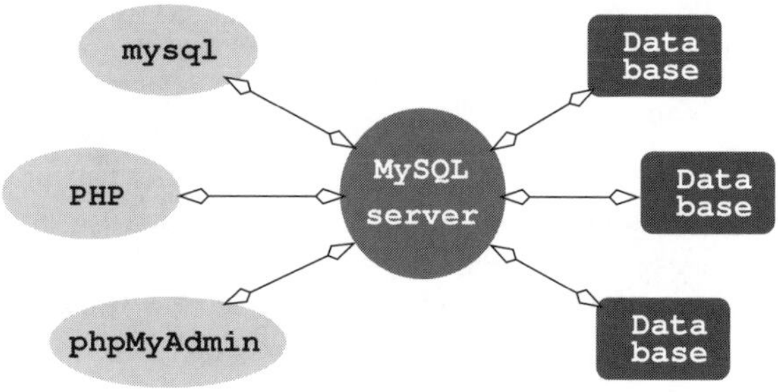

10.25.2 MySQL Run-Time Configuration

As **mysqld** (the database server) starts, it reads configuration values in `my.cnf` (usually kept in `/etc` or `/etc/mysql`). Specified are the data folder, the socket location, the userid of **mysqld**, and possibly many other settings. Edit `my.cnf`, and delete the line `bind-address = 127.0.0.1` if present.

It is also recommended that you consider running a local-access-only MySQL server rather than one that is network enabled. The latter allows MySQL clients to access the server via a network, which can mean security problems. The former will limit access to MySQL clients on the same host, making it much more secure. To do this, add the configuration setting

```
skip-networking
```

to both the `[mysqld]` and the `[mysqld_safe]` sections in `my.cnf`. You need to restart **mysqld** after making changes to the configurations. See the MySQL documentation for details about MySQL configuration.

It is a good idea to run the Linux command

mysql_secure_installation

to improve the security of your MySQL installation.

After starting **mysqld**, you can use **netstat**, a command to display networking status and activity on your system, to double-check. Run the command

```
netstat -tap | grep  mysqld
```

If you see a display, it means **mysqld** is allowing network access. If you see no display, then only local clients are allowed access. The `-tap` option tells **netstat** to display all information related to TCP with names of programs involved.

10.25.3   Administering MySQL

MySQL protects databases by requiring a userid and password, and, depending on what privileges the user has, various operations/accesses are allowed or denied.

At the beginning, MySQL has an administrator (**root**) and a blank password. The very first administrative task is to set a password for **root**[4].

mysqladmin -u root password *new_password*

The option -u specifies the MySQL userid **root** and the admin operation is password setting. Make sure you save the password for future use. Let's assume the **root** password is **foobar**.

The MySQL **root** is the user who can create new databases, add users, and set privileges for them. The command **mysqladmin**

mysqladmin -h localhost -u root -p*password* create dwph5

takes the hostname, userid, and *password* information and creates the new database **dwph5**.

Now we can add **pwang** as a user with all privileges to use **dwph5**. One way is to use the **mysql** tool, which is a command-line interface to the MySQL database server. Give the command

mysql -h localhost -u root -p*password* dwph5

Then you are working within **mysql**, and you may enter SQL queries. Do the following (**Ex: adduser.sql**):

```
mysql> USE mysql;                   (setting database name to mysql)
mysql> SHOW TABLES;                 (listing names of tables)
+-----------------+
| Tables_in_mysql |
+-----------------+
| columns_priv    |
| db              |
| func            |
| host            |
| tables_priv     |
| user            |
+-----------------+
mysql> INSERT INTO user (Host, User, Password, Select_priv)
    ->   VALUES ('', 'pwang', password('thePassword'), 'Y');
mysql> FLUSH PRIVILEGES;
mysql> GRANT ALL PRIVILEGES ON dwph5.* TO pwang
    ->   IDENTIFIED BY 'thePassword';
mysql> FLUSH PRIVILEGES;
mysql> quit
```

---

[4]Not to be confused with the Linux super user, which is also **root**.

Then inform user **pwang** about his userid, password, and database name. See the MySQL documentation for more information on setting user privileges. To reset the password for **pwang**, use the SQL

```
mysql> USE mysql;

mysql> update user set Password=PASSWORD('newOne')
    -> WHERE User='pwang';
```

Because PHP is usually available on the same host, the free *phpMyAdmin* tool (**phpmyadmin.net**) is often also installed to enable MySQL administration over the Web. *PhpMyAdmin* (Section 10.26) supports a wide range of operations with MySQL. The most frequently used operations are supported by the Web browser supplied GUI (managing databases, tables, fields, relations, indexes, users, permissions, and so on). Other operations are always doable via direct SQL statements. Both the **root** user and any user for a specific database can do database administration through *phpMyAdmin* from anywhere on the Web.

MySQL Workbench is a separate administration tool, and *MySQL Query Browser* is a tool for creating, executing, and optimizing queries. You can download and install these from **mysql.com**.

10.25.4   Resetting the MySQL Root Password

It is important to not forget the MySQL root password. However, if you find yourself in a such a situation, you can reset it. As Linux root, first stop the **mysqld**:

service mysqld stop

Then run **mysqld** in safe mode without security checking:

/usr/bin/mysqld_safe --skip-grant-tables &

Then run **mysql** on the default database **mysql**:

mysql -u root mysql

Then update the password for **root**:

```
mysql> update user set Password=PASSWORD('anything')
    -> WHERE User='root';
Query OK, 2 rows affected (0.04 sec)
Rows matched: 2  Changed: 2  Warnings: 0

mysql> flush privileges; exit;
```

Now kill the **mysqld_safe** process and restart the **mysqld**.

10.26 Installing phpMyAdmin

First, download the latest version from `phpmyadmin.net` and unpack in your Web document root folder (usually `/var/www/html`). For example (**Ex: myadmin.install**),

```
cd /var/www/html
tar jxvpf phpMyAdmin-3.3.4-english.bz2
rm phpMyAdmin-3.3.4-english.bz2
mv phpMyAdmin-3.3.4-english phpMyAdmin
```

The resulting `phpMyAdmin` folder is now in place under the Web document root and you can display installation instructions and other documentation with the URL

```
http://localhost.localdomain/phpMyAdmin/Documentation.html
```

To install `phpMyAdmin`, you only need to do a few things. In the `phpMyAdmin` folder, create a configuration file `config.inc.php` by copying and editing the sample file `config.sample.inc.php`.

It is recommended that you pick the `cookie` authentication method and set up a *control user*, as indicated by the sample configuration file, on your MySQL server so anyone who has a MySQL login can use `phpMyAdmin` to manage databases accessible to that particular user. See the `phpMyAdmin` documentation for configuration details.

After installation, the URL

```
http://host/phpMyAdmin
```

reaches the on-Web MySQL admin tool for any valid user to manage the database server. (Figure 10.13).

FIGURE 10.13: phpMyAdmin Tool

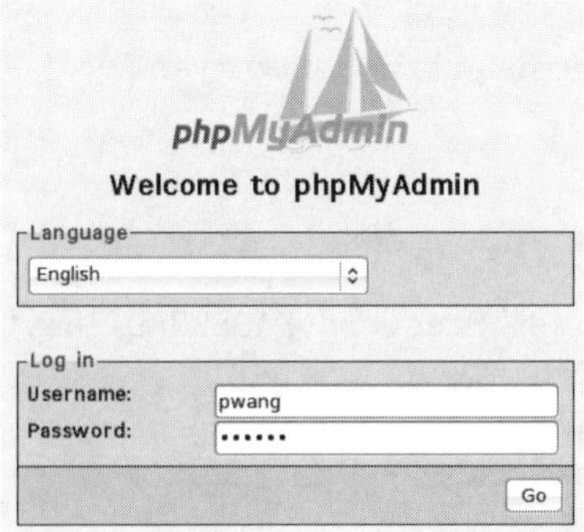

Be sure to install the latest version of phpMyAdmin, earlier versions have known security problems.

10.27 Installing MySQL

MySQL comes with most Linux distributions. In case there is a need, the Linux package management system makes installation/update easy. For CentOS/Fedora, do as root one of

```
yum install mysql-server mysql
yum update mysql-server mysql
```

For Ubuntu/Debian, do one of

```
sudo apt-get install mysql-server
sudo apt-get update mysql-server
```

Now proceed to edit the my.cnf file (Section 10.25.2) and then start/restart the **mysqld** (Section 10.25.1).

If you wish to install/update Apache+PHP+MySQL to achieve LAMP all at once, use these commands:

CentOS/Fedora:

```
yum install httpd php  mysql-server mysql
yum update httpd  php  mysql-server mysql
```

Ubuntu/Debian:

```
sudo apt-get install apache2 php5 \
                 libapache2-mod-php5 mysql-server
sudo apt-get update apache2 php5 \
                 libapache2-mod-php5 mysql-server
```

It may be even easier if you install *XAMPP for Linux*, which is LAMP made easier to install on Linux. Remember that these installations are very nice as developmental systems, but not secure enough as production systems. Enterprise editions of Linux will most likely include a production Web server with LAMP and more. What you learn here will apply directly to such production servers.

Refer to dev.mysql.com/downloads/ at the MySQL site for manual installation.

The MySQL Workbench is a GUI tool for administering databases and working with tables, queries, and other aspects of individual databases with a convenient graphical user interface. It is freely available and works on all major platforms. On Linux, the command

mysql-workbench

launches the GUI tool. See mysql.com/products/workbench for more information.

10.28   Linux Package Management

A *package management system* automates the installation and maintenance of software applications for any given operating system. For Linux we have two major systems: the *Advanced Packaging Tool* (**apt**) for the Debian family and the *Yellow dog Updater, Modified* (**yum**) for the Red Hat family.

Using the package management tools, you can easily install/remove, configure, and update packages made available by developers in online *repositories*. The checking of software dependencies and placement/replacement of files and commands are performed automatically.

We have used YUM and APT commands to install/update packages, especially in this chapter. Let's give a brief summary of package management commands here.

10.28.1   YUM and RPM

On CentOS/Fedora, the **yum** command is used for package management. It is basically a front end for the lower-level *rpm* tool.

- **yum** `install` *package-name* ...—Installs the specified packages along with any required dependencies.

- **yum** `groupinstall` *group-name* ...—Installs the specified package groups along with any required dependencies.

- **yum** `erase` *package-name* ...—Removes the specified packages from your system.

- **yum** `search` *string*—Searches the list of packages for names and descriptions that contain the fixed *string* and displays the matching package names, with architectures and a brief description of the package contents.

- **yum** `deplist` *package-name*—Displays a list of all libraries and modules on which the given package depends.

- **yum** `check-update`—Checks and lists available updates to installed packages.

- **yum** `info` *package-name*—Displays the name, description, version, size, and other useful information of the software.

- **yum** `reinstall` *package-name* ...—Removes and then installs a new copy of each given package.

- **yum** `localinstall` *local-rpm-file*—Installs without having to download.

- **yum** `update` *package-name* ...—Downloads and installs all updates including bug fixes, security releases, and upgrades, as provided by the distributors of your Linux. If no package name is given, all packages will be updated.

- **yum** `groupupdate` *group-name* ...—Downloads and installs all updates for the named group.

- **yum** `upgrade`–Upgrades all packages installed in your system to the latest release.

10.28.2 APT

On Ubuntu/Debian, use the **apt-get** command for package management.

- **sudo apt-get** `install package-name` ...—Installs the given packages, along with any dependencies.

- **sudo apt-get** `remove package-name` ...—Removes the packages specified, but does not remove dependencies.

- **sudo apt-get** `autoremove`—Removes any dependencies that remain installed but are not used by any applications.

- **sudo apt-get** `clean`–Removes downloaded package files for software already installed.

- **sudo apt-get** `purge package-name` ...—Combines the functions of `remove` and `clean` for specified packages. Also removes their configuration files.

- **sudo apt-get** `update`—Reads the `/etc/apt/sources.list` file and updates the system's database of packages available for installation. Run this after editing `sources.list`.

- **sudo apt-get** `upgrade`—Upgrades all packages if there are updates available. Run this after the command **apt-get** `update`.

The **aptitude** command is an interactive command-line front end for **apt-get** and can be more convenient to use.

10.29 For More Information

- Complete information for the Apache Web server can be found at `httpd.apache.org/`.

- The latest releases and documentation for PHP are at `php.net/index.php`.

- The site `www.mysql.com` contains all current releases and other information for MySQL.

- There is also a site for building LAMP servers at `www.lamphowto.com`.

- Linux package repositories can be easily found on the Web. For example, see `http://rpm.pbone.net` for RPM packages and `http://www.debian.org/distrib/packages` for APT packages.

- For a textbook on Linux we recommend *Mastering Linux*, by Paul S. Wang (`ml.sofpower.com`)

10.30 Summary

The combination of Linux, Apache, MySQL, and PHP (LAMP) forms a popular and powerful Web hosting environment. The freely available LAMP makes a great developmental system but should not be used directly as a production Web server for security reasons.

A Web server follows HTTP to receive requests and send responses. Its main function is to map incoming URIs to files and programs in the document space designated for the Web.

The Apache **httpd** Web server supports dynamic module loading and runtime configuration, making it very easy to customize and fit the requirements of a wide range of Web hosting operations. Configuration directives can be placed in central files and in *access files* (`.htaccess`) under individual folders within the document space.

In addition to controlling features and behaviors of **httpd**, Apache configurations can specify access limitations to parts of the document space and can require login with HTTP Basic or Digest Authentication. Apache supports HTTP 1.1 features and offers name-based and IP-based virtual hosts, automatic file deflation, and HTTP cache control.

PHP can be installed as an Apache module and will interpret embedded PHP scripts as the Apache **httpd** delivers a response page. When PHP starts, it sets options by reading configuration directives contained in the `php.ini` file. The value of certain PHP options may also be set in user scripts, in `.htaccess`, or in `httpd.conf`.

PHP has good database support and works well with MySQL database systems. MySQL supports multiple databases protected by userid and password. Different database users may have different access privileges and can be managed easily using Linux commands (**mysqladmin**, **mysql**, and so on), the Web-based phpMyAdmin tool, or the GUI tool MySQL Workbench.

The YUM/APT package management tools are handy for installing and maintaining software packages on Linux Web hosts.

Exercises

10.1. Assuming your Linux is running the Apache Web server, find the version of Apache server, the `httpd.conf` file, and the document root folder.

10.2. How do you go about finding out if your Apache Web server supports per-user Web space?

10.3. How do you find out if your Apache has PHP support? If so, where is the file `php.ini` and for what purpose?

10.4. Install your own Apache server with PHP support under your home directory (Hint: Use a non-privileged port). After installation, start your own **httpd** and test it.

10.5. Set up your Apache to automatically deflate `.html`, `.css`, and `.js` files.

10.6. What is a real host? A virtual host? An IP-based virtual host? A name-based virtual host?

10.7. Configure your Apache to require a password on some Web folder. Create some valid users and test your setting to make sure that it works.

10.8. Install your own MySQL under your home directory. You will be the root database user. Create a new test database and some tables using the **mysql** tool.

10.9. Install the phpMyAdmin tool. Use it to manage your MySQL database.

10.10. Set up some database tables for the Web in your MySQL using your phpMyAdmin tool. Test your setup with PHP code in a webpage.

10.11. Is it possible to deflate output of PHP pages? If so, how?

10.12. Is it possible to take advantage of HTTP caching for PHP-generated pages? If so, what type of generated pages, and how?

Chapter 11

XML, AJAX, and Web Services

HTML supplies elements designed to markup webpages. We see how useful HTML is to structure and organize content for the Web. The markup makes webpages easy to process by computers and that is a huge advantage.

Now, what about applying the same markup technique to structure and organize other types of textual information? What if we desire to have a set of tags for book catalogs, news feeds, college course listings, shipping rates, TV shows, sports, vector graphics objects, or even mathematical formulas?

In fact, we can create a set of elements for any such content we care to so organize. The technology is called XML, the *eXtensible Markup Language.*

XML is used widely in many areas including the Web. For example, XHTML is an XML-compliant version of HTML and polyglot HTML5 (XHTML5) is also compliant with XML (Section 2.6). Furthermore, SVG (*Scalable Vector Graphics*) and MathML (*Mathematical Markup Language*), both XML-defined languages, are part of HTML5.

The purpose of XML is to make documents easy to communicate among heterogeneous platforms and efficient to process by programs. The set of XML tools is extensive. Web browsers and server-side scripting languages such as PHP all provide built-in support for XML processing.

AJAX (*Asynchronous JavaScript and XML*) and *Web Services* are important and widely used Web technologies that often rely on XML for data representation. AJAX can update information on a webpage, by making HTTP requests back to the server, without navigating to a new page. Web services can make useful computations and data available to others through HTTP, returning results in XML or some other well-defined format.

Information and techniques presented in this chapter can make any website more functional, dynamic, efficient, and effective.

11.1 What Is XML?

XML is not a markup language with a set of tags (elements) specific to structuring a certain kind of document. In fact, XML defines no tags at all. Instead, XML is a set of rules by which you can invent your own set of tags (elements) to organize and structure any specific type of textual documents you wish. Thus, XML is a way to define your own markup language. Each XML-defined markup language is known as an *XML application.* Documents so marked up are called XML documents. XML is also a W3C standard and there are many tools, mostly free, to create, process, and display XML documents.

There are many XML applications in addition to SVG and MathML. For example, RSS and Atom are standard Web news syndication formats based on XML (Section 11.3). SMIL (*Synchronized Multimedia Integration Language*) is an XML application to display multimedia contents in a time-synchronized manner. SMIL is often considered PowerPoint™ for the Web. The list of publicly used XML applications is long (See the wikipedia). And then there are non-public markups used within organizations and even by individuals. XML makes it possible and easy for anyone to create a new markup language.

11.2   XML Document Format

As shown in Figure 11.1, an XML document begins with processing instructions, followed by a *root element* that may contain other elements. The elements are defined by individual XML applications following rules given by XML.

FIGURE 11.1: An XML Document

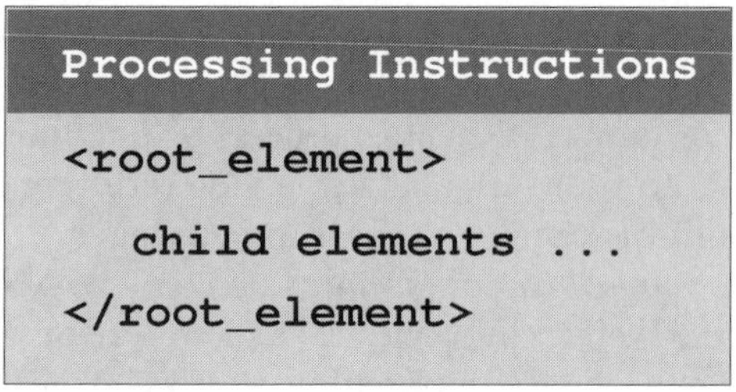

For example, here is a simple XML document:

```
<?xml version="1.0" encoding="UTF-8"?>          (1)
<address>                                        (2)
    <street>432 Main Street</street>
    <city>Arbville</city>
    <state>Ohio</state>
    <zip>02437</zip>
</address>
```

The `<?xml` processing instruction (line 1) is required as the first line of any XML document. It provides the XML version and the character encoding used for the document. Other processing instructions follow the first line and can supply style sheets, definitions for the particular type of XML document, and other meta information for the document. Coming after the processing

instructions is the root element that encloses the document content. The root element in this example is the `address` element (line 2), which constitutes the remainder of the XML document.

11.3   News Syndication Formats

XML is used widely in practice. Perhaps the best-known example is news syndication on the Web. The W3C standard RSS (*Really Simple Syndication*) is an XML application for *Web feeds*, frequently updated content such as blog entries and news headlines. An RSS document can supply an abstract, author, organization, publishing dates, and a link to the full story or article. Using a Web browser or an RSS reader, end users can *subscribe* to RSS feeds and easily see the changing headlines.

Here is a simple RSS example (**Ex: RSS**), the `books.rss` file:

```
<?xml version="1.0" encoding="ISO-8859-1" ?>
<rss version="2.0"><channel>
 <title>Sofpower Book Sites</title>
 <link>http://sofpower.com</link>
 <description>Companion websites for textbooks</description>
 <item>
   <title>Mastering Linux</title>
   <link>http://ml.sofpower.com</link>
   <description>Resources for "Mastering Linux",
   a highly recommended Linux book (2010)</description>
 </item>
 <item>
   <title>Dynamic Web Programming and HTML5</title>
   <link>http://dwp.sofpower.com</link>
   <description>Resources for "Dynamic Web Programming and
   HTML5", an in-depth Web programming book</description>
 </item>
</channel></rss>
```

The file lists two information/news items from the sofpower.com (channel) each linking to a website.

Atom is another XML-based news syndication standard from the IETF (Internet Engineering Task Force).

Web browsers and news reader applications understand RSS and Atom formats. Clicking on a link leading to an RSS or Atom file causes a Web browser to ask you if you wish to subscribe to the news feed. If you do subscribe, your browser will bookmark the feed and monitor any new headlines from the feed.

News feed links are usually marked by an orange-colored icon. See `cnn.com/services/rss` for RSS feeds from CNN, or

`hosted2.ap.org/APDEFAULT/APNewsFeeds`

for Atom feeds from AP.

11.3.1   XML Document Syntax Rules

- An XML document consists of an XML declaration followed by a *root element*. All other elements are contained inside the root element.

- XML tag and attribute names are case sensitive.

- XML elements must have both open and close tags.

- Elements must be properly nested.

- Element content (textual data) is placed between the open and close tags. *Entities* can be used for special characters in element content. The five predefined entities are `&lt;`, `&gt;`, `&amp;`, `&apos;` (`'`), `&quot;` (`"`). If an XML element encloses no data (empty), then it can use the short-hand form `<tagName ... />`.

- An XML attribute must always be given with a value and the value must always be quoted (with either single `'` or double `"` quotes).

- XML preserves white space. There is no automatic collapsing of white space.

- XML uses *lf* (`\n`) as line terminator, and any CR LF is converted to LF.

- XML comments are given inside `<!--` and `-->`.

An XML application defines its own elements. Element names must follow these rules:

- A name is a sequence of letters, digits, and other characters.

- A name must not contain white space.

- A name must not begin with a digit, punctuation, `XML`, or `Xml`.

- Avoid using : in a name. This is reserved for XML namespaces.

An XML element may have attributes supplying metadata for the element: `id`, `style`, `class`, `title`, and so on. Values for attributes are quoted strings (single or double quotes). Allowed/recognized attributes can be defined in a DTD (*Document Type Definition*) or XML Schema.

11.3.2 Validity of an XML Document

An XML document is *well-formed* if it follows the XML syntax rules. An XML document is *valid* if it is well-formed and conforms to a document definition specified in a DTD or XML Schema and attached to the XML document via a processing instruction. We may use XML documents without first developing any document definitions. In many situations, a well-formed XML document is enough.

Browsers can load and validate XML documents. The W3C and others make XML validators available on the Web.

11.4 Browser Display of XML Documents

XML documents are used for many purposes. In particular, they are also understood by Web browsers. Take, for example, the following `catalog` we invented (in file `catalog.xml`) to illustrate XML usage (**Ex: Catalog**):

```
<?xml version="1.0" encoding="UTF-8"?>
<catalog>
   <course>
      <title>Web Design and Programming I</title>
      <instructor>
         <name>Tony Samangy</name>
         <department>VCD</department>
         <email>tsamangy@adelphia.net</email>
      </instructor>
      <instructor>
         <name>Paul Wang</name>
         <department>CS</department>
         <email>pwang@cs.kent.edu</email>
      </instructor>
      <semester>Fall</semester>
   </course>
<!-- more courses -->
</catalog>
```

The root element `catalog` may contain multiple `course` elements. Each `course` has a title, one or more instructors, and a semester designation. The name, department, and email are specified for each instructor.

What if we put such a `catalog` in a file, say `catalog.xml`, and put it on the Web? A Web server would deliver such a file with the content type `text/xml`. As long as the file is well-formed, a Web browser will display the source code in a nicely indented manner showing the parent-child element containment structure.

Now if we attach a CSS style sheet to `catalog.xml` (**Ex: CatalogCss**):

```
<?xml version="1.0" encoding="UTF-8"?>
<?xml-stylesheet type="text/css" href="catalog.css"?>
<catalog>
   . . .
</catalog>
```

where `catalog.css` is

```
/*****    A CSS file for course.xml    *****/
catalog
{ display: list; margin: 2em;
  background: #def; line-height: 130% }
course { display: list-item; margin-bottom: 20px; }
name, department, email { display: inline; color: blue; }
instructor, semester { display: block; }
semester { font-weight: bold;  }
title { display: block; font-weight: bold; }
email { font-family: courier; }
```

The XML file now displays as an itemized list (Figure 11.2).

FIGURE 11.2: CSS-Styled XML Display

In practice, it is often more useful to display XML supplied data inside a webpage rather than as a separate page. We can embed XML supplied data easily with a `object` tag (**Ex: XmlData**)

```
<h2>Our Course Listing</h2>
<object style="border: thin solid black" width="400"
      height="150" data="catalogcss.xml" /></object>
```

resulting in a display shown in Figure 11.3.

FIGURE 11.3: object-Embedded XML

11.5   Transforming XML Documents with XSL

XML documents make it easy to communicate data, for many different purposes, among heterogeneous computers and programs. When an XML document is received by a program, it often needs to extract the information contained in the XML document and use the extracted data in some way. The W3C standard XSL (*Extensible Stylesheet Language*) is designed for defining arbitrary data extraction and transformations for XML documents. With XSL we potentially are able to take the same XML *source document* and put it to use in many directions through different XSL-defined transformations.

XSL has two main components: XSLT (*XSL Transform*), a language for defining rewrite rules to transform an XML document, and XPath, a notation to select parts inside the source XML document. We can think of an XML document as a data source (like a database) and the XSL as a query language to extract information from an XML document. For Web programming, we will focus on generating XML and HTML code using XSL.

11.5.1   A First XSL Transform

As our first example (**Ex: CatalogOL**), let's write an XSL program (called an XSL style sheet) that generates an HTML ordered list by transforming the source XML document `catalog.xml` (Section 11.4). The resulting HTML code produces a display shown in Figure 11.4.

The XSL style sheet `catalogOL.xsl` is written in XML:

```
<?xml version="1.0" encoding="ISO-8859-1"?>
<xsl:stylesheet version="1.0" xmlns:xsl=          (A)
    "http://www.w3.org/1999/XSL/Transform">
<xsl:template match="/">                          (B)
<html><body>                                      (C)
<h2>Course Listing (from XSL)</h2>
  <ol>                                            (D)
    <xsl:for-each select="catalog/course">        (E)
      <li><p><xsl:value-of select="title"/>:       (F)
```

FIGURE 11.4: Ordered List Display

```
        <xsl:value-of select="semester"/></p></li>    (G)
    </xsl:for-each>                                    (H)
  </ol>
</body></html>
</xsl:template>
</xsl:stylesheet>
```

The root XSL element is `stylesheet` as given on line A, where `xsl:` is a *namespace prefix* indicating elements belonging to XSL. A `stylesheet` element can provide one or more *template rules* (`xsl:template`) to specify rewrite rules on the source XML document. A `stylesheet` can use `xsl:include` and/or `xsl:import` to incorporate other XSL files. A `stylesheet` element may also define global values `xsl:param` and `xsl:variable` used in templates. The `xsl:param` element also enables parameters to be passed into a style sheet.

The result produced by a template rule contains fixed parts and computed parts that are selected from the XML source file. The single template (line B) in our `catalogOL.xsl` example specifies fixed HTML code such as lines C through D and computed parts on lines E through F.

The `match` attribute (line B) indicates, using XPath notation (Section 11.8), a place in the source XML where the template applies. Here, `/` indicates the root element of the source XML document, namely `<catalog>`.

The `for-each` element (lines E–H) contains code applied to each child `course` of `catalog` in the source XML. With the `value-of` element, the value for `title` (line F) and value for `semester` (line G) from the XML source document are combined with the fixed parts to produce the desired result, as follows:

```
<html><body>
<h2>Course Listing (from XSL)</h2>
<ol>
<li><p>Web Design and Programming I:
        Fall</p></li>
```

```
<li><p>Web Design and Programming II:
         Spring</p></li>
<li><p>Web Design and Programming Studio:
         Spring</p></li>
</ol>
</body></html>
```

Note that the *value of an XML element* is all the non-tag strings contained in the element and its descendant elements. You can use the `copy-of` element to include the XML tags. Thus, the value of the first instructor element in `catalog.xml` is

```
Tony Samangy
VCD
tsamangy@adelphia.net
```

whereas the copy of the same element would be

```
<name>Tony Samangy</name>
<department>VCD</department>
<email>tsamangy@adelphia.net</email>
```

To apply `catalogOL.xsl` to `catalog.xml`, we can add the processing instruction

```
<?xml-stylesheet type="text/xsl" href="catalogOL.xsl"?>
```

to `catalog.xml`. If we do that, then accessing `catalog.xml` from any browser will produce the display shown in Figure 11.4.

11.5.2 XSL Transforms

We have seen the `for-each` element of XSLT for looping over a set of selected elements in the source XML. The `if` and `choose` elements give you the ability to apply transformations conditionally.

For example, the XSL file (`if.xsl`) uses the following template

```
<xsl:template match="/catalog">
<html><body><h2>Fall Courses</h2>
  <ol><xsl:for-each select="course">
    <xsl:if test="semester ='Fall'">                    (I)
      <li><p><xsl:value-of select="title"/>
          </p></li>
    </xsl:if>
  </xsl:for-each></ol>
</body></html>
</xsl:template>
```

which when applied to `catalog.xml` produces the result

```
<html><body>
<h2>Fall Courses</h2>
<ol><li><p>Web Design and Programming I</p></li></ol>
</body></html>
```

The value of the `test` attribute (line I) is an XPath expression (Section 11.8) checking if the `semester` in a `course` is equal to `Fall`. Child elements of an `if` only take effect when the if condition is true.

The `choose` element is used for multi-branch conditionals and can contain one or more `when` child elements followed by an `otherwise` element.

```
<xsl:choose>
  <xsl:when test="expr1"> ...  </xsl:when>
  <xsl:when test="expr2"> ...  </xsl:when>
  ...
  <xsl:otherwise> ...  </xsl:otherwise>
</xsl:choose>
```

11.5.3   XSL Templates

XSL transform builds a result tree by processing each node in the *current node list* containing specific nodes from the source XML tree. At the beginning, the current node list contains one node, the root node of the source XML tree (Figure 11.5).

FIGURE 11.5: XSL Transform

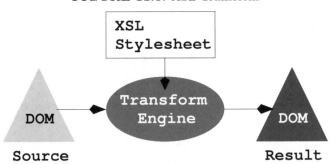

An XSL file contains one or more templates. The `match` expression defines to which source nodes a template may apply. To process a node, XSL picks the best template from applicable templates and *instantiates* it with the node (the node becoming the template's *current node*) and applies the template. After applying the template, the node is removed from the current node list. When the current node list becomes empty, XSL processing terminates.

To process a node on the current node list without applicable templates, the node is replaced by its child nodes on the current node list and node processing continues. To process a text node without applicable templates, the value of the text node is inserted in the result tree. Thus, XSL processing performs a depth-first tree traversal of the source tree looking for applicable templates to apply to the current node. When a template is applied, the template code specifies how the subtree rooted at the current node is handled. This traversal is known as *apply-templates* on the root node.

At the end of XSL processing, a result tree (DOM) is built. Let's look at (`apply.xsl`), which uses several templates to transform the document `catalog.xml` (**Ex**: XSLTemplate).

```
<xsl:stylesheet>
 <xsl:template match="/">                      (1)
  <html><body><h2>Course Listing</h2>
     <ol><xsl:apply-templates /></ol>          (2)
  </body></html>
 </xsl:template>

 <xsl:template match="course">                 (3)
  <li><p style="color: blue">
   <xsl:value-of select="title"/>:
   <xsl:value-of select="semester"/> Semester
   (<xsl:apply-templates select="instructor"/>) (4)
  </p></li>
 </xsl:template>

 <xsl:template match="instructor[1]">          (5)
   <b><xsl:value-of select="name" /></b>
 </xsl:template>

 <xsl:template match="instructor">             (6)
   <xsl:text>, </xsl:text>                      (7)
   <b><xsl:value-of select="name" /></b>
 </xsl:template>
</xsl:stylesheet>
```

Processing starts with the application of template (line 1) to the root node (the current node). The `apply-templates` (line 2) finds a template to apply to the current node and all its descendants. Thus, the `course` template (line 3) is applied to each `course`.

The `apply-templates` on line 4 finds a template to apply to each `instructor` and its descendants. The template rule on line 5 is selected to apply to the first `instructor` because it is more specific than the template on line 6, which is applicable to all `instructors`. Note the use of `xsl:text` to generate a comma followed by a SPACE.

The result tree generated is as follows, with a visual display shown in Figure 11.6.

```
<html><body><h2>Course Listing</h2>
<ol><li><p style="color: blue">Web Design and
  Programming I: Fall Semester
  ( <b>Tony Samangy</b>, <b>Paul Wang</b> )</p></li>
  <li><p style="color: blue">Web Design and
  Programming II: Spring Semester
  ( <b>Sanda Katila</b>, <b>Paul Wang</b> )</p></li>
  <li><p style="color: blue">Web Design and
  Programming Studio: Spring Semester
  ( <b>Sanda Katila</b>, <b>Paul Wang</b> )</p></li>
</ol></body></html>
```

FIGURE 11.6: Display of `apply-templates` Result

Course Listing

1. Web Design and Programming I: Fall Semester
   (**Tony Samangy**, **Paul Wang**)

2. Web Design and Programming II: Spring
   Semester (**Sanda Katila**, **Paul Wang**)

3. Web Design and Programming Studio: Spring
   Semester (**Sanda Katila**, **Paul Wang**)

Here is how `apply-templates` works. It selects all child nodes of the current node or a set of nodes specified by its `select` attribute. For the selected nodes and all their descendants, `apply-templates` directs the XSLT processor to find an appropriate template to apply. Template applicability is determined by the match attribute. If multiple templates are applicable, the one most specific or has the highest priority is chosen. If no template is applicable to a node, `apply-templates` processing continues to its child nodes. If the node is a text node and no template is applicable, the text is placed in the result tree.

11.6 Named Templates and Template Calls

Instead of using `apply-templates`, which can have unforeseen effects, many prefer to write XSL templates with names, and perhaps also parameters, and call them explicitly. The situation is similar to defining and calling subroutines in procedural programming.

Let's take the *apply.xsl* and modify it slightly to become `call.xsl` (**Ex: XSLCall**. The changed parts are shown here:

```
<xsl:template match="/catalog">
<html> <body> <h2>Course Listing</h2>
    <ol>
    <xsl:for-each select="course">              (10)
    <xsl:call-template name="course"/>
    </xsl:for-each>                             (11)
    </ol>
</body> </html>
</xsl:template>


<xsl:template name="course">                  (12)
<!-- same code as before -->
</xsl:template>
```

The `apply-templates` code (line 2) is replaced by a `for-each` element (lines 10 and 11), calling the named template `course` (line 12) on each course. The `course` template is the same as the one on line 3 except the `match` is changed to `name`. The files `apply.xsl` and `call.xsl` produce the same result. When calling a template, the current node becomes the current node of the called template.

It is possible to pass parameters in a template call. For example, the template `abc`,

```
<xsl:template name="abc">
  <xsl:param name="x" />
  <xsl:param name="y">default value for y</xsl:param>
  ...
</xsl:template>
is called with the code
\begin{verbatim}
<xsl:call-template name="abc">
  <xsl:with-param name="x">value for x</xsl:with-param>
  <xsl:with-param name="y">value for y</xsl:with-param>
</xsl:call-template>
```

The received parameter value for x can be referenced as $x in `select` XPath expressions. For example,

```
<xsl:value-of select="$x" />
```

and in the output, for example,

```
<h3 class="{$x}">Abstract</h3>
```

You may also place a `param` element before the first template to establish a global parameter usable in all the templates. Values for such global templates can be set after the XSL style sheet has been loaded into an XSL transform engine.

For example, in the style sheet `param.xsl`, the global `semester` parameter has the default value `'Fall'` (line 13). See Section 11.7 for passing a different value to `semester` from a PHP XSLT object.

```
<xsl:param name="semester">Fall</xsl:param>         (13)
<xsl:template match="/catalog">
  <h2><xsl:value-of select="$semester"/>
     Courses</h2>
    <ol><xsl:for-each select="course">
    <xsl:if test="semester = $semester">
      <li><p><xsl:value-of select="title"/>
           </p></li>
    </xsl:if>
    </xsl:for-each></ol>
</xsl:template>
```

See Section 11.10 for an example where the XSL parameter value is set in JavaScript.

Similar to parameters, you can also define variables in XSL:

```
<xsl:variable name="x" select="expr"/>
<xsl:variable name="x">value</xsl:variable>
```

A variable is global if it is defined as a child of the root element. Otherwise, the variable is local within its context.

11.6.1   Sorting in XSL

The XSL `sort` element can be used within `for-each` or `apply-templates` to order the results produced. For example,

```
<xsl:for-each select="catalog/course">
 <xsl:sort select="semester"/>
   <li><p><xsl:value-of select="title"/>:
       <xsl:value-of select="semester"/>
       Semester</p></li>
</xsl:for-each>
```

produces a list sorted by the name of the semester. The `select` attribute indicating the sort key is a required attribute for `sort`. Optional attributes include: `order` (ascending or descending), `data-type` (for key, text, or number), and `case-order` (upper-first or lower-first).

11.7   XML Processing in PHP

PHP has excellent support for XML processing. A `SimpleXMLElement` object can be used to represent and manipulate any XML document. You can load a valid XML string or file into a `SimpleXMLElement` object and then process it using PHP property selectors and array iterators. The structure of a `SimpleXMLElement` object is simple. Its property name is an XML element name. Its property value is a string, a `SimpleXMLElement` object, or an array of `SimpleXMLElement` objects.

For example (**Ex: SimpleXMLDemo**), we can load `catalog.xml` into a `SimpleXMLElement` object:

```php
<?php    $xmlsource="catalog.xml";
if (file_exists($xmlsource))
{  $cat = simplexml_load_file($xmlsource); }    // (a)
if ( $cat )
{  print_r($cat);                               // (b)
   echo $cat->course[1]->title . "\n";          // (c)
}
?>
```

The function `simplexml_load_file` loads a file (local or remote) and returns a `SimpleXMLElement` object `$cat` (line a). Use `simplexml_load_string` to load an XML string. The structure of `$cat` is easily revealed by the `print_r` call (line b). The code on line c displays the title of the second course under the `catalog`.

`SimpleXMLElement` methods such as

```
children()
attributes()
addChild($sxmlObj)
addAttribute($name,$value)
```

can read or set the XML structure. The constructor call

```
new SimpleXMLElement($XMLstr)
```

creates an object from the given string.

The method `asXML()` gives you the XML string of the object. PHP also provides a `DOM` class that implements the W3C standard XML DOM interface. See the PHP manual for more information on `SimpleXMLElement` and `DOM`.

11.7.1 Making XSL Transforms in PHP

Follow these steps to apply any XSL style sheet to any source XML document:

1. Load the style sheet and the source XML document into separate `SimpleXMLElement` elements (lines d and e).

2. Create an `XsltProcessor` (line f), import the loaded style sheet (line g), and set any desired parameters (line h).

3. Apply the transform processor to the loaded XML source document and obtain the result XML (line i).

```php
<?php /////    xslProc.php     /////
$transform="param.xsl";
$xmlsource="catalog.xml";

if (file_exists($xmlsource))
{  $xml = simplexml_load_file($xmlsource); }      // (d)
if (file_exists($transform))
{  $xsl = simplexml_load_file($transform); }      // (e)
if ( $xsl && $xml )
{  $xslt = new XsltProcessor();                    // (f)
   $xslt->importStylesheet($xsl);                  // (g)
   $xslt->setParameter("", 'semester', 'Spring');// (h)
   if ($result = $xslt->transformToXML($xml))      // (i)
   {  echo $result; }
   else
   { trigger_error('XSL transformation failed.',
     E_USER_ERROR);
   } // if
}
?>
```

Running `xslProc.php` produces the following output:

```
<?xml version="1.0"?>
<h2>Fall Courses</h2><ol><li>
<p>Web Design and Programming I</p></li></ol>
```

Change the first two lines in `xslProc.php` to apply any XSL style sheet to any XML file you like (**Ex: PHPXsl**).

11.8   XPath

XPath is a W3C standard notation used in XSL to specify expressions. In XSLT template rules, XPath *location* notations are used to select a set of nodes on the source XML tree for processing. The notation is interpreted relative to the *context node* or the *current node* being transformed. Table 11.1 shows many common location notations where we use the abbreviations: ch (child node), chn (child nodes), attr (attribute), and doc (document).

   Node sets can be combined with the | operator. For example, if the current node is /catalog, then the expression

```
course[semester="Fall"]/instructor |
    course[semester="Spring"]/instructor
```

TABLE 11.1: XPath Location Notations

Location	Selects	Location	Selects
.	the node itself	xy[1]	1st xy ch
..	parent	xy[last()]	last xy ch
xy	all xy chn	*/xy	all xy grandchn
*	all chn	text()	all text node chn
.//xy	xy descendants	/a/b[2]/c[3]	3rd c of 2nd b of a
@name	the name attr	xy[@id="me"]	all xy with id="me" attr
../@red	red attr of parent node	xy[@id="me"][5]	the 5th of xy chn with id="me" attr
chap[au]	chap chn with <au> ch	xy[5][@id="me"]	the 5th xy ch only if it has an id="me" attr
chap//xy	xy descendants of chap chn	chap[au="Li"]	chap children with <au>Li</au> child
//xy	all xy in doc	//inst/name	all inst/name in doc
@*	all attrs	xy[@red and @blue]	xy chn with both attrs

gives all instructor nodes for courses in the Spring or Fall semester.

XPath expressions may use the usual arithmetic and logical operators such as +, -, *, div (division), >, <, >=, <=, and, or, not, and mod (remainder).

XPath also provides many useful functions. String functions include

- string(*arg*)—returns the string representation of the argument.

- concat(s1, s2, ...)—returns the concatenated string.

- starts-with(s, prefix)—returns true or false.

- contains(s, substring)—returns true or false.

- substring-before(s, substring)—returns the substring in s.

- substring-after(s, substring)—returns the substring in s.

- substring(s, pos, len)—returns the substring in s from the given position of the given length. len is optional. (1-based positioning).

- string-length(*str*)—returns number of characters. If *str* is not given, the string rep of the context node is used.

- normalize-space(*str*)—returns a **normalized** string by stripping leading and training white spaces and multiple white spaces into one.

- translate(s, from, to)—returns a translated string in the spirit of the UNIX **tr** command.

Number functions include

- **number(** *arg* **)**—returns the argument converted to a floating-point number.

- **floor(** *number* **)**—returns the floor.

- **ceiling(** *number* **)**—returns the ceiling.

- **round(** *number* **)**—returns the rounded number.

11.9   AJAX

AJAX (*Asynchronous JavaScript and XML*) is not a language or protocol. It is a technique to dynamically update information on a webpage using client-side JavaScript, without reloading the page. The actions usually take place *asynchronously* or concurrently with other browser/GUI activities.

Perhaps the most widely used AJAX effect is dynamic text completion (or suggestions) for user input. As a user inputs for a search box or other form field, suggestions can be retrieved via AJAX from the server side and displayed immediately. Many users find this very helpful and convenient. Other useful AJAX applications include updating sports scores and stock prices, user control of map display, weather report/status, games, client-side user input validation, user-friendly file upload, and many, many more. With AJAX we can also develop Web applications that work like traditional offline applications. Basically, AJAX enables us to create a browser-supplied GUI that can employ all the capabilities provided by JavaScript on the client-side and Web services on the Internet. With such a GUI, users are not limited to browsing the Web in the traditional sense of visiting different webpages.

AJAX requires the following JavaScript capabilities:

- Making HTTP requests (see **Ex: JSHttpGet** and **Ex: JSHttpPost**)

- Receiving HTTP responses

- Processing response data

- Manipulating HTML and XML DOM trees

and combines them to achieve intended effects. Our first AJAX example demonstrates dynamic text suggestions.

11.9.1   Auto Completion with AJAX

Let's demonstrate AJAX by a simple example of completing names for Ohio counties. As the user enters the county name, the partial text entered is sent to a server-side program which sends back all Ohio county names with that prefix. The list of county names is displayed in a suggestion box. Clicking on a suggestion enters it into the input box. The suggestion box is updated on a per-keystroke basis (Figure 11.7). The HTML code for this example (**Ex:**

FIGURE 11.7: AJAX Auto-Complete Example

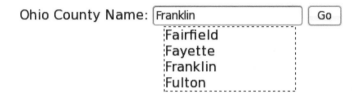

AutoComplete) is

```
<h3>AJAX Auto-complete Example</h3>
<form action="showpost.php" method="post">
Ohio County Name: <input type="text" id="county"
  onkeyup="getSuggestion(this.value)"              (A)
  name="county"/>  <input type="submit" value=" Go "/>
</form>
<div class="suggestion" id="suggestions"></div>     (B)
```

For each character entered, the JavaScript function getSuggestion is called with the current input value (line **A**). Any suggestions obtained will be displayed in the suggestion box (line **B**).

The JavaScript function getSuggestion

```
function getSuggestion(str)
{ if ( !req ) init();                                // (C)
  str=str.replace(/^\s+||\s+$/g,'');                 // (D)
  if (str==="") { sbox.style.display="none"; return; }
  req.open("GET", "suggest.php?qs="+str, true);      // (E)
  req.send();                                        // (F)
}
```

calls init() for a one-time setup of global values, including the req request object. White spaces are removed from the user input string (line D) and sent in a GET request. The open method (line E) sets the method, URL, and async properties for the request. The send method (line F) actually makes the connection and sends the request over the network. The third argument for req.open (true or false) makes the HTTP request *asynchronous* or *synchronous*. In asynchronous mode (line E), the req.send() call will return immediately without waiting for the response, which can be time consuming.

For security reasons, a JavaScript HTTP request may only be sent back to the *page host*, the Web server from where the webpage containing or referencing the JavaScript comes. Thus, a webpage cannot make an AJAX HTTP request to any Web server except the page host.

The `init()` call sets up an XMLHTTP object (using browser-dependent code), the suggestion box (`sbox`), and an event handler `display_suggestion` (line G) called when the HTTP response becomes ready.

```
var req=false, sbox;
function init()
{ req = (window.XMLHttpRequest)
      // for IE7+,Firefox,Chrome,Opera,Safari
      ? new XMLHttpRequest()
      // code for IE6, IE5
      : new ActiveXObject("Microsoft.XMLHTTP");
  sbox=document.getElementById("suggestions");
  req.onreadystatechange=display_suggestion;         // (G)
}


function display_suggestion()
{ var sg;
  if (this.readyState==this.DONE                      // (H)
      && this.status==200)
  { if ( (sg=req.responseText) && sg !="")            // (I)
    { sbox.innerHTML=sg;
      sbox.style.display="block";
    }
    else { sbox.style.display="none"; }
  }
}
```

The `onreadystatechange` handler is called each time the `readyState` property (the status of the response) of the `XMLHttpRequest` object changes. The `readyState` (line H) may have the following predefined constant values:

```
UNSENT = 0;   OPENED = 1; HEADERS_RECEIVED = 2;
LOADING = 3;  DONE = 4;
```

After the request is done, the result can be obtained in

- `req.responseXML`—An XML DOM object that can be accessed and manipulated from JavaScript following the appropriate XML DOM API.

- `req.responseText`—The server response as a text string (line I).

The server-side PHP script computes the suggestions by matching (line K) the incoming string with entries in an array of Ohio county names (line J). Each matching county name results in a clickable suggestion (line L).

```
<?php  /////    suggest.php    /////
$a=array("Adams", "Allen", "Ashland", "Ashtabula",      // (J)
"Athens", "Auglaize", "Belmont", "Brown", "Butler",
  . . .
"Washington", "Wayne", "Williams", "Wood", "Wyandot");

//get the qs parameter from URL
$names="";    $q=$_REQUEST["qs"];

if ($q !== "")
{ $q=strtolower($q); $len=strlen($q);
  foreach($a as $name)
  { if (stristr($q, substr($name,0,$len)))               // (K)
      $names .= "<span onclick='fill(this)'>            // (L)
              $name</span>\n";
  }
}
echo $names; ?>
```

The JavaScript function `fill` puts the suggested string in the input box:

```
function fill(node)
{ document.getElementById('county').value=
        node.innerHTML;
}
```

11.10   Soup of the Day

As our second example (**Ex:** SODJs), we will use AJAX to find the soup-of-the-day information kept on the server side in an XML document (`soups.xml`):

```
<?xml version="1.0" encoding="UTF-8"?>
<souplist>
  <Monday>Home Made Chicken Noodle Soup</Monday>
  <Tuesday>Beef and Vegetable Soup</Tuesday>
  <Wednesday>Cream of Mushroom Soup</Wednesday>
  <Thursday>French Onion Soup</Thursday>
  <Friday>Hot and Sour Soup</Friday>
  <Saturday>New England Clam Chowder</Saturday>
  <Sunday>Italian Wedding Soup</Sunday>
</souplist>
```

Our three-step approach is

1. Load the XML document into an XML DOM under JavaScript.

2. Find the soup-of-the-day string from the XML DOM tree.

3. Update the page display through the HTML DOM (Figure 11.8).

FIGURE 11.8: Soup Of the Day

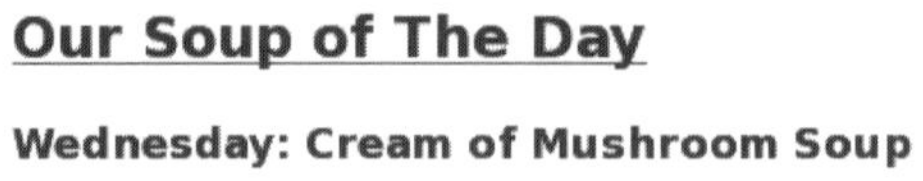

Our Soup of The Day

Wednesday: Cream of Mushroom Soup

The function displaySOD is called with the URL of the XML document and the id of the element where to display the soup information. The cross-browser xmlDoc loads the given URL (line 3) asynchronously (line 1) with the onreadystatechange event handling set (line 2).

```
var xmlDoc, target;
function displaySOD(url, t)
{ target=document.getElementById(t);
  if (window.ActiveXObject)   // Windows IE
  { xmlDoc=new
      ActiveXObject("Msxml2.DOMDocument.6.0");
  }
  else if (document.implementation // other browsers
      && document.implementation.createDocument)
  { xmlDoc = document.implementation
          .createDocument("", "soup", null);
  }
  xmlDoc.async=true;                          // (1)
  xmlDoc.onreadystatechange=getsoup;          // (2)
  xmlDoc.load(url);                           // (3)
}
```

The event handler getsoup obtains the soup information from the loaded xmlDoc (line 4) by a tagname matching today's day computed by whcihDay():

```
function getsoup()
{  var day=whichDay();
   soup = xmlDoc.getElementsByTagName(day).item(0);  // (4)
   target.innerHTML = day +": "+
                       soup.firstChild.nodeValue;
}
function whichDay()
{  var date=new Date();
   var today=date.getDay(); // 0-6 0=Sun
   var days = ["Sunday", "Monday", "Tuesday",
           "Wednesday", "Thursday",
```

```
          "Friday", "Saturday" ];
    return days[today];
}
```

Another approach for this example uses XSL to extract the soup information from the XML document for any particular day. The soup.xsl file contains a parameter day (line 5) that can be set when soup.xsl is invoked.

```
<?xml version="1.0" encoding="ISO-8859-1"?>
<xsl:stylesheet version="1.0" xmlns:xsl=
  "http://www.w3.org/1999/XSL/Transform">
<xsl:output method="html" />
<xsl:strip-space elements="*"/>
<xsl:param name="day">Monday</xsl:param>          (5)
<xsl:template match="/souplist/*">
 <xsl:if test="name(.)=$day">
   <span id="{$day}"><xsl:value-of select="$day"/>:
   <xsl:value-of select="." /></span>
 </xsl:if>
</xsl:template>
</xsl:stylesheet>
```

See **Ex**: XSLOutput for the effect of the xsl:output processing instruction.

The revised displaySOD function sets up two XML document objects: one (xml) for load soups.xml and one (xsl) for loading soup.xsl. After the two documents have been loaded asynchronously, an action function, which is different for IE and non-IE browsers, is called.

```
function displaySOD(x, s, t)
{ xmlURL=x;   xslURL=s;
  xmlFlag=false; xslFlag=false;
  target=document.getElementById(t);
  if (window.ActiveXObject)
  { xml=new ActiveXObject("Msxml2.DOMDocument.6.0");
    xsl=new ActiveXObject
          ("Msxml2.FreeThreadedDOMDocument.6.0");
    action=transformIE;
  }
  else if (document.implementation &&
          document.implementation.createDocument)
  { xml=document.implementation
        .createDocument("", "", null);
    xsl=document.implementation
        .createDocument("", "", null);
    action=transformFF;
  }
  xml.async=true; xsl.async=true;
```

```
  xml.onreadystatechange=xmlReady;
  xsl.onreadystatechange=xslReady;
  xml.load(xmlURL); xsl.load(xslURL);
}
```

The event handlers sets a ready flag and calls `action()`:

```
function xmlReady() { xmlFlag=true; action(); }
function xslReady() { xslFlag=true; action(); }
```

For non-IE browsers, the action is `transformFF`, which creates an XSLTProcessor (line 6), loads the XSL style sheet (line 7), sets the XSL parameter day (line 8), and then applies the transformation to th XML document (line 9). The result is an HTML DOM fragment that can be placed as the child of the `target` display element (line 10).

```
function transformFF()
{ if ( ! xmlFlag || ! xslFlag ) return;
  var xsltProcessor = new XSLTProcessor();      // (6)
  xsltProcessor.importStylesheet(xsl);          // (7)
  xsltProcessor.setParameter(null,
                    "day", whichDay());         // (8)
  var frag=xsltProcessor
     .transformToFragment(xml, document);       // (9)
  if ( target.hasChildNodes() )                 // (10)
     target.replaceChild(frag, target.firstChild);
  else
     target.appendChild(frag);
}
```

See the DWP website (**Ex:** SODXsl) for the code for `transformIE` which is somewhat different.

11.11   AJAX GET and POST Requests

To make an HTTP request from JavaScript, we first create an `XMLHttpRequest` object

```
req=new XMLHttpRequest();
```

and then use it to make an HTTP GET or POST request. Before sending an asynchronous request, we need to set an event handler that is called when the HTTP reply comes back:

```
req.onreadystatechange=event_handler;
```

To send a GET request to a desired *URL*, use

```
req.open("GET", URL, async);
req.send();
```

where *async* can be `true` or `false` to indicate an asynchronous or synchronous request.

We have seen the use of an AJAX GET request in **Ex: AutoComplete** (Section 11.9.1).

A browser may cache the result of a previously made HTTP GET request. This is especially true for Internet Explorer. To make sure an AJAX request gets a fresh response from the server side, we can append an extra query-string *key=value* to the end of the *URL*, making it different from any previous request. The value used can be a random number or the current time.

To make an HTTP POST request from JavaScript, we first set the request method to POST, assign values to HTTP request headers, and then send the request body. For example,

```
req.open("POST", url, true);
var d = new Date();
body="time="+d+"&browser="+navigator.userAgent+
    "&number="+Math.random() +
    "&formula="+encodeURIComponent("x=a+b");    // (11)
req.setRequestHeader("Content-Length",
                        body.length);
req.setRequestHeader("Content-Type",
          "application/x-www-form-urlencoded");
// sends HTTP POST request async
req.send(body);
```

It is important to make sure that any GET query string or POST request body is properly URL encoded (line 11).

The `XMLHttpRequest` object is being standardized by the W3C. Despite its name, it can be used to make requests in protocols other than HTTP (including FILE and FTP). Also, it can receive HTTP responses in any content type, not just XML. See the resources page on the DWP website for links to XMLHttpRequest documentation.

HTTP requests from JavaScript can be made easier by AJAX objects such as the light-weight *The Ultimate Ajax Object* or in large JavaScript libraries, *JQuery* for example.

The remainder of this chapter contains more AJAX examples.

11.12   Web Services

A *Web service* is a resource or program on the Web that can be invoked with an HTTP request. A Web service usually computes a result based on the request input and sends back a result in a well-defined format. The most widely used

Web service result formats are XML and JSON (*JavaScript Object Notation*). A Web service is like a remote procedure, it runs on a remote computer and provides specific results useful for clients. But, unlike remote procedures, Web services will always use HTTP for request and response (Figure 11.9).

FIGURE 11.9: Web Service Overview

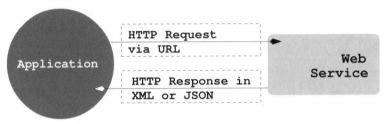

In the early days of Web services, the SOAP (Simple Object Access Protocol) was widely used. Contemporary Web service and client developers prefer REST (*REpresentational State Transfer*), which is an academic term for URLs, query strings, and request/response bodies used in HTTP. In other words, REST-based Web services receive HTTP POST/GET requests and send back results, often in XML or JSON, in an HTTP response body.

Application programs running on a computer connected to the Internet can easily make requests to Web services and obtain results to be used in the application. For example, a news reader application can obtain news feeds (the simplest kind of Web service) via HTTP requests and then use the results (RSS documents in XML) for displaying and retrieving news articles. For websites, we can use a server-side script such as PHP to include information obtained from Web services dynamically. The server-side scripts can also be invoked via AJAX from the client-side, making it possible for a webpage to employ any Web services available.

Technologies related to Web services also include UDDI (Universal Description, Discovery and Integration) and WSDL (Web Service Description Language). They help describe individual services and their automatic discovery. These are outside the scope of this textbook.

11.12.1 Example Web Services

There are many Web services throughout the world. Here are some examples:

- Google Maps API Web services—for requesting Maps API data to be used in your own Maps applications; available to registered users.

- PayPal Web services—for automating various aspects of PayPal payment processing; available for PayPal account holders.

- National Oceanic and Atmospheric Administration's National Weather

Service Web service—for hourly updated weather forecasts, watches, warnings, and advisories; available to the public.

- FAA airport status Web service—for US airport status, delays, and airport weather; available to the public.

- United Parcel Service (UPS) Web services—for address verification, shipping rates, tracking, and other services; available for registered customers.

Online listings and directories for Web services are available. The *Web Services Directory* at `programmableweb.com` is an example.

11.13 Accessing Web Services from PHP

We have seen how PHP can be used to generate webpages dynamically using client data (from the browser or form input) or databases. Accessing Web services is another significant way PHP can contribute to the functionality and dynamism of websites.

Information obtained from Web services can be injected into a webpage directly on the server side or returned to the client side in response to AJAX requests.

As an example, let's look at the airport status Web service provided by the FAA (the United States Federal Aviation Administration). It is a REST service that can respond in XML or JSON. To access this FAA service, use the following Web service URL:

```
http://services.faa.gov/airport/status/code
        ?format=application/format
```

where *code* is the standard airport code[1] (SFO, for example) and the *format* is `xml` or `json`.

Running the PHP code

```php
<?php $url="http://services.faa.gov/airport/"          // (A)
        ."status/SFO?format=application/json";
    $handle = fopen($url, "r");
    if (!$handle)
    {echo "{error: 'failed to open $url'}"; exit; }
    $contents = stream_get_contents($handle);
    fclose($handle);  echo $contents;
?>
```

produces the following JSON result:

---

[1] By the IATA, the International Air Transport Association.

```
{"name":"San Francisco International","ICAO":"KSFO",
 "state":"California",
 "status":{"avgDelay":"","closureEnd":"", "closureBegin":"",
          "type":"","minDelay":"","trend":"",
          "reason":"No known delays for this airport.",
          "maxDelay":"", "endTime":""},
 "delay":"false","IATA":"SFO","city":"San Francisco",
 "weather":{"weather":"Fair",
          "meta":{"credit":"NOAA's National Weather Service",
                 "url":"http://weather.gov/",
                 "updated":"9:56 AM Local"},
          "wind":"North at 5.8mph","temp":"63.0 F (17.2 C)",
          "visibility":"10.00"}
}
```

The code is a JavaScript object with *property*: *value* pairs separated by commas.

We can turn the preceding PHP code (line **A**) into a server-side Web service proxy for AJAX:

```php
<?php /////    getWS.php    /////
header("Content-Type: application/json");
$url=$_REQUEST['url'];
$handle = fopen($url, "r");
if (!$handle)
{  echo "{error: 'failed to open $url'}"; exit; }
$contents = stream_get_contents($handle);
fclose($handle);  echo $contents;
?>
```

We can now call `getWS.php` using AJAX, passing it a Web service URL, and splice the resulting JSON information into the webpage on-the-fly (Figure 11.10).

FIGURE 11.10: Airport Status Web Service via AJAX

The HTML code is

```
<h2>Airport Status</h2>
<p><input id="apc" placeholder="3-letter Airport Code"
   required=""/> <input type="button" onclick=
   "showAS(document.getElementById('apc').value,'status')"
value="Get Airport Status" /></p>
<div id="status"></div>
```

And the JavaScript function `showAS` sets up the FAA url to be used by the PHP Web service proxy (`getWS.php`) and calls `getResponse` with it. The result retrieved will be displayed in `disp_node`:

```
var req, disp_node;

function showAS(airport_code, id)
{ if (airport_code )
  { url="http://services.faa.gov/airport/status/" +
      airport_code + "?format=application/json";
    disp_node=document.getElementById(id);
    getResponse(url);
  }
}
```

After creating an `XMLHttpRequest` object, a query string is prepared (line 1) for the proxy `getWS.php`.

```
function getResponse(url)
{ req = (window.XMLHttpRequest)
    ? new XMLHttpRequest()
    : new ActiveXObject("Microsoft.XMLHTTP");
  prx="getWS.php?url="+encodeURIComponent(url);   // (1)
  req.onreadystatechange=doResult;                // (2)
  req.open("GET", prx, true);
  req.send();
}
```

The response event handler (line 2) displays HTML code computed by `airportDisplay` (line 3):

```
function doResult()
{ if (this.readyState==this.DONE && this.status==200)
  { json=this.responseText;
    disp_node.innerHTML=airportDisplay(json);   // (3)
  }
}
```

The JSON content received by the function `airportDisplay` is evaluated to become a JavaScript object (line 4), which allows easy assess to the information content received (lines 5 through 6).

```
function airportDisplay(js)
{ alert(js);
  eval("json=" + js + ";");                    // (4)
  code="<ul>";
  code += "<li>"+json.name+"("+json.IATA+"), "  // (5)
         +json.city+", "+json.state+"</p></li>";
  code += "<li>"+(json.status.type?
              json.status.type+', ':"")
              +json.status.reason+"</p></li>";
  code += "<li>"+json.weather.weather+", "
         +json.weather.temp+"</p></li>";        // (6)
  code += "</ul>"; return code;
}
```

The Web service proxy arrangement allows us to use AJAX to retrieve data from any Web service without the client-side restrictions. Try this example (**Ex:** `AirportStatus`) at the DWP site.

11.14   Address Verification Web Service

As another example, let's look at the UPS (United Parcel Service)[2] address verification (AV) service. To use this particular Web service, you need to register and obtain an access code. It is free for potential UPS customers. Requests and responses to the UPS AV service are in predefined XML formats. In our example, we will apply the UPS AV service to validate the city, state, and zip part of an address.

The required content posted to the UPS address verification service at

```
$AVurl='https://wwwcie.ups.com/ups.app/xml/AV';
```

consists of two XML documents: an `AccessRequest` for the registered access code, userid, and password; and an `AddressValidationRequest` giving the city, state, and zip to be verified. The file `AVRequestXML.php` constructs the required XML string.

```
<?php   /////   AVRequestXML.php   /////
$reqXMLstr=
'<?xml version="1.0" encoding="UTF-8"?>
<AccessRequest>
  <AccessLicenseNumber>'.$access.'</AccessLicenseNumber>
  <UserId>' . $userid . '</UserId>
  <Password>' . $passwd . '</Password>
</AccessRequest>
```

---

[2]UPS, the UPS brand mark, is a trademark of United Parcel Service of America, Inc. All Rights Reserved.

```
<?xml version="1.0" encoding="UTF-8"?>
<AddressValidationRequest>
 <Request>
   <TransactionReference>
     <CustomerContext>' .$myctx. '</CustomerContext>
   </TransactionReference>
   <RequestAction>AV</RequestAction>
 </Request>
 <Address>
   <City>' . $city . '</City>
   <StateProvinceCode>' .$state. '</StateProvinceCode>
   <PostalCode>' . $zip . '</PostalCode>
 </Address>
</AddressValidationRequest>';
?>
```

The function getResponse formulates a POST request (lines A through B),
opens an HTTP connection to the designated UPS Web service URL (line C),
and returns the response content (lines D through E).

```
function getResponse(&$AVurl, &$reqXMLstr)
{ try
  { $form = array                                 // (A)
    ( 'http' => array
      ( 'method' => 'POST',
        'header' => 'Content-type: application'
            . '/x-www-form-urlencoded',
        'content' => "$reqXMLstr"
      )
    );
    $req = stream_context_create($form);          // (B)
    $conn = fopen($AVurl, 'rb', false, $req);      // (C)
    if(!$conn) throw new Exception
            ("{'error':'Connection failed'}");
    $response = stream_get_contents($conn);        // (D)
    fclose($conn);
    if($response == false) throw new
            Exception("{'error': 'Bad data'}");
  } catch(Exception $ex) { echo $ex; }
  return $response;                               // (E)
}
```

The correct response is in XML, and it can be passed to the function verify,
which checks the result of address verification.

```
function verify(&$response)
{ global $city, $state, $zip;
```

```
try {
$resp = new SimpleXMLElement($response);
$status=$resp->Response->ResponseStatusDescription;
$quality=$resp->AddressValidationResult->Quality;
$vcity=$resp->AddressValidationResult->Address->City;
$vstate=$resp->AddressValidationResult->
          Address->StateProvinceCode;
if ( $status=="Success" && $quality > 0.96       // (F)
     && $vcity == strtoupper($city)
     && $vstate == strtoupper($state)
   ) echo'{"AV": 1}';                             // (G)
else echo '{"AV": 0}';
} catch(Exception $ex)
  { echo "{'error':'" . $ex . "'}"; }
}
```

Using a `SimpleXMLElement`, we easily get to the information contained in the XML response. We make the appropriate check (lines **F** through **G**) and output the JSON `{"AV": 1}` for success and `{"AV": 0}` for failure.

With these functions, we can create a flexible `AVClient.php` that can be invoked from the command line or by an HTTP request:

```
<?php /////    AVClient.php    /////
header("Content-Type: application/json");        // (H)
//Configuration begin
  $access = "1C957AEB4365B1A8";
  $userid = "wangtong";
  $passwd = "ajaxU2012";
  $AVurl = 'https://wwwcie.ups.com/ups.app/xml/AV';
  $myctx="Testing1234";
//Configuration end

// functions  definitions

if ( !empty($argv[1]) )
{ $city=$argv[1];$state=$argv[2];$zip=$argv[3]; }
else if ( !empty($_REQUEST['c']) &&
          !empty($_REQUEST['s']) &&
          !empty($_REQUEST['z']) )
{ $city=$_REQUEST['c']; $state=$_REQUEST['s'];
  $zip=$_REQUEST['z'];
} else {  exit;  }

require("AVRequestXML.php"); // sets $reqXMLstr
verify(getResponse($AVurl, $reqXMLstr));
?>
```

For example, the command

```
php  AVClient.php Kent OH 442420001
```

outputs {"AV": 1}, a valid JSON string, as indicated by the Content-Type header used (line H).

11.15   AJAX: Form Address Verification

With AVClient.php in place, we can use it in an AJAX application that checks the city, state, and ZIP entries (Figure 11.11) in any HTML form (**Ex: AJAXav**).

FIGURE 11.11: Address Verification

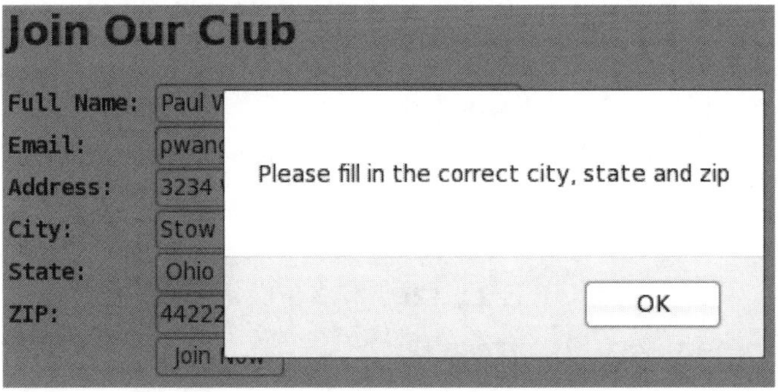

The form checking JavaScript function is as follows:

```
function verify_csz()
{ var req, city, state, zip, ans;
  req = (window.XMLHttpRequest)
    ? new XMLHttpRequest()
    : new ActiveXObject("Microsoft.XMLHTTP");
  city=document.getElementById('city').value;
  state=document.getElementById('state').value;
  zip=document.getElementById('zip').value;
  if ( city && state && zip )
  { q_str="c="+city+"&s="+state+"&z="+zip;
    req.open("GET","AVClient.php?"+q_str,false);    // (I)
    req.send();
    if (req.readyState==req.DONE && req.status==200)
    { eval("ans=" + req.responseText +";");
      if ( ans.AV == 1 ) return true;
    }
  }
```

```
window.alert("Please fill in the correct city,
             state and zip");
return false;
}
```

Note: We used a synchronous GET call for `AVClient.php` (line I).

11.16   AJAX File Upload

In Chapter 7 we discussed a file picker (Section 7.13) and a file dropzone (Section 7.16) for improving the client-side handling of files. We mentioned that the resulting list of files can be processed by JavaScript for many purposes. One important application is to upload these files to the server side.

In this section, we combine the user interface ideas from the above-mentioned two sections and add AJAX code for file uploading (**Ex: AJAXUpload**). The files are uploaded in turn, but each individual file is uploaded asynchronously. A progress bar displays upload status. Completion of one file upload triggers the next. After all files have been processed, a user gets an in-page confirmation and may use the same tool to upload more files (Figure 11.12).

FIGURE 11.12: AJAX File Upload

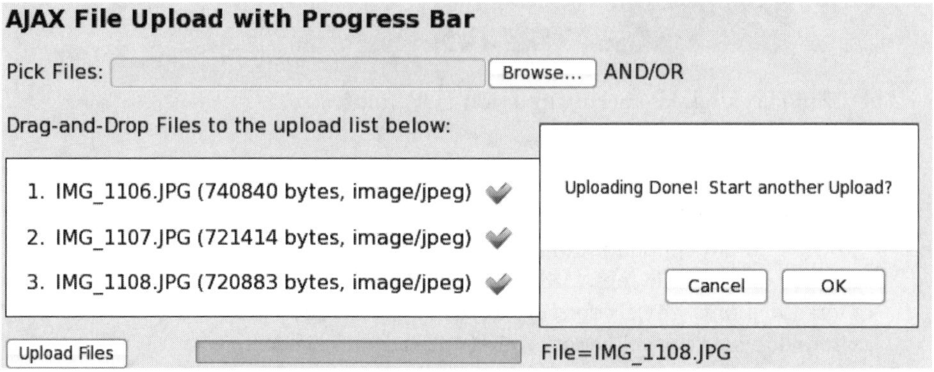

We took the HTML for file DnD from Section 7.16 and added a file picker:

```
<label>Pick Files: </label><input id="picker"
   size="35" type="file" multiple=""
   onchange="listFiles(this.files,              (A)
         document.getElementById('filelist'))"
   accept="image/jpg" />
```

Files picked via the picker are added to the list of files dropped to the file dropzone by calling `listFiles` (line A). The file list is kept in a global array, `the_files`. You can find `listFiles` in the file `pickfiles.js`, which is basically the same as `dropfiles.js` from Section 7.16.

We also added a progress bar:

```
<progress id="pb" value="0" style="width: 260px;
        background:lightgreen" max="100"></progress>
```

The file `upload.js` implements AJAX file upload. To start file uploading, call `ajaxUploadNext(0)`:

```
function ajaxUploadNext(i)
{ if (! p_bar)
  { p_bar=document.getElementById('pb');
    f_name=document.getElementById('filename');
  }
  p_bar.value=0;
  if (i < the_files.length)                      // (B)
  { file=the_files[i];
    if ( file != null ) fileUpload(file,i);      // (C)
  }
  else
  { if(i==0) {alert("No files to upload."); return; }  // (D)
    if(confirm("Upload Done! Start another Upload?"))   // (E)
    { upload_reset();  }
  }
}
```

If there is an ith file and it is not `null` (lines B and C), we call `fileUpload`. Otherwise, we report file upload results (lines D and E).

The function `fileUpload` uploads the given file `f` if it is of the right type and size (line F), resets the progress bar, and adds a file name after it (line G).

```
function fileUpload(f, i)
{ var req = new XMLHttpRequest();
  if (req.upload && f.size<=f_max && check_type(f))  // (F)
  { p_bar.value=0; f_name.innerHTML="File="+f.name;  // (G)
    req.upload.addEventListener("progress",          // (H)
          makeProgress, false);
    req.upload.addEventListener("load",              // (I)
          loaded, false);
    req.onreadystatechange=nextFile;
    req.open("POST", url, true);                     // (J)
    req.setRequestHeader("X_FILENAME", f.name);      // (K)
    p_bar.value=5;  req.send(f);                     // (L)
  }
  else { if (req.upload) ajaxUploadNext(i+1); }      // (M)
}
```

After setting event handlers, we initiate an asynchronous HTTP POST request (line J), sending the filename under the X_FILENAME header (line K) and the file string as the POST message body (line L). If the test on line F fails, we proceed to the next file (line M).

The progress event listener (line H)

```
function makeProgress(e)
{ if (e.lengthComputable)
  { var pc = Math.round((e.loaded*100)/e.total);
    p_bar.value=pc;                                    // (N)
  }
}
```

updates the progress bar to indicate the percentage of the file uploaded (line N).

The load listener is called after a file is done uploading:

```
function loaded() { p_bar.value=100;  }
```

and readystatechange event listener nextFile

```
function nextFile()
{ var result;
  if (this.readyState==this.DONE)  // HTTP request is done
    if (this.status==200)  // HTTP request normal
    { result=this.responseText;
      if ( result.indexOf("upload complete") )         // (O)
      { p_bar.value=100;
        document.getElementById("cd_"+(file_i+1))
.src="check_small.png"; alert(result);          // (P)
        ajaxUploadNext(file_i+1);                      // (Q)
      }
      else { alert(result + "Aborting."); }
    }
}
```

checks the upload response (line O), replaces the red delete symbol (x) by a green checkmark for the file uploaded (line P), and initiates the uploading of the next file (line Q).

Global variables used include

```
var p_bar=null, f_name=null,  start_count=0,
    done_count=0,  done=false,
    c_type="image/jpeg",    // file content type
    url="uploadfile_2.php",  // upload receiver url
    file_i,                  // current file index
    f_max=2097152;           // 2MB per file size limit
```

The function `upload_reset` wipes the slate clean for a new round of file uploading.

```
function upload_reset()
{  the_files=new Array();
   while (drop_zone.hasChildNodes())
      drop_zone.removeChild( drop_zone.firstChild );
   document.getElementById('pb').value=0;
   document.getElementById('filename').innerHTML="";
}
```

On the server side, the receiving program for file upload is `uploadfile_2.php` which is in the form of a Web service.

```
$target="up_dir";  // upload files folder

function upload_file($fname)
{  global $target;
   file_put_contents("$target/$fname",       // save file
      file_get_contents('php://input'));
}

$fname= isset($_SERVER['HTTP_X_FILENAME'])  // file name
   ? $_SERVER['HTTP_X_FILENAME'] : false;

if ( $fname && upload_file($fname) )
   echo "$fname upload complete";
else
   echo "$fname upload failed";
```

Note: We have put to good use the PHP functions

```
file_put_contents
file_get_contents
```

We have combined user interface design, AJAX, HTML5 File API, and PHP to construct this fileupload tool. It can be very helpful for uploading multiple files that are time consuming. At the DWP website, a fully working example (**Ex: AJAXUpload**) can be found that also provides an alternative implementation using `multipart/form-data` upload instead of the single-part post request body method (lines K and L).

11.17   XML Namespaces

XML enables everyone to invent his/her own markup elements and give them names and attributes. It is a very good thing because XML has been applied extensively. Ad-hoc markups have been used in many self-contained applications. And there are also many XML applications standardized for public use. These include SMIL, SVG, MathML, RSS, XSL, XHTML5, and others.

A natural desire then is to use different XML *vocabularies* together in one document. Because different vocabularies may define elements with the same names, we need to solve this name conflict problem before allowing mixing markups together in one document. The solution, as in many other programming contexts, is the use of *namespaces.*

An XML namespace is a collection of all element and attribute names defined in a particular XML vocabulary. When mixing XML vocabularies in one document, all we need to do is to clearly indicate which name comes from which namespace and the name conflict problem is solved.

The attribute

```
xmlns="URI
```

defines the *default XML namespace* for an element and all its descendants. Any unqualified element or attribute name is considered local or in the default namespace. The *URI* must lead to the DTD or Schema that defines the XML vocabulary. The URI can be a Web URL such as

```
http://www.w3.org/1999/xhtml            (for HTML)
http://www.w3.org/1999/XSL/Transform    (for XSL)
http://www.w3.org/1999/xlink            (for XLink)
http://www.w3.org/1998/Math/MathML      (for MathML)
http://www.w3.org/2000/svg              (for SVG)
http://www.rssboard.org/rss-specification (for RSS)
```

or a *Uniform Resource Name* (URN) such as

```
urn:ISBN:0-395-36341-6
urn:mpeg:mpeg7:schema:2001
```

For example, the `<html>` attribute

```
xmlns="http://www.w3.org/1999/xhtml"
```

puts the entire HTML code in the XHTML namespace. And the `<rss>` attribute

```
xmlns="http://www.rssboard.org/rss-specification"
```

puts the whole RSS in its proper namespace.

In an XML document, you can use a *namespace prefix* to place a name in another (non-local) namespace:

prefix: `name`

The prefix is known as a *name qualifier* and is defined with the attribute

```
xmlns:xyz="URI"
```

where *xyz* is the desired prefix. In an XSL document, we routinely mix XSL names with names not in XSL. For example,

```
<xsl:stylesheet version="1.0"
  xmlns:xsl="http://www.w3.org/1999/XSL/Transform">
<xsl:template match="/">
<html><body>

</html>
</xsl:template>
</xsl:stylesheet>
```

We can use the HTML img tag in an XML document as follows:

```
<?xml version="1.0"?>
<catalog xmlns:html="http://www.w3.org/1999/xhtml">
...
  <html:img src="..." alt="picture" ... />
...
</catalog>
```

When a namespace URI is defined by a W3C Schema, as opposed to a DTD, a more complicated way of referencing the Schema URI is needed:

```
<?xml version="1.0"?>
<BaseballTeam xmlns="http://www.mlb.com"
  xmlns:xsi="http://www.w3.org/2001/XMLSchema-instance"
  xsi:schemaLocation=
    "http://www.mlb.com xs/BaseballTeam.xsd">
<team name="Cleveland Indians">

  <!-- More XML Content -->
</team>
</BaseballTeam>
```

HTML5 now allows the direct use of MathML and SVG elements inside HTML5 code without bothering with namespaces. This is very convenient as we will see in Chapter 12.

11.18 For More Information

XML, XSL, and XML DOM standards can be found at the W3C website.

JavaScript support for making HTTP requests and for XML processing (parsing and serializing XML) can be found at the Mozilla Developer Network and at the Mircrosoft Developer Network.

PHP support for XML can be found at

`www.php.net/manual/en/refs.xml.php`

11.19   Summary

XML is a collection of technologies that allows the development of markup languages for different purposes. Server-side and client-side tools are available to retrieve, load, parse, represent (XML DOM), and display XML documents.

Web services are programs made available on the Web that can be called by their URL (REST), and that returns results mostly in XML or JSON.

AJAX combines JavaScript and XML to perform computations and access the Web asynchronously while the user continues to interact with a webpage. AJAX can make Web applications behave in quite the same way as regular non-Web applications and make webpages much more functional and interactive. Examples given include auto-completion of user input, accessing FAA airport Web service, and address verification using the UPS Web service.

XSL is an XML application used to define how to rewrite XML documents into other XML/HTML or text documents. XSL programming involves writing XSLT templates to specify transformations and using XPath expressions to select elements and data from XML structures. XSL transforms are supported by JavaScript and by PHP.

A file upload example (**Ex: AJAXUpload**) with progress bar combines HTML5 File API, user interface design, and AJAX to demonstrate the usefulness of the materials covered.

Exercises

11.1. What is a well-formed XML document? A validated XML document?

11.2. When using Javscript to make HTTP requests from the client side, what restrictions apply, and why?

11.3. Find an interesting Web service and describe how it works.

11.4. Here is a totally empty XSL style sheet:

```
<?xml version="1.0" encoding="ISO-8859-1"?>
<xsl:stylesheet version="1.0" xmlns:xsl=
    "http://www.w3.org/1999/XSL/Transform">
</xsl:stylesheet>
```

What do you expect if we apply it to our `catalog.xml` in Section 11.4? Try it and then explain the result.

11.5. Look up the W3C standard on `XMLHttpRequest`. What arguments can the send method of this interface support? What is the purpose of the XMLHttpRequest responseType property, and what values can you assign to this property? What is the purpose of the XMLHttpRequest upload property? Where in this chapter did we use this property?

11.6. Look at the DTD file catalog.dtd:

```
<!ELEMENT catalog (course+)>

<!ELEMENT course (title, instructor+)>
<!ELEMENT title (#PCDATA)>
<!ATTLIST course semester (Fall|Spring|Summer|All) #REQUIRED>

<!ELEMENT instructor (name,department,email)>
<!ELEMENT name (#PCDATA)>
<!ELEMENT department (#PCDATA)>
<!ELEMENT email (#PCDATA)>
```

Take the catalog.xml in Section 11.4 and modify it to conform to the above DTD. Add to the file the processing instruction:

```
<!DOCTYPE catalog PUBLIC "-//my/DTD CAT 1.0"
  "http:full-url-to/catalog.dtd">
```

Then try validating the resulting XML document using the W3C validator http://validator.w3.org/.

11.7. PHP supports making HTTP requests in multiple ways. Look into and explain (a) Simply use the PHP fopen or fsocketopen functions; (b) Use PEAR HTTP Request2; (c) Use the HTTP PHP extension.

11.8. Look into PHP support of XML and explain the different ways it provides for XML processing.

11.9. Take our address verification example (**Ex: AJAXav** in Section 11.15) and make it return a result in JSON or XML depending on an indication provided in the request URL.

11.10. Look into address validation Web services provided by someone other than UPS. Make our address verification example (**Ex: AJAXav** in Section 11.15) work for that service.

11.11. Build a AJAX file downloader that can download one or more files from the server side and display download progress with a progress bar. Allow the user to stop/resume the downloading.

11.12. Follow the code of our AJAX fileupload example (Section 11.16) and build a manage pictures tool that uses AJAX techniques to allow a user to add, delete, change name, add caption for a set of pictures kept on the server side.

Chapter 12

SVG and MathML

An interesting and important aspect of HTML5 is the fact that SVG (Scalable Vector Graphics) and MathML (Mathematical Markup Language), both XML applications, have been imported into HTML5. SVG supports drawing, transforming and animating two-dimensional (2D) vector graphics. MathML provides encodings for the display (*Presentation MathML*) and semantics (*Content MathML*) of mathematical expressions.

In addition to displaying the occasional vector graphics and mathematical expressions, SVG and MathML add significant scientific dimensions to HTML5 in dealing with 2D geometry and manipulating mathematical formulas. Their impact on technical communication and education will surely grow with time.

In this chapter, we begin by showing how HTML5 incorporates SVG and MathML with some simple examples. Then SVG and MathML will each be introduced. JavaScript control of and user interaction with SVG and MathML DOMs are explained with realistic examples.

12.1 Including SVG and MathML in HTML5

The HTML5 standard calls for inline inclusion of SVG and MathML and most browsers support both *natively* without loading plug-ins. This means you would add SVG in HTML5 inside the `svg` element

```
<svg>

      proper SVG code here

</svg>
```

and that you would add MathML in HTML5 inside the `math` element

```
<math>

      proper MathML code here

</math>
```

Each element and all its child elements will automatically be placed in the default SVG or MathML namespace as appropriate. An inline `svg`/`math` element is treated as an inline element (phrasing/flow content) by HTML5.

For example (**Ex:** `Bullseye`), the following code

```
<!DOCTYPE html>
<html xmlns="http://www.w3.org/1999/xhtml"
      lang="en" xml:lang="en">
<head> <meta charset="utf-8"/>
<title>Inline SVG</title></head>
<body style="background-color: #def">
<h3>HTML5: Inline Inclusion of SVG Code</h3>
<p>Look at this bullseye
<svg xmlns="http://www.w3.org/2000/svg"
     style="vertical-align:middle; width:30px; height:30px">
 <circle cx="15" cy="15" r="10"
    fill="none" stroke="rgb(128,0,0)" stroke-width="3px"/>
 <circle cx="15" cy="15" r="4" fill="rgb(128,0,0)" />
</svg> that is coded in SVG.</p>
<p>Works in all major browsers.</p></body></html>
```

produces the display shown in Figure 12.1. Two circles are given in SVG

FIGURE 12.1: Bullseye with Inline SVG

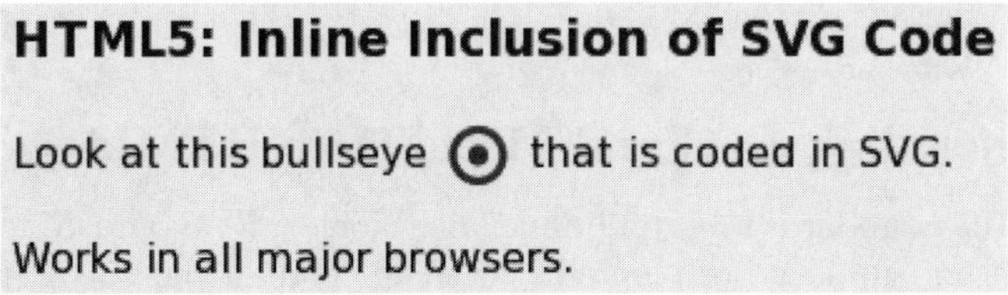

with center coordinates (`cx`, `cy`) and radius `r`. Note that the explicit XML namespace specification

```
xmlns="http://www.w3.org/2000/svg"
```

for the `svg` element is added as an insurance. HTML5-compliant browsers should set that namespace implicitly.

Although not recommended, it is also possible to place the SVG code in a separate file and include it in a webpage using the `img`, `embed`, or `object` tag (**Ex:** `BullseyeEmbed`).

Here is `bullseye.svg` as a separate file:

```
<?xml version='1.0' standalone="no"?>
<!DOCTYPE svg PUBLIC "-//W3C//DTD SVG 1.1//EN"
 "http://www.w3.org/Graphics/SVG/1.1/DTD/svg11.dtd">
```

```
<svg viewBox="0 0 30 30"
    xmlns="http://www.w3.org/2000/svg">
 <circle cx="15" cy="15" r="10"
    fill="none" stroke="rgb(128,0,0)" stroke-width="3px" />
 <circle cx="15" cy="15" r="4" fill="rgb(128,0,0)" />
</svg>
```

Please note how the separate SVG file is constructed. The first 3 lines are the XML processing instructions, and the root svg element must have the default namespace set correctly.

The file bullseye.svg can then be the src of an image:

```
<img src="bullseye.svg" width="100" height="100"
    alt="SVG bullseye" />
```

the src of an embed, or the data of an object tag.

```
<embed src="bullseye.svg" type="image/svg+xml"
    style="vertical-align:middle;
    width:30px;height:30px"/>
```

```
<object data="bullseye.svg" type="image/svg+xml"
    style="vertical-align:middle;
    width:30px;height:30px"/>
```

The embed and object are equivalent and will embed bullseye.svg and produce the display shown in Figure 12.2.

FIGURE 12.2: Bullseye with Embedded SVG

Similarly, MathML code can be included inline in HTML5 (**Ex:** MathMLDemo). For example,

```
<h3>HTML5: Inline Inclusion of MathML Code</h3>
<p>A quadratic equation
<math xmlns="http://www.w3.org/1998/Math/MathML">
 <mrow>
    <mrow>
      <mi>a</mi><mo>&InvisibleTimes;</mo>
        <msup><mi>x</mi><mn>2</mn></msup>
    </mrow>
```

```
 <mo>+</mo>
 <mrow>
   <mi>b</mi><mo>&InvisibleTimes;</mo><mi>x</mi>
 </mrow>
 <mo>+</mo><mi>c</mi>
</mrow>
   <mo>=</mo><mn>0</mn>
</math></body></html>
```

produces the display in Figure 12.3.

FIGURE 12.3: Quadratic Equation with Inline MathML

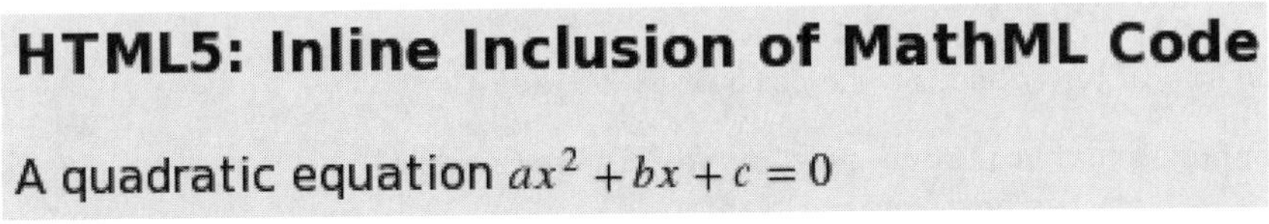

HTML5: Inline Inclusion of MathML Code

A quadratic equation $ax^2 + bx + c = 0$

The preceding Presentation MathML marks up identifiers (`mi`), operators (`mo`), numbers (`mn`), superscripts (`msup`), and groupings (`mrow`).

It is recommended that SVG and MathML code be included inline in HTML5, and we focus only on the inline coding method in this chapter.

12.2   SVG

Scalable Vector Graphics (SVG) is an XML application for 2D vector graphics and a W3C standard. SVG uses text to describe points, lines, and curves. Many say SVG is to graphics what HTML is to text. Compared to the proprietary Adobe Flash, SVG uses an open and standard textual format as opposed to a closed binary format. Using vector graphics, SVG files are much smaller in size than using raster graphics such as PNG or JPG. The same SVG graphics can be scaled up/down without losing sharpness or becoming pixelized. Also, SVG is much easier for applications to transform and manipulate and has great potential on mobile devices. With HTML5 and native support from major browsers, SVG will surely see increasingly wider applications.

12.2.1   Shape Elements

SVG elements for basic geometric shapes include `circle`, `rect` (rectangle with/without rounded corners), `line` (straight line segment), `polyline` (multisegment line), `polygon` (closed multisegment line), and `ellipse`:

```
<circle cx="100" cy="100" r="50" />
```

```
<rect x="170" y="110" width="80" height="40" />

<rect x="250" y="50" rx="6" ry="6"
            width="80" height="40" />

<line x1="330" y1="150" x2="430" y2="50" />

<polyline points="60 300 75 350 100 250
                  125 350 150 250 165 300" />

<polygon points="250 250 297 284 279 340 220 340
                 202 284" />

<ellipse cx="400" cy="300" rx="72" ry="40" />
```

Figure 12.4 shows a display of these elements (**Ex: SVGshapes**) using the SVG style attributes.

```
fill="none" stroke="rgb(128,0,0)" stroke-width="2px"
```

FIGURE 12.4: SVG Shapes

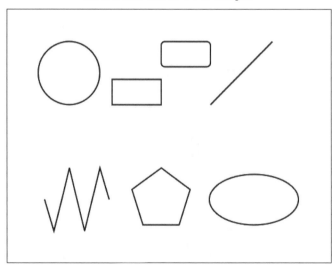

12.2.2 Including Raster Images

The image element allows you to include a raster image as part of your SVG picture:

```
<image width="100" height="100" xlink:href="imgURL">
```

```
<image width="100" height="100"
    xlink:href="data:image/jpg;base64, dataString">
```

The *imgURL* points to the desired raster image. In the second version, we encode the raster image file as a base64 string *dataString* and include it in the image tag. This way, the image element is self-contained.

12.2.3   The path Element

The SVG path element can be used to draw almost any lines and shapes. You simply specify a sequence of points and what to draw from one point to the next. The points can be given in absolute or relative (to the previous point) coordinates, and the drawing commands are given in simple one-character codes. The path element is typically given in the form

```
<path d="pathdata" />
```

where the *pathdata* provides one or more points and drawing commands. For example, a filled green triangle (Figure 12.5) can be given as (**Ex: SVGTriangle**)

```
<path d="M 20 40 L 80 40 L 50 10 z"
      fill="green" stroke="green" />
```

FIGURE 12.5: SVG Triangle

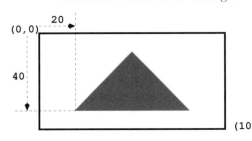

The pathdata means "Move to (20,40), draw a line to (80,40), then a line to (50,10), then close the path with a line to the first point." The concise datapath can be further compacted to

```
d="M20 40H80L50 10z"
```

Available pathdata commands include M (moveto), L (lineto), H (horizontal lineto), V (vertical lineto), A (elliptic arcto), Q or T (quadratic Bézier arcto), C or S (cubic Bézier arcto), and Z (closepath, lineto initial point). Each command is followed by one or more points as required arguments. The points are in absolute or relative coordinates, depending on whether a command is given in upper- or lower-case. A command letter can be omitted if it is the same as the previous command. Unnecessary white space in pathdata can always be omitted.

12.2.4 The `text` Element

You can place text in SVG at any given starting point (x,y) with a desired style:

```
<text x="20" y="50" fill="blue" font-weight="bold"
    font-style="normal" font-size="38px">Hello SVG</text>
```

Text can also be rendered along a given path (Figure 12.6) as in (**Ex:** TextOnPath)

```
<path id="stamp" d='M20 50 A30 30 0 1 1 80 50
                    A30 30 0 1 1 20 50'
 fill="none" stroke="black" stroke-width="0.2px"/>
<text fill="black" font-size="60%">
  <textPath xlink:href="#stamp">Dynamic
Web Programming by Wang</textPath></text>
```

FIGURE 12.6: Text along a Path

Here, the path is specified by elliptic arcs. See the next section for information on the path command `A`.

12.2.5   Curved Arcs

We explained the meaning of Bézier spline curves in Section 7.17.2 where drawing on the HTML5 `canvas` element was first discussed. In SVG pathdata, we use the notations

- `Q x1 y1 x y`—A quadratic Bézier spline from the current point to (x,y), with control point $(x1,y1)$.

- `T x y`—A quadratic Bézier spline from the current point to (x,y), with control point the reflection, with respect to the current point, of the previous control point.

- `C x1 y1 x2 y2 x y`—A cubic Bézier spline from the current point to (x,y) with control points $(x1,y1)$ and $(x2,y2)$.

- `S x2 y2 x y`—A cubic Bézier spline from the current point to (x,y) with the first control point being the reflection, with respect to the current point, of the previous control point and the second control point being $(x2,y2)$.

Elliptical arcs are specified with

`A rx ry tilt large-arc-flag sweep-flag x y`

The curve leads from the current point to (x,y) with the given radii `rx` and `ry`, and a `tilt` degree rotation with respect to the x-axis. But, because there can be four such elliptical arcs, we must use the flag values to choose among them. The four arcs in Figure 12.7 are

FIGURE 12.7: Elliptic Arcs

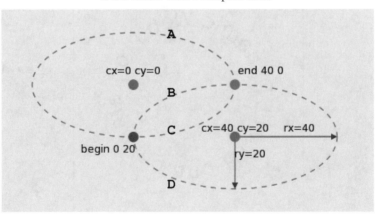

```
d='M0 20 A40 20 0 1 1 40 0'        (arc A)
d='M0 20 A40 20 0 0 1 40 0'        (arc B)
```

```
d='M0 20 A40 20 0 0 0 40 0'         (arc C)
d='M0 20 A40 20 0 1 0 40 0'         (arc D)
```

12.3   SVG Markers, Patterns, and Transforms

SVG graphics are placed on a 2D *canvas* using the usual computer graphics coordinate system, where the y-axis goes downward and angles are measured clockwise with respect to the x-axis.

The width and height of the root `svg` element establishes the initial *viewport*, which also defines the size of the browser display area. The `viewbox` attribute of the `svg` element (`viewBox="x0 y0 width height"`) specifies the part of the canvas to be displayed in the display area (scaling to fit). For example,

```
<svg width='480' height='360' viewBox='-60 -45 120 90'>
```

By changing the location or size of the viewBox, you can show different parts of the canvas; and by mapping the display area to a smaller/larger viewBox, you can zoom in/out.

We have seen coordinate transforms in Section 6.20 and Section 7.17.3. Individual or groups of SVG elements (with the `<g>` element) can also set the `transform` attribute to one or a sequence of the following transforms (**applied last first**):

- `translate(x, y)`—Adds x and y to all points.

- `rotate(t, x0, y0)`—Rotates clockwise about (x0,y0) t degrees.

- `scale(sx, sy)`—Multiplies x and y by the given factors. If they are the same, you may give just one.

- `skewX(t)`—Sets x=x+y*tan(t); the larger the y, the more the x gets pushed to the right.

- `skewY(t)`—Sets y=y+x*tan(t); the larger the x, the more the y gets pushed downward.

Figure 12.8 shows a clock hand rotated 30 degrees (**Ex: SVGclock**). The `svg` element maps a 300×300 square area at $(-150, -150)$ to a viewport of the same size (line 1).

```
<svg id='root' width='300' height='300'
     viewBox='-150 -150 300 300'>                      (1)
<defs>
 <marker fill='red' id='Triangle' viewBox='0 0 10 10' (2)
    refX='0' refY='5' markerWidth='8' markerHeight='6'
    orient='auto'>
```

FIGURE 12.8: Rotated Clock Hand

```
   <path d='M 0 0 L 10 5 L 0 10 z'/></marker>
</defs>
<rect x='-150' y='-150' width='100%' height='100%'
     fill='#def'/>
<g fill="none' stroke="rgb(0,0,128)" stroke-width="3.5px">
 <circle cx='0' cy='0' r="120"/>
 <path id="clockhand"  transform='rotate(30,0,0)'        (3)
    d='M0 0 L0 -76' marker-end='url(#Triangle)' />
 <text font-size='130%' x='0' y='-96'
     text-anchor='middle'>12</text>
</g></svg>
```

The clock hand is drawn from the clock center at $(0,0)$ to $(0,-76)$ with a red arrowhead marker and a 30-degree rotation about the point $(0,0)$ (line 3). The arrowhead marker (line 2) is defined within a `defs` element and placed at the end of a path with the `marker-end` attribute (line 3).

Graphics defined in `defs` are for later use in the code, often repeatedly.

A marker can be attached to one or more vertices of a `path`, `line`, `polyline`, or `polygon` element with one or more of the attributes `marker` (all vertices), `marker-start` (start point), `marker-mid` (all mid points), and `marker-end` (end point). A marker is usually defined as a self-contained view-box with a reference point (`refx,refy`) for attaching to a given vertex. The orientation is automatic (`orient='auto'`) or a specified degree (`orient='deg'`).

A pattern is a graphic that can be used as a fill pattern or as a stroke pattern. Here is a `checker` pattern with two white and two black squares:

```
<defs><pattern id="checker" x="0" y="0" width="20"
```

```
  height="20" patternUnits="userSpaceOnUse">
<rect x="0" y="0" width="10"
      height="10" fill="white" />
<rect x="10" y="0" width="10"
      height="10" fill="black" />
<rect x="0" y="10" width="10"
      height="10" fill="black" />
<rect x="10" y="10" width="10"
      height="10" fill="white" />
</pattern></defs>
```

We can use it to draw a checkerboard (Figure 12.9, left) with (**Ex: Checkerboard**)

```
<rect fill="url(#checker)"  x="10" y="10"
          width="80" height="80" />
<rect x="7.5" y="7.5" width="85" height="85"
  fill="none" stroke="grey" stroke-width="5px" />
```

Or a square with a checker border ((Figure 12.9, right) with (**Ex: CheckerBorder**)

```
<defs><pattern id="checkerstk" xlink:href="#checker"
      patternTransform="scale(0.2, 0.2)" /></defs>
<rect stroke="url(#checkerstk)"
   stroke-width="4" fill="none" x="10"
   y="10" width="80" height="80" />
```

FIGURE 12.9: Checker Pattern Usage

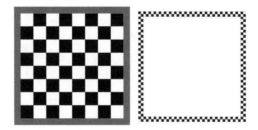

12.3.1 Coordinate Transformation Matrix

From mathematics, 2D coordinate transformations can be stated as a *coordinate transform matrix* (CTM).

Let

$$CTM = \begin{pmatrix} a & c & e \\ b & d & f \\ 0 & 0 & 1 \end{pmatrix}$$

Then, any point (x, y) is transformed to (x', y') by CTM as follows:

$$\begin{pmatrix} x' \\ y' \\ 1 \end{pmatrix} = CTM \cdot \begin{pmatrix} x \\ y \\ 1 \end{pmatrix}$$

In terms of the CTM,

`transform="translate(h, v)"` is

$$\begin{pmatrix} 1 & 0 & h \\ 0 & 1 & v \\ 0 & 0 & 1 \end{pmatrix}$$

`transform="scale(sx, sy)` is

$$\begin{pmatrix} sx & 0 & 0 \\ 0 & sy & 0 \\ 0 & 0 & 1 \end{pmatrix}$$

`transform="rotate(t)` is

$$\begin{pmatrix} \cos(t) & -\sin(t) & 0 \\ \sin(t) & \cos(t) & 0 \\ 0 & 0 & 1 \end{pmatrix}$$

and rotation about other points is a combination of this rotation with translations.

`transform="skewX(t)` is

$$\begin{pmatrix} 1 & \tan(t) & 0 \\ 0 & 1 & 0 \\ 0 & 0 & 1 \end{pmatrix}$$

`transform="skewY(t)` is

$$\begin{pmatrix} 1 & 0 & 0 \\ \tan(t) & 1 & 0 \\ 0 & 0 & 1 \end{pmatrix}$$

A sequence of these transformations $(M_1, M_2, M_3, ...)$ is a matrix product $(M_1 \cdot M_2 \cdot M_3 \, ... \,)$ of these basic matrices. You can efficiently set the `transform` attribute to `matrix(a,b,c,d,e,f)` to specify the CTM.

Let's apply CTM to obtain the mirror image of the `webtong.com` logo (Figure 12.10, left) defined as an SVG `symbol`:

```
<defs><symbol id="webtong" viewBox="0,0 1051,951">
  <path d="M750 0 L150 650 825 650 z" fill="red" />
  <path d="M675 700 L1050 700 750 950 z" fill="blue" />
  <path d="M0 700 L300 700 50 950 z" fill="blue" />
</symbol></defs>
```

To obtain its mirror image we need to multiply by -1 the x-coordinates of all points. In other words, we need the CTM

$$mirror = \begin{pmatrix} -1 & 0 & 0 \\ 0 & 1 & 0 \\ 0 & 0 & 1 \end{pmatrix}$$

To translate the mirror image to the right side of the logo by 250, we need the CTM

$$transX = \begin{pmatrix} 1 & 0 & 250 \\ 0 & 1 & 0 \\ 0 & 0 & 1 \end{pmatrix}$$

And the combined CTM is

$$transX \cdot mirror = \begin{pmatrix} -1 & 0 & 250 \\ 0 & 1 & 0 \\ 0 & 0 & 1 \end{pmatrix}$$

Thus, the SVG code (**Ex:** `Mirror`)

```
<use x="0" y="0" width="100" height="100"
     transform="matrix(-1,0,0,1,250,0)"
     xlink:href="#webtong" />
```

produces the mirror image on the right side of Figure 12.10.

FIGURE 12.10: Mirror Image by CTM

12.4 SVG Animation

SVG follows the SMIL (Synchronized Multimedia Integration Language) standard in defining animation of graphical objects. Most attributes and style properties can be animated. With animation, it is easy to produce motion, zoom/pan, fade-in/out, size and color changes, and other dynamic effects. Major browsers provides increasingly better support for SVG animation. At this point (Fall 2012), the best are Opera 12, Chrome 18, Firefox 15, and Safari 6. IE 9 has some support and IE 10 under Windows 8 should become much better.

To specify animation, you simply place *animation elements* as child nodes in the element being animated. SVG animation elements specify animation on their parent element and they include

- `animate`—Animates general attributes, properties.

- `set`—Sets attributes to discrete values at predetermined times.

- `animateTransform`—Performs coordinate transformations.

- `animateMotion`—Specifies movement along a specified path.

- `animateColor`—Makes color changes.

For example, we can move the clock hand (Figure 12.11) in **Ex: SVGclock** (Section 12.3) by animating the `rotate` transform

FIGURE 12.11: Animated Clock

```
<path id="clockhand" d='M0 0 L0 -76' marker-end='url(#Triangle)'>
<animateTransform attributeName="transform"
    type="rotate" from="0,0,0" to="360,0,0"
    dur="60s" begin="0.5s" repeatCount="indefinite" />
</path>
```

The `animateTransform` element here says the `transform` attribute for the `path` (the clock hand) is to rotate from 0 to 360 degrees about the point (0,0)

every 60 seconds, starting at 0.5 seconds (after element loading). The animation is repeated indefinitely. See **Ex: AnimClock** for the actual animation at the DWP website.

The next example (**Ex: SVGScroll**) shows scrolling text (Figure 12.12).

FIGURE 12.12: Scrolling Text

```
<svg width="980" height="100" viewBox="0 0 980 100"
    style="background-color: #def">
<text x="60" y="60" transform="translate(980,0)"        (A)
  fill="red" style="font-size:3em;font-family:Arial">
Dynamic Web Programming and HTML5
<animateMotion repeatCount="1" dur="5s"
    path="M0 0 -980 0" fill="freeze" />                 (B)
</text></svg>
```

Initially, the text is translated to just outside the viewing area (line A). It scrolls to the left 980 pixels following a path for 5 seconds and the graphics is frozen at the end of the animation (line B). The same technique can move any SVG element along a path. By adding other animations, the object may change size, color, orientation, and so on as it moves.

Here is a red ball that moves and changes size at the same time:

```
<circle cx="0" cy="0" r="8" fill="red">
  <animateMotion dur="6s" path="M0 0 L 200 200"
      repeatCount="indefinite" />
  <animate attributeName="r" begin="0s"
      dur="6s" values="8;16;8"
      keyTimes="0;0.5;1"
      repeatCount="indefinite" />
</circle>
```

Note that radius **r** changes from 8 to 16, then back to 8 again at the **keyTimes** specified start, halfway, and end of each 6-second cycle.

The value of the attribute being animated is computed based on these attribute settings:

```
from="start-value"       to="end-value"
by="incr-value"          values="list"
```

You may use **from-to** animation, **from-by** animation, **by** animation, and **to** animation. If **values** is given, **from**, **to**, and **by** will be ignored. The value of the animated attribute starts with the `from` value unless you specify `additive='sum'`, in which case it starts from the attribute's current value plus the `from` value. For example, the following two animations are the same:

```
<path id="clockhand" d='M0 0 L76 0' marker-end='url(#Triangle)'>
<animateTransform attributeName="transform"
   type="rotate" from="0,0,0" to="360,0,0" additive="sum"
   dur="60s" begin="0.5s" repeatCount="indefinite" />
</path>


<path id="clockhand" d='M0 0 L76 0' marker-end='url(#Triangle)'>
<animateTransform attributeName="transform"
   type="rotate" to="360,0,0"
   dur="60s" begin="0.5s" repeatCount="indefinite" />
</path>
```

Each moves the second hand from the original (3 o'clock) position.

12.4.1 Clipping

When you place a *clip mask* (called a clip path in SVG) over one or a group of graphics elements, only the part inside the masking shape shows through. Other parts will not be rendered. The `clipPath` element is used to define clip masks:

```
<defs> <clipPath id="spot">
<circle cx="0" cy="50" r="48"></circle>
</clipPath></defs>
```

The `clip-path="url(#spot)"` attribute associates a clip mask to an element to be clipped. A moving mask can be obtained using a shape that moves.

Figure 12.13 shows a snapshot of the following clipping animation (**Ex:** Spotlight) whose code begins by defining `symbols` and a `clipPath` needed.

FIGURE 12.13: Moving Spotlight

```
<svg style="width:700px; height:100px;
            background-color: black">
<defs>
 <symbol id="back">
   <rect x="0" y="0" width="700" height="100" />
```

```
  </symbol>
  <symbol id="dwp">
    <text x="350" y="60" style="text-anchor:middle;
         font-size:35px;font-family:Arial">
      Dynamic Web Programming and HTML5</text>
  </symbol>
  <clipPath id="spot">
    <circle cx="0" cy="50" r="48">
    <animateMotion dur="6s" path="M-50 0 750 0"
          repeatCount="indefinite" />
    </circle>
  </clipPath>
</defs>
```

We display the base sign (line 1) and then a moving spotlight within which
the background is white and the text is red (line 2):

```
<use xlink:href="#dwp" fill="yellow"           (1)
     fill-opacity="0.6"/>
<g clip-path="url(#spot)">                      (2)
  <use xlink:href="#back" fill="white"/>
  <use xlink:href="#dwp" fill="red" />
</g></svg>
```

Try ths example (**Ex: Spotlight**) at the DWP website.

12.5   SVG User Interaction and Scripting

Major browsers support scripting in SVG in much the same ways as in HTML.
The SVG `script` tag is used to supply JavaScript code either directly in the
SVG file or through a URL in a separate file. The former takes the form

```
script type="text/javascript">
<![CDATA[

    JavaScript code here

]]></script>
```

and the latter takes the form

```
<script type="text/javascript" xlink:href="url" />
```

The namespace prefix `xlink` needs to be defined unless the SVG code is inline
inside HTML5.

For SVG code placed inline in HTML5, the following are true:

- SVG JavaScript code can also be introduced with the HTML `script` element, just as for HTML.

- HTML and SVG JavaScript code can call each other directly.

- HTML and SVG JavaScript code can respond to SVG and HTML events.

- Each inline `svg` element becomes the root of an SVG DOM tree that is attached to the HTML DOM of the containing webpage.

- HTML and SVG share the same HTML `document` object, their `ownerDocument`.

The preceding points are not true when SVG is embedded with `img`, `object`, or `embed`.

As an example (Figure 12.14), let's add scripting to the animated clock (Section 12.4) so the second hand can be started, stopped, and resumed by mouse-clicks.

FIGURE 12.14: Interactive Clock

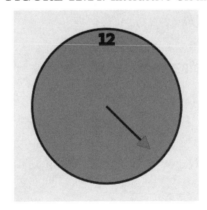

We add code to be performed onload (**Ex:** StopClock):

```
<svg height='300' width='300' onload="svgInit(this)"
     id='root' viewBox='-150 -150 300 300'>
<script type="text/javascript" xlink:href="clock.js" />
```

and a click handler on the clock face

```
<circle cx='0' cy='0' r="120" fill="#0d0" onclick="ss()" />
```

Note the clock face is filled with `#0d0`, a dark green color. If we don't fill the clock face, clicks inside the clock face will not be delivered to the `circle` element; only clicks on the circle stroke (the perimeter of the clock face) will be received by the `circle` element. With the fill, clicks anywhere on the clock face will be delivered to the `circle` element. The SVG style property `pointer-events` can be set to control when pointer events are delivered to SVG elements. The setting `pointer-events: none` makes an element not the target of mouse events. In the above example, we could use the following for the same effects:

```
<circle cx='0' cy='0' style="pointer-events:visible"
        r="120" onclick="ss()" />
```

There are other settings for `pointer-events`, and this style property is making its way into the general HTML CSS specification.

The `clock.js` file is

```
var svgElement, hand, started;

function svgInit(e)
{ svgElement=e;                                // (A)
  hand=document.getElementById('shand');       // (B)
  started=false;
}

function ss()
{ if ( ! started )
  { hand.beginElement(); started=true; return; } // (C)
  svgElement.animationsPaused()
  ? svgElement.unpauseAnimations()             // (D)
  : svgElement.pauseAnimations();              // (E)
}
```

When the `svg` code finishes loading, our onload handler `svgInit` attached to the `svg` element is called. The argument passed, recorded in the global variable `svgElement`, is the root `svg` element, which implements the `SVGSVGElement` interface of the W3C SVG DOM API (line **A**).

Because the SVG is inline, the function has direct access to the HTML `document` object (line **B**). When a user clicks on the clock for the first time, the handler `ss` starts the clock by calling the `beginElement()` of the animated second hand (line **A**). For this to work, we also must use the `begin="indefinite"` attribute setting for the animation element.

Subsequent clicks on the clock face will either pause (line **E**) or resume (line **D**) the animation. The methods `svg` element

```
animationsPaused()
pauseAnimations()
unpauseAnimations()
```

affect all animations inside the root `svg` element. They are specified by the `SVGSVGElement` interface.

We can also make the clockface change color on mouseover easily using the following code (**Ex:** `RollClock`).

```
<circle cx='0' cy='0' r="120" fill="#def" onclick="ss()"
    onmouseover="this.setAttribute('fill', '#0a0')"
    onmouseout="this.setAttribute('fill', '#def')" />
```

As another example (**Ex:** Zoom), let's animate zoom in and zoom out where the user can click any figure on the full display (Figure 12.15, left) and the display window will zoom in on that figure through an animated transition (Figure 12.15, right). Clicking on the figure again zooms out to the original full display.

FIGURE 12.15: SVG Zoom

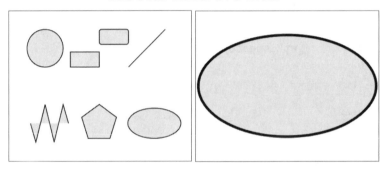

Here is the SVG code:

```
<svg id="svgroot" width="500" height="400"
  viewBox="0 0 500 400" onload="svginit()"
  style="border:thin solid black">
<script type="text/javascript" xlink:href="azoom.js" />
<animate id="azoom" attributeName="viewBox" values=""      (a)
  begin="undefined" dur="0.75s" fill="freeze" />
<rect x="0" y="0" width="500" height="400" fill="white"/>
<g id="group" fill="#def" stroke="rgb(0,0,128)"
                    stroke-width="2px">
<circle cx="100" cy="100" r="50" onclick="zoom(evt)" />   (b)
<!-- more figures coded the same way   -->
<ellipse cx="400" cy="300" rx="72" ry="40"                (c)
        onclick="zoom(evt)"   />
</g></svg>
```

The animation is performed on the `viewBox` attribute of the root `svg` element (line a). Each figure calls `zoom(evt)` when clicked (lines b and c).

The JavaScript file `azoom.js` starts with initialization code:

```
var azoom, svgBox, inNode=null, inBox;

function svginit()
{ /* root svg element  */
  svgBox = bbStr(document.getElementById("svgroot"),0);
  /* azoom is animate element */
  azoom = document.getElementById("azoom");
}
```

The function bbStr is important. It obtains the bounding box of a given SVG node, adds a desired margin m to it, and returns a string representation:

```
function bbStr(node, m)
{  var box = node.getBBox();
   return (box.x-m) + " " + (box.y-m) + " "
     + (box.width+m+m)+" "+(box.height+m+m);
}
```

In our example, svgBox is the string "0 0 500 400" and the bounding box string for the ellipse, with m=2, is "326 258 148 84". Thus, we will set the values (line a) for the viewBox animation to

0 0 500 400;326 258 148 84

for zoom-in on the ellipse and to

0 0 500 400;326 258 148 84

for zoom-out. That is what the event handler zoom does.

```
function zoom(evt)
{ if ( inNode == evt.target )  /* zoom out */
  { azoom.setAttribute("values", inBox+';'+svgBox);
    inNode=null;
  }
  else    /* zoom in */
  { inNode=evt.target; inBox= bbStr(inNode, 2);
    azoom.setAttribute("values", svgBox+';'+inBox);
  }
  azoom.beginElement();
}
```

12.5.1 SVG DOM Manipulations

Just like the HTML DOM, the SVG DOM is also a specialization of the XML DOM API. Thus, your knowledge of the HTML DOM will be very helpful here. For inline SVG code, an svg element represents a node on the HTML DOM tree, and the top node of an SVG DOM tree implements the SVGSVGElement interface.

For example, the onclick handler,

```
function rm(evt)
{  rn=evt.target; pn=rn.parentNode;
   pn.removeChild(rn);
}
```

when attached to any target SVG or HTML element, can remove the element onclick (**Ex:** Remove and **Ex:** Catch). And the function svgBubble creates and returns a new bubble, a white circle centered at canvas position (x,y) that disappears onclick, ready to be placed on the canvas as child of the root svg element.

```
function svgBubble(x,y)
{ var c=document.createElementNS(svgns,"circle");     // (1)
  c.setAttribute("fill","white");
  c.setAttribute("onclick","rm(evt)");
  c.setAttribute("cx",x); c.setAttribute("cy",y);
  r=10+24*Math.random();   c.setAttribute("r",r);
  return c;
}
```

Note that we call createElementNS instead of createElement (line 1) because the latter would create an element in the default namespace, which in this case is HTML. The global variable svgns must be already set to the correct SVG namespace URI.

Putting rm and svgBubble to use, we now create a little game where if a user clicks on a canvas and a white bubble appears at the click location (Figure 12.16). If the click is on a bubble, it disappears.

FIGURE 12.16: Click Bubble Game

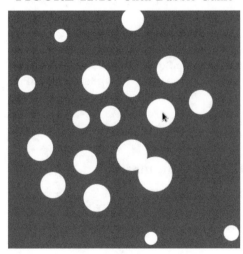

The HTML part of the example (**Ex:** Bubble) is

```
<p>Click to create and burst bubbles.<br /></br /></p>
<div><svg id="svgroot" width="500" height="500"
    viewBox="0 0 500 500"
    onload="svginit(evt)" onclick="makeBubble(evt)"
    style="background-color: green"></svg></div>
```

The JavaScript file `bubble.js` attached to the HTML head is

```
var svgns, SVGRoot, dx, dy;

function svginit(evt)
{ SVGRoot=evt.target;
  svgns=SVGRoot.namespaceURI;
  window.onresize = offset; offset();
}

function offset()                                   // (2)
{ dx=xyPosition(SVGRoot.parentNode);
  dy=dx.y; dx=dx.x;
}

function makeBubble(evt)
{ var b=svgBubble(evt.clientX-dx+window.scrollX,   // (3)
                  evt.clientY-dy+window.scrollY);
  SVGRoot.appendChild(b);                          // (4)
}

function xyPosition(nd)
{ var x=0, y=0;
  while ( nd && nd.offsetParent )
  { x += nd.offsetLeft; y += nd.offsetTop;
    nd = nd.offsetParent;
  }
  return {'x':x, 'y':y};
}

// plus functions rm and svgBubble
```

The offset position of the `svg` canvas display area (line 2), the scroll lengths, and the mouse-click position are used to compute the center position of a new bubble (line 3) which is placed on the canvas right away (line 4). Try the ready-to-run example (**Ex: Bubble**) on the DWP website.

12.6 Knocking Balls Simulation

As another SVG example, we will combine SVG with scripting to simulate the often-seen knocking ball toy (Figure 12.17).

FIGURE 12.17: Knocking Ball Toy Simulation

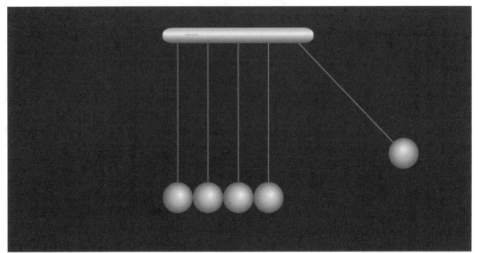

Our implementation (**Ex: Knocking**) begins with SVG definitions for a teathered ball, which apply a radial color gradient to highlight the silver/chrome ball with a 20-pixel radius and a vertical silver line attached (`id="ball"`).

```
<defs><radialGradient id="thegrad"
  gradientUnits="userSpaceOnUse"
  cx="60" cy="200" r="22" fx="64" fy="196">
  <stop offset="0" stop-color="rgb(250,250,250)"/>
  <stop offset="0.55" stop-color="silver"/>
  <stop offset="1" stop-color="rgb(110,110,110)"/>
</radialGradient>
<symbol id="ball">
  <line x1="60" y1="0" x2="60" y2="180" stroke="silver"/>
  <circle cx="60" cy="200" r="20" fill="url(#thegrad)"/>
</symbol></defs>
```

The beam from which the balls hang is a rectangle with rounded corners and a linear color gradient plus a green/red light on it (`id="beam"`).

```
<defs><linearGradient id="lg"
  gradientUnits="userSpaceOnUse"
  x1="80" y1="0" x2="80" y2="20">
```

```
<stop offset="0" stop-color="silver"/>
<stop offset="0.4" stop-color="rgb(250,250,250)"/>
<stop offset="1" stop-color="rgb(120,120,120)"/>
</linearGradient>
<symbol id="beam">
<rect x="0" y="0" width="200"
  height="20" rx="14" ry="8" fill="url(#lg)"/>
<rect id="light" x="30" y="8" width="18" height="4"
      rx="2" ry="1" fill="lightgreen" />
</symbol></defs>
```

Against a black background, the beam and five balls are displayed as follows:

```
<use xlink:href="#beam" x="80" y="30" onclick="stop()"/>
<!-- ball one here -->
<use xlink:href="#ball" x="80" y="50" />
<use xlink:href="#ball" x="120" y="50" />
<use xlink:href="#ball" x="160" y="50" />
<!-- ball five here -->
```

Balls one and five have rotational animations and event handling attached to them. Here is ball one:

```
<use xlink:href="#ball" x="40" y="50" onclick="start('t1s')" >
<animateTransform id="t1s" attributeName="transform"
  attributeType="XML" type="rotate" onend="next('t1d');"
  values="0,100,50;70,100,50" begin="indefinite"
  dur="1.5s" additive="replace" fill="freeze" />
<animateTransform id="t1u" attributeName="transform"
  attributeType="XML" type="rotate" onend="next('t1d')"
  values="0,100,50;20,100,50;55,100,50;70,100,50"
  keyTiems="0,0.1,0.6,1" begin="indefinite"
  dur="0.4s" additive="replace" fill="freeze" />
<animateTransform id="t1d" attributeName="transform"
  attributeType="XML" type="rotate" onend="next('t5u')"
  values="70,100,50;55,100,50;20,100,50;0,100,50"
  keyTiems="0,0.4,0.9,1" begin="indefinite"
  dur="0.4s" additive="replace" fill="freeze" />
</use>
```

The t1s animates pulling the ball to start the knocking motion. The t1d animates the ball coming down and hitting the second ball (stopping). The t1u animates the first ball rising as a result of ball five striking ball four. Ball five has entirely similar animations.

The onend event handler next is called at the end of one animation to start a specific next one:

```
function next(id)
{ var anim = document.getElementById(id);
  if ( ! end || id.charAt(2)=="d" )
    anim.beginElement();
  else
  {   end = false;
      started=false;
      lt=document.getElementById('light');
      lt.setAttribute("fill", "lightgreen");
  }
}
```

Clicking on ball one or five invokes the `start` handler (line A).

```
var started = false, end = false, lt;

function start(id)                                    // (A)
{ var anim = document.getElementById(id);
  started=true;    anim.beginElement();
}

function stop()                                       // (B)
{ if ( started )
  { end = true;
    lt=document.getElementById('light');
    lt.setAttribute("fill", "orangered");
  }
}
```

Clicking on the beam invokes the `stop` handler (line B). Play with this simulated toy (**Ex: Knocking**) at the DWP website.

12.7   SVG versus HTML5 Canvas

In HTML5, SVG and the Canvas API are two ways to draw graphics interactively. Both support drawing points, lines, and simple shapes, coordinate transformations, and raster image inclusion. The question is when to use which one.

The Canvas is a drawing area completely controlled from JavaScript. It is best for creating and rendering raster graphics such as plotting mathematical functions (curves and surfaces), raster image manipulation, and editing such as red-eye removal. Canvas graphics and user interactions often involve heavy JavaScript programming. But it has poor support for displaying text.

SVG is an XML application, a language with a DOM structure. It is also resolution independent because of its scalability. Animations can be declared and applied; events can be attached to graphical elements. Thus, SVG is suitable for displaying vector graphics and performing animations with user interaction. Sophisticated animations with user control can often be implemented with a small amount of JavaScript code. SVG is ideal for implementing systems for teaching and learning plane geometry. It is also very suitable for mobile devices.

12.8   MathML

MathML is an XML application that enables "inclusion of mathematical expressions in webpages" according to the *W3C Math Home* site. The W3C Math Working Group has been developing *Presentation MathML* for rendering of mathematical expressions and *Content MathML* for capturing the semantics of mathematical expressions since the late 1990s. W3C recommendations for MathML have moved from MathML 1.0 to 2.0 and now 3.0. Presentation MathML is important for browser display and printing. Content MathML makes it possible to communicate mathematical expressions between computer programs for processing such as evaluation, simplification, derivation, integration, ans so on. MathML 3.0 introduced important refinements, including support for CSS styling and for hyperlinks. As the Web begins to have an impact on education at many levels, MathML will see increasing use in improving mathematics education.

Firefox is the first to support Presentation MathML natively and is still the best to use for MathML. Other browsers are catching up. The MathPlayer plug-in enables Internet Explorer to display MathML nicely. The display quality of MathML also depends on the fonts available. By allowing inline coding of MathML, HTML5 will make cross-browser mathematics finally a reality.

Try the MathML examples in this book on Firefox or IE with MathPlayer installed. See the "fonts for MathML" page on the Mozilla Developer Network (MDN) if you need to install fonts.

12.8.1   Presentation versus Content MathML

To understand the difference between Presentation MathML and Content MathML, consider the expression *lengthsup*2. But what does it mean? Does it mean "the second footnote for `length`" or "*length squared*"?

The Presentation MathML for it is

```
<msup><mi>length</mi><mn>2</mn></msup>
```

The Content MathML for the same is

```
<apply>
    <power/><ci>length</ci><cn type="integer">2</cn>
</apply>
```

The former is enough for rendering but it takes the latter to capture the computational semantics of the expression.

12.9　Infix to MathML Conversion

To illustrate the usage of MathML, we will look at a demo that converts mathematical expressions given in the familiar infix notation to MathML code (Figure 12.18).

FIGURE 12.18: Infix to MathML Converter

Infix to MathML Conversion

To try the demo, enter an infix expresson and click the convert button.

```
sqrt(-x+3)/4 + 100*x - PI/9 = x^3
```

| Get MathML Code | Render MathML |

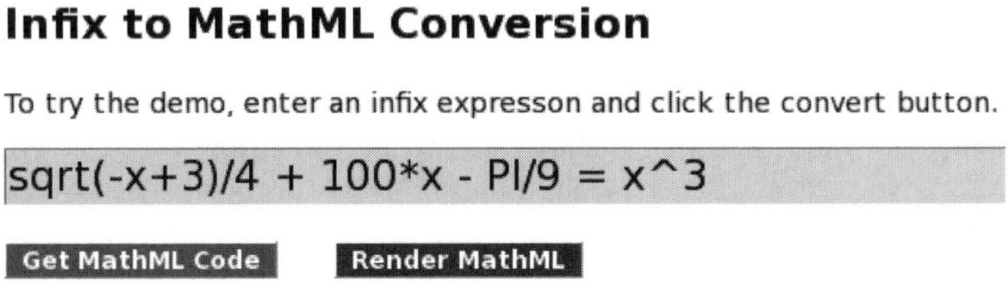

This infix to MathML converter (**Ex: Infix2Mathml**) involves a Web service[1] that does the actual parsing of the infix string and converting it to MathML. An HTML5 page collects user input, the desired infix string, and makes AJAX calls to the Web service to obtain the conversion result (MathML code). The resulting MathML source code and the browser rendering of it are displayed on user command.

Figure 12.19 shows the actual Firefox display of the resulting MathML code.

FIGURE 12.19: Sample MathML Display

$$100x + \frac{\sqrt{3-x}}{4} - \frac{\pi}{9} = x^3$$

The actual MathML code returned by the Web service consists of the Presentation MathML:

```
<mrow>
    <mrow>
        <mrow><mn>100</mn><mo></mo><mi>x</mi></mrow>
        <mo>+</mo>
        <mfrac>
            <mrow><msqrt><mrow>
```

---

[1]Provided by the author's research at Kent State University.

```
            <mn>3</mn><mo>-</mo><mi>x</mi>
        </mrow></msqrt></mrow>
        <mrow><mn>4</mn></mrow>
    </mfrac>
    <mo>-</mo>
    <mfrac><mrow><mi>&pi;</mi></mrow>
        <mrow><mn>9</mn></mrow></mfrac>
  </mrow>
  <mo>=</mo><msup><mi>x</mi><mn>3</mn></msup>
</mrow>
```

and the Content MathML

```
<apply><eq/>
    <apply><plus/>
        <apply><times/>
            <cn type="rational">-1<sep/>9</cn>
            <cn type="constant">&pi;</cn>
        </apply>
        <apply><times/>
            <cn type="rational">1<sep/>4</cn>
            <apply><power/>
                <apply><plus/>
                    <cn type="integer">3</cn>
                    <apply><times/>
                        <cn type="integer">-1</cn><ci>x</ci>
                    </apply>
                </apply>
                <cn type="rational">1<sep/>2</cn>
            </apply>
        </apply>
        <apply><times/>
            <cn type="integer">100</cn><ci>x</ci>
        </apply>
    </apply>
    <apply><power/>
        <ci>x</ci><cn type="integer">3</cn>
    </apply>
</apply>
```

Content MathML uses the *prefix notation*

`<apply>` *operator* *operand*$_1$ *operand*$_2$... `</apply>`

where the *operator* can be any MathML-defined operator element, such as
`<plus/>`, `<times/>`, `<power/>`, and so on. Each of the one or more operands
are valid Content MathML number (`<cn>`), identifier (`<ci>`), or prefix notation.

The complete MathML markup for this example is in *mixed* MathML as follows:

```
<math xmlns="http://www.w3.org/1998/Math/MathML">
<semantics>
    <!-- Presentation MathML Here -->
 <annotation-xml encoding="MathML-Content">
    <!-- Content MathML Here -->
 </annotation-xml>
 <annotation-xml encoding="Text-Infix">
    sqrt(-x+3)/4 + 100*x - PI/9 = x^3
 </annotation-xml>
</semantics></math>
```

Thus, the code uses the `<semantics>` element to bind one or more semantics (meaning) to the presentation code. This way, the code not only can be displayed easily but also can be used in computations by programs.

The AJAX code for this demo is in `infix2mathml.js`, which begins with initializing global variables:

```
/////    infix2mathml.js    /////
var infixStr="", code=null,
req = (window.XMLHttpRequest) ? new XMLHttpRequest()
      : new ActiveXObject("Microsoft.XMLHTTP"),
proxyURL="getMathML.php?expr=",
```

The PHP proxy `getMathML.php` will send the user input infix expression to the correct server at icm.mcs.kent.edu.

The function `infix2MathML` is invoked by clicking the Get MathML Code button:

```
function infix2MathML(infix)
{ if (infix == infixStr)
  { document.getElementById('code').value=
       document.getElementById('math_display').innerHTML;
    return;
  }
  infixStr=infix;
  var url = proxyURL+encodeURIComponent(infix);
  req.onreadystatechange=showCode;
  req.open("GET", url, true);  // upper-case GET
  req.send(null);
}
```

And the resulting MathML code is displayed:

```
function showCode()
{  if (this.readyState==this.DONE)
```

```
    {  if (this.status==200 && this.responseText)
         document.getElementById('code').value =
             this.responseText;
       else
         alert("Problem retrieving response data");
    }
}
```

The `displayMath` displays the MathML code in the output `math_display` by setting its `innerHTML` either from the `code` output data or by calling the AJAX proxy:

```
function displayMath(infix)
{ if (infix == infixStr)
  {  document.getElementById('math_display').innerHTML=
         document.getElementById('code').value;
     return;
  }
  infixStr=infix;
  var url = proxyURL+encodeURIComponent(infix);
  req.onreadystatechange=displayMathML;
  req.open("GET", url, true); req.send(null);
}


function displayMathML()
{ if (this.readyState==this.DONE)
  { if (this.status==200 && this.responseText)
       document.getElementById('math_display').innerHTML=
           this.responseText;
    else
      alert("Problem retrieving MathML Code");
  }
}
```

The PHP proxy is

```
<?php   /////   getMathML.php   /////
header("Content-Type: application/mathml+xml");
// an ajax proxy

$serviceURL="http://icm.mcs.kent.edu/cgi-bin/"
  ."infix2mathml.cgi?format=mathml&simplify=yes&expr=";
$expr=$_REQUEST['expr'];
$url=$serviceURL . urlencode($expr);

$handle = fopen($url, "r");
if (!$handle)
```

```
{  echo "{error: 'failed to open $url'}"; exit; }
$contents = stream_get_contents($handle);
fclose($handle);  echo $contents;
?>
```

The converter is a handy tool for obtaining MathML code for expressions one can easily input using infix notation. Try **Ex: Infix2Mathml** on the DWP website.

12.10   MathML for Mathematics Education

Support in HTML5 for MathML will help unleash the huge potential of its application in Web-based mathematics education at all levels. Let's look at an example where a student can enter the derivative of a given expression and the answer can be properly displayed and also checked for correctness. Figure 12.20 shows such a page where the student can practice on computing the derivative for any given expressions.

Clicking the mathematics icon displays an input tool (*MathEdit*) for entering and editing mathematical expressions. This tool is written in JavaScript and works in all major browsers.

After composing the answer (the derivative), the student can then ask that the answer be checked. The page may allow several tries before disclosing the correct answer. The MathEdit tool and the answer checking Web service are part of the R&D efforts by the WME (Web-based Mathematics Education) group at Kent State University.

The original expression and derivative given by the student, both represented in MathML, are sent to the answer checker, which computes the derivative and compares that with the student answer. The answer checker returns a JSON result, indicating the correctness of the student answer and, if necessary, the actual correct answer.

The working demo for this example (**Ex: Derivative**) can be found at the DWP website.

12.11   Combining SVG with MathML

An instruction page can apply SVG and MathML together productively. For example, for any two arbitrary points the user may pick on an SVG-implemented xy-plane (Figure 12.21),
the following actions can be performed automatically (**Ex: LineEq**):

- A line is drawn through the two points.

- Point coordinates are displayed.

FIGURE 12.20: Derivative Checking

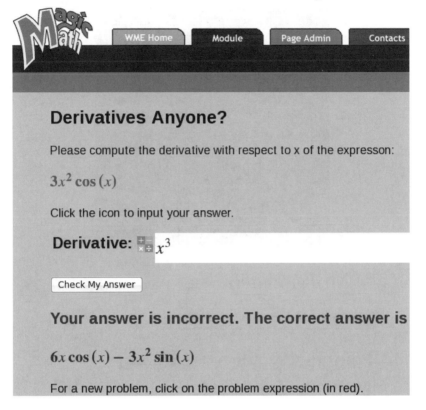

- The equation for the straight line is computed in MathML and displayed (Figure 12.22).

The SVG xy-plane is

```
<div><svg id="svgroot" width="500" height="500"
    viewBox="-250 -250 500 500"
    onload="svginit(evt)" onclick="makePoint(evt)"
    style="background-color: green">
<g id="coord" stroke="white" stroke-width="1px">
<line x1="-250" y1="0" x2="250" y2="0" />
<line x1="0" y1="-250" x2="0" y2="250" /></g>
</svg></div>
```

Clicking on the xy-plane calls the event handler `makePoint` which keeps a point count in the global `p_cnt` and resets the SVG canvas on the third click (line A). It calls `svgPoint` to create a 3-pixel circle for each point at the mouse-click position (lines B through C).

FIGURE 12.21: X–Y Plane

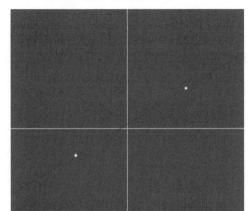

FIGURE 12.22: Line Equation

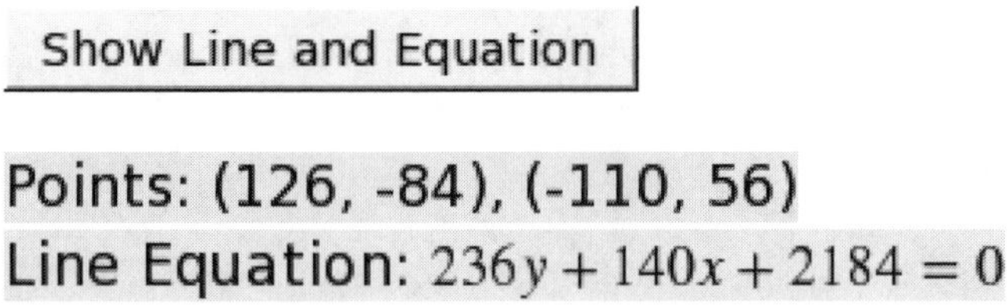

Show Line and Equation

Points: (126, -84), (-110, 56)
Line Equation: $236y + 140x + 2184 = 0$

```
function makePoint(evt)
{ if ( p_cnt > 1 ) reset();                        // (A)
  var b=svgPoint(evt.clientX-dx+window.scrollX,    // (B)
            evt.clientY-dy+window.scrollY);
  if ( b )
  {  p_cnt ? p1 = b : p2 = b; p_cnt++;
     SVGRoot.appendChild(b);                        // (C)
  }
}

function reset()
{  p_cnt=0;
   if ( p1 ) { SVGRoot.removeChild(p1); p1=null; }
   if ( p2 ) { SVGRoot.removeChild(p2); p2=null; }
   if ( line )  { plane.removeChild(line); line=null; }
   document.getElementById('linediv').innerHTML="";
}
```

```
function svgPoint(x,y)
{ var c=document.createElementNS(svgns,"circle");
  x=Math.floor(x/2.0)*2-250;
  y=Math.floor(y/2.0)*2-250;
  if ( p_cnt==2 && p1.getAttribute("cx")==x &&        // (D)
       p1.getAttribute("cy")==y)
  { alert("The second point must be different");
    return null;
  }
  c.setAttribute("fill","white");
  c.setAttribute("onclick","tooClose(evt)");          // (E)
  c.setAttribute("cx",x); c.setAttribute("cy",y);
  c.setAttribute("r",3); return c;
}


function tooClose(evt)
{ evt.stopPropagation();
  alert("points are too close"); return false;
}
```

The two points are kept from being too close to each other (lines D and E).

The button "Show Line Equation" triggers makeline(), which draws the straight line (line F) and displays the line equation with MathML (line G).

```
function makeLine()
{ var x1, y1, x2, y2;
  if ( p_cnt < 2 ) return;
  x1=p1.getAttribute("cx"); y1=p1.getAttribute("cy");
  x2=p2.getAttribute("cx"); y2=p2.getAttribute("cy");
  displayLine(x1,y1,x2,y2);                           // (F)
  lineEquation(x1,y1,x2,y2);                          // (G)
}
```

To display the line through the given two points, displayLine adds a red line that extends to the border of the visible xy-plane (lines H through I).

```
function displayLine(x1,y1,x2,y2)
{ var a=y1-y2, b=x2-x1, c=x1*y2-x2*y1;                // (H)
  var slope = Math.abs(a/(b*1.0));
  if ( slope < 1.0 )
  { x3=250; x4=-250; y3=(-c-a*x3)/(b*1.0);
    y4=(-c-a*x4)/(b*1.0);
  }
  else
  { y3=250; y4=-250; x3=(-c-b*y3)/(a*1.0);
    x4=(-c-b*y4)/(a*1.0);
```

```
  }
  line=document.createElementNS(svgns,"line");
  line.setAttribute("x1",x3); line.setAttribute("y1",y3);
  line.setAttribute("x2",x4); line.setAttribute("y2",y4);
  line.setAttribute("stroke", "red");
  plane.appendChild(line);                              // (I)
}
```

To obtain the MathML for the line equation, `lineEquation` constructs the infix form of the equation (lines J through K).

```
function lineEquation(x1,y1,x2,y2)
{ var a=y1-y2, b=x2-x1, c=x1*y2-x2*y1;                  // (J)
  var infix;
  if ( a < 0 && b < 0 ) { a = -a; b=-b; c=-c; }
  if ( a == 0 && c == 0) infix="y=0";
  else if ( b==0 && c==0 ) infix="x=0";
  else infix = ""+a+"*x+"+b+"*y+"+c+" = 0";             // (K)
  points="<p>Points:   (" + x1 + ", " + y1 +"), (" +
         x2 + ", " + y2 +")</p>Line Equation: ";        // (L)
  displayMath(infix);                                   // (M)
}
```

The `displayMath` function calls the infix-to-MathML conversion service and then displays the resulting MathML code together with the point coordinates computed on line L.

```
function displayMath(infix)
{ var url = proxyURL+encodeURIComponent(infix);
  req.onreadystatechange=displayMathML;
  req.open("GET", url, true);
  req.send(null);
}


function displayMathML()
{ if (this.readyState==this.DONE)
  {   if (this.status==200 && this.responseText)
      document.getElementById('linediv').innerHTML =
          points + this.responseText;
      else
        alert("Problem retrieving MathML Code");
  }
}
```

This example only begins to scratch the surface of what can be done with HTML5, SVG, and MathML for mathematics, science and engineering on the Web.

Please try the full example (**Ex: LineEq**) on the DWP website.

12.12   For More Information

For the most current SVG language standard please visit/search the W3C site which also has up-to-date SVG DOM API documentation. The Mozilla Developer Network (MDN) also has complete coverage of SVG at developer.mozilla.org/en/SVG. The SVG authoring guidelines at jwatt.org/svg/authoring/ can be very helpful.

Visit w3.org/Math/ for the new MathML 3.0 standard. The Planet MathML site provides news and activities regarding MathML. The Web-based Mathematics Education (WME) site at Kent State University (wme.cs.kent.edu) is an interesting project applying SVG and MathML, among other techniques, for school mathematics education.

See the Design Sciences site for MatyPlayer, an IE plug-in for MathML. See MDN for fonts for displaying MathML.

12.13   Summary

Scalable Vector Graphics (SVG) enables textual markup of 2D vector graphics, including `circle`, `rect` (rectangle with/without rounded corners), `line` (straight line segment), `polyline` (multisegment line), `polygon` (closed multisegment line), `ellipse`, arbitrary `path`, and text. Also supported are scaling, color gradient, clipping, filtering, coordinate transformation, and animation.

Presentation MathML enables rendering of mathematical formulas by browsers and other devices. *Content MathML* captures the meaning (operations) of mathematical expressions so it can be used in computations by programs. Together they make communicating mathematics for viewing and for processing possible on the Web.

Both SVG and MathML can be included **inline** in HTML5. The included `<svg>`/`<math>` element becomes a self-contained root node of an SVG/MathML DOM tree on the HTML5 DOM. JavaScript for SVG/MathML can be attached either to the child document or to the parent HTML5 document within the same browsing context. Within the child DOM, follow the SVG DOM API or MathML DOM API. For creating new elements in the child DOM, use `document.createElementNS(...)` with the correct namespace URI.

The combination of SVG, MathML, and HTML5 has great potential in learning, teaching, and communicating technical materials on the Web.

Exercises

12.1. List, with examples, the different ways SVG can be embedded in a webpage.

12.2. What is the namespace URI for SVG? For MathML? What is the root element for SVG? For MathML?

12.3. Explain the difference between Presentation and Content MathML. Why do we say Content MathML uses a prefix notation?

12.4. Attach an `onclick` event handler to an SVG triangle that changes its stroke color. Do you need to fill the triangle? Why?

12.5. Use SVG animation and scripting to make a bouncing ball display.

12.6. Add a minute hand to the clock animation in Section 12.5.

12.7. Add a digital readout to the clock.

12.8. Make an SVG display with several objects; each can be repositioned with the mouse.

12.9. Take the straight line example from Section 12.11 and make it possible to drag either point that defines the line to a different position. The display is animated to follow the dragging operation.

12.10. Find out about SVG support for filtering.

12.11. Follow the demos in Section 12.8 and write a tool that takes two polynomials entered in infix format by the user and displays their product (expanded) using MathML.

12.12. Explain how the line equation example in Section 12.11 computes the point position on the SVG canvas from the mouse-click position.

Chapter 13

HTML5 and Mobile Websites

Mobile computing devices, smartphones and tablets in particular, experienced explosive growth since the early 2000s and the trend is predicted to continue. Apple® introduced the iPhone® in 2007 and Google released the first Android® OS in late 2008. Worldwide sales of smartphone units for 2011 was *"472 million or 31% of mobile communication device sales,"* according to Gartner. A Pew Internet Project survey from May 2011 found 35% of US adults owned a smartphone.

Analysis of global Q2 2012 smartphone sales by Nielsen shows the Google Android OS dominating market share (48.5%), with Apple's iOS® a distant second (32.0%, Figure 13.1):

FIGURE 13.1: Q2 2012 Smartphone OS Shares

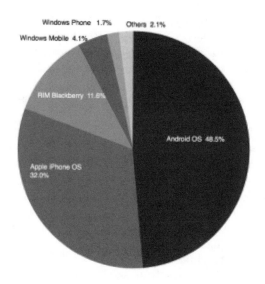

Morgan Stanley Research estimates that sales of smartphones will exceed those of PCs in 2012, and IMS Research expects smartphone annual sales to reach 1 billion in 2016 (half the mobile device market).

Major browsers for smartphones and tablets include Android Browser (Google), iPhone Safari® (Apple), BlackBerry® Storm Browser, Nokia® S60

Touch Browser, Microsoft IE Mobile®, Opera Mini®, Chrome® for Android, and Mobile Firefox® (Fennec). Many of these provide good support for standard Web technologies, especially HTML5, which is designed to work well on desktops, laptops, tablets, and smartphones. Mobile operating systems and browsers are expected to evolve at a rapid pace. But the trend is moving toward standard HTML5, CSS, JavaScript, and away from mobile/wireless markup languages.

Given these trends, it becomes obvious that Web developers need to understand the mobile browsing environment, the difference between browsing on mobile and desktop/laptop devices, how to design websites targeted for mobile devices, how to make/transform existing websites to be *mobile friendly*, and how to take advantage of unique or convenient features of mobile devices such as touch-screen, momentum scrolling, panning, zooming, orientation switching, phone dialing, texting (SMS), picture/video taking, speech input, and geolocation.

13.1   Mobile Web Browsing

The most notable charateristic of smartphones is their palm-fitting size. Smartphone screen sizes range from less than 3 inches to about 5 inches, measured diagonally. A 2002 study by Strategy Analytics found, most smartphone users felt 4 to 4.5 inches to be the ideal screen size for holding in the palm. Tablets offer much larger, typically 7 to 10-inch, screens.

Screen resolutions for modern mobile devices are constantly increasing. They typically range from 240×320 to 480×800 pixels. The iPhone 5® is 640×1136 (326 ppi) and the Samsung Galaxy® III is 720×1280 pixels (306 ppi). As you can see, mobile phones tend to have much higher pixel densities, resulting in clear fine details for fonts and pictures, if you have great vision to see them. Thus, the smaller screen size is still a primary design consideration for mobile phones if much less so for tablets.

Perhaps just as notable is the touch-screen. Mobile devices have no mouse, no windows that can resize or have scrollbars, and usually no physical keyboard. Users touch the screen for scrolling, panning, zooming, navigating, as well as text input and editing. Usually, the user clicks with a finger tap, right-clicks by holding the finger down, pans and scrolls by flicking a finger, zooms in/out by double-tapping, or pinch open/close. The gestures may have some equivalents on laptop touchpads but are generally not available on desktops. Smartphone users are also very agile in scrolling the screen in either direction, whereas horizontal scrolling is usually a bad idea for desktop/laptop browsing. The `onmouseover` rollover effects so widespread on desktop/laptop browsing environments are impossible or not needed on touch-screen devices.

Mobile devices are generally slower in speed. Thus it is advisable to make mobile pages concise, to the point, and without much decorating graphics, or large pictures. Because a hand-held device usually offers a *portrait* and a *land-*

scape orientation, mobile webpages must account for that. This means a fluid design that reacts well to screen resizing would be much more mobile friendly than ice designs. In fact, an ice-design page will not react to orientation change and will frustrate users.

To make fluid designs for mobile browsing, it is advisable to avoid HTML elements with significant widths such as larger pictures, tables, or videos. The goal is to have the content flow and line wrap as much as possible to better fit smaller screens. Even though smartphone users generally do not mind scrolling, it is wise to reduce the white space between lines and paragraphs to save precious screen space. Sizable head banners and footers, desirable on desktop/laptop environments, are to be avoided on smartphones.

13.2   Mobile Website Strategies

So what ways are there to deliver a website to mobile devices?

For larger companies with the resources and manpower, a dedicated *mobile app* can be developed to provide information and intended services. Such an app would be written in a language dictated by the mobile OS (Objective C for iOS and Java for Android, for example) and would be downloaded and installed by each mobile customer, usually from an *app store*. Such native apps enjoy full OS support, including access to files and devices, and UI features. The disadvantages include high development cost, the need for customer installation, the app presenting a different and unfamiliar user interface, and cost of maintenance and update of the app as a separate piece of software that must also be kept consistent and in sync with the company's website. To reach all mobile customers, a company taking this approach must develop such an app for each different mobile OS and keep them compatible as new mobile OSs are released.

Instead of developing native apps, a business may decide to use the Web to reach mobile customers as well. This makes sense from many viewpoints:

- The Web, especially with HTML5, is standard technology, allowing you to reach any customer with a Web browser, mobile or not.

- Customers are familiar with the Web browser interface and will not have to switch to a different UI when on mobile devices.

- There is nothing to download, install, or update.

- Users have a better understanding of security afforded by the Web under `https` and may not feel at ease with a downloaded app.

- It can be much easier for a business to keep Web content on mobile and nonmobile sites consistent.

To reach customers on the mobile Web, a company may develop a completely separate website just for them. However, it makes sense to have the mobile and nonmobile sites share content and code as much as possible. With ever increasing speed, size, and screen resolution, the difference between laptops and tablets are disappearing. Thus, more often than not, it is reasonable for organizations to have a mobile site for smartphones and a regular site for larger devices.

In this chapter, we explore how to apply standard technologies HTML5, CSS3, PHP, and JavaScript to make websites mobile friendly and approaches to deliver them to the mobile Web.

13.3   Mobile Layout and Navigation

To make a mobile website fit well with various small screen sizes and resolutions is a challenge. It is advisable to keep a simple one-column layout with no margins, borders, or paddings. This is usually true for the `body` and top-level container elements. For the main content, it is still good to have a small margin, say 2 to 3 pixels, on the left and right, making text easier to read and reassuring users that nothing is cut off at the edges. Avoid verbosity and unnecessary images. Some vertical scrolling of a page is often unavoidable. However, mobile users are accustomed to scrolling both horizontally and vertically. The positive tactile feedback can be engaging.

Any top banner with its elaborate graphics, top navbar, and left/right navigation panels must usually be eliminated. Instead, a one-row top banner can display website identity and double as a navigation menu. For example, a top banner and navbar for the birdsong farm regular website (Figure 13.2)

FIGURE 13.2: Birdsong Farm Regular Site Entry

can be simplified to a single row consisting of a logo (line A) and a navigation menu coded with the `select` element (lines B through C):

```
<nav class="mobile">
<img src="bslogo.png" alt="birdsong farm logo"          // (A)
```

```
      id="mlogo" />
<select size="1" id="mnav" onchange="menuAction(this)"> // (B)
  <option value="index.html" selected="">Birdsong Farm
  </option>
  <option value="produce.html" >Our Produce</option>
  <option value="csa.html" >CSA Shares</option>
  <option value="news.html" >News</option>
  <option value="faq.html" >FAQ</option>
  <option value="contact.html" >Contact Us</option>     // (C)
</select></nav>
```

Figure 13.3 (left) shows the mobile version of the Birdsong Farm site entry.
The top-bar displays two inline blocks, one for the logo (line D) and one for
the select menu (line E) as follows:

FIGURE 13.3: Space-Saving Banner/Navbar

```
nav.mobile
{ display: block;   width: 100%;
  background-color: #005027;
  color: white; margin: 0px; padding: 10px;
  border-bottom: 2px solid #ffa500;
}

nav.mobile > img#mlogo                               /* (D) */
{ width: 40px; display: inline-block;
  vertical-align: middle;
}
```

```
nav.mobile > select                              /* (E) */
{ width: 85%;  display: inline-block;
  text-align: center; background-color: darkgreen;
  padding: 4px; color: white; font-weight: bold;
  font-size: x-large; vertical-align: middle;
}
```

Figure 13.3 (right) shows the pop-up navigation menu display on the same Android phone. Selecting an option triggers the **onchange** handler menuAction.

```
function menuAction(sel)
{   var index = sel.selectedIndex;
    var url = sel.options[index].value;
    sel.selectedIndex = 0;
    if ( url != -1)
    {   window.location=url; }
}
```

On the mobile screen, it is best not to have small hyperlinks or links packed too close to one another. Users will have a hard time clicking those links with their thumbs.

See **Ex:** BirdsongMobile at the DWP site.

13.4   CSS Media Queries

The beauty of HTML5 and CSS has always been the separation of document structure and presentation. With a variety of mobile phones, tablets, laptops, and desktops, it is more important than ever that we keep webpages structured well and present them with CSS for different target devices.

The CSS3 *media query* feature makes it possible to more precisely label styles under different media conditions. With media query, you can specify styles that apply only for certain screen resolutions, color or black/white, window sizes, aspect ratios, portrait or landscape orientation, and so on.

For example, the Birdsong Farm (birdsongfarmohio.com) site delivers the following style sheet for desktop, laptop, and 10-inch tablet browsers (Figure 13.2):

```
/**** desktop, laptop, 10-inch tablet screen  ****/
@charset "UTF-8";
body
{ font-size: 100%;
  font-family: Helvetica, Arial, sans-serif;
  line-height: 120%; background-color: #005027;
```

```
}
nav.mobile {display: none;}                              /* 1 */

.container
{ width: 960px; padding: 10px; margin: 0 auto;
  background-image: url(birdsongFarm-header.jpg);        /* 2 */
  background-repeat: no-repeat;
  background-position: center top; height: 500px;
}
nav.header
{ padding: 10px; margin: 310px 10px 0px 10px;
  position: relative;
}
nav.header .tabs                                         /* 3 */
{ margin: 0px -4px 7px 1px;
  float: left; }
a { text-decoration: none; }
nav.header .tabs a { color: #FFF; padding: 10px; }

nav.header .tabs a#tab1:hover                            /* 4 */
{ background-image: url(images/tab1.png);
  background-repeat: no-repeat;
  background-position: center top;
}
  /*   ... more ...   */
```

It removes the mobile header (line 1) and gives us a nice graphical top banner (line 2), and navigation tabs (line 3) with rollover links (line 4).

To style the same Birdsong Farm site for smartphones, one approach is to add media-dependent styles through the CSS3 media query feature. The following style sheet applies only to devices satisfying the given media queries (line 5), namely, devices with a maximum physical screen width and height of 640 pixels. The precise pixel number may vary but the intention is to use media queries to rule out desktops, laptops, and 10-inch tablets and to rule in most smartphones.

```
/**** m.birdsong.css ****/
@charset "UTF-8";
@media only screen and (max-device-width:640px)
               and (max-device-height:640px)           /* 5 */
{ nav.header {display: none;}                           /* 6 */

  nav.mobile                                            /* 7 */
  { display: block; width: 100%;
    background-color: #005027; color: white;
    margin: 0px; padding: 10px, 10px, 5px, 5px;
```

```
    border-bottom: 2px solid #ffa500;
  }

  nav.mobile > img#mlogo                          /* 8 */
  { width: 40px; display: inline-block;
    vertical-align: middle; padding-left:3px;
    padding-right:3px;
  }

  nav.mobile > select                             /* 9 */
  { width: 84%;  display: inline-block;
    text-align: center; background-color: darkgreen;
    padding: 4px; color: white; font-weight: bold;
    font-size: x-large; vertical-align: middle;
  }
  /*  ... more ... */
}
```

This mobile style sheet removes the regular header/navbar (line 6) and adds a single-row topbar that doubles as a site banner and navigation menu (lines 7 through 9).

The Birdsong Farm site is a case study of making the same site serve both regular and mobile customers using various techniques in combination with media queries, as we shall explain (**Ex:** BirdsongMobile).

For example, the header code has two parts: one for regular and one for mobile browsers. Which one to display is controlled by CSS with media queries.

```
<nav class="mobile">
<img src="apple-touch-icon.png" alt="birdsong logo" id="mlogo" />
<select size="1" id="mnav" onchange="menuAction(this)">
<option value="index.html" selected="">Birdsong Farm</option>
<!-- more options --></nav>
<nav class="header"><div class="tabs">
<a href="index.html" id="tab1"
   class="active1">Birdsong Farm</a></div>
<!-- more tabs -->
</nav>
```

The smartphone style sheet m.birdsong.css is attached to each page as follows:

```
<link href="m.birdsong.css" media="screen          (10)
          and (max-device-width: 640px)
          and (max-device-height: 640px)"
      rel="stylesheet" type="text/css" />
```

Note the added media attribute of the link element (line 10) tells a browser the media query conditions before it loads the style sheet. The style sheet does not apply to any device whose physical screen is wider or higher than 640 pixels. That should exclude desktops, laptops, and 10-inch tablets.

The media query feature is part of the CSS3 standard. A media at-rule (Section 4.22) starts with @media, followed by one or more comma separated *media queries*. Each media query is zero or one media type followed by zero or more media condition expressions.

media_type AND *expression* AND *expression* ...

You may also add a keyword only or not before the *media_type*. The only keyword is redundant and can be omitted but it can unconfuse browsers not supporting media queries. See Section 4.22 for a list of media types.

Each *expression* is in the form of

(*media_feature*: *value*)

An @media rule is applied only if all its components evaluate to true on a particular device.

Media features include width (of display window), height, device-width (of physical screen), device-height, aspect-ratio, device-aspect-ratio, color, orientation, and so on. Many accept min/max prefix. On smartphones and touch-screen tablets, the device width/height and the display window width/height are usually the same.

With media queries, we can easily specify style rules that apply under various media features and conditions. For example,

```
<link href="smartphone.css" rel="stylesheet" type="text/css"
 media="screen and (min-width:320px) and (max-width:480px)"/>
<link href="tablet.css" rel="stylesheet" type="text/css"
 media="screen and (min-width:600) and (max-width:1280px)"/>
<link href="desk.css" rel="stylesheet" type="text/css"
 media="screen and (min-width:1281px)"/>
```

where smartphone.css applies on most smartphones in portrait orientation, tablet.css applies on smartphones in landscape orientation and on touch-screen tablets, whereas desk.css provides the larger-screen devices.

Browser support for CSS3 media queries is improving but not uniform yet.

13.5   Mobile Webpages

While multiple media-query controlled style sheets can deliver the same webpages to different devices, it is always possible to specialize pages for delivery to different types of devices. Let's look at the author's homepage as a case study on delivering separate regular and smartphone versions of webpages.

For Paul's homepage, the layout for regular (desktop/laptop) browsers is a centered, two-column, fluid layout with header and footer as illustrated by Figure 13.4.

FIGURE 13.4: Two-Column Layout

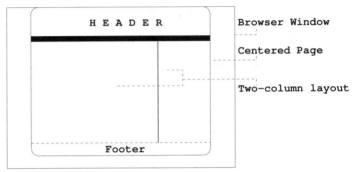

The same pages also work well on 10-inch touch-screen tablets in landscape orientation. For tablets in portrait orientation, we add the following style:

```
@media only screen AND (max-width:800px)
{ div#centerpage
   { margin-left: auto; margin-right: auto; width: 98%;
     border: 2px solid darkblue; border-radius: 16px;
     overflow-x: auto;
   }
}
```

which changes the `width` for the centered page from the default 75% to 98%. Figure 13.5 shows the beginning of this regular homepage.

Paul's homepage consists of five sections: a Brief Bio, Research, Consulting, Web Development, and Books. For a smartphone, a multi-pane layout (Figure 13.6) is used, with one pane for each section.

Each pane has a width and height equal to that of the physical display screen. Each pane can be scrolled vertically. If all panes are made visible, they would form a horizontal strip of panes. Initially, only the first pane is shown. The top navigation bar provides arrows to pan from one pane to the next, back and forth like reading pages in a magazine. Smartphone users are familiar with such layouts and find panning through pages pleasant and natural. Figure 13.7 shows the first pane, a middle pane, and the last pane for Paul's homepage as seen on an Android phone.

Different URLs are use for the regular and mobile homepages:

http://www.cs.kent.edu/~pwang   (regular version)
http://www.cs.kent.edu/~pwang/m   (smartphone version)

FIGURE 13.5: Author's Homepage

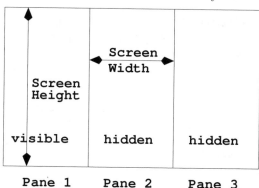

The HTML and CSS are different for the regular and the smartphone versions to achieve the desired layout and UI effects. But the actual content delivered are included from the same PHP files (Figure 13.8), each for one of the five sections for Paul's homepage:

```
bio.php      research.php  consulting.php
webtong.php  books.php
```

The same five PHP files are loaded by the regular homepage and by the smartphone version. The latter loads each of the preceding files into a (page*n*.php) page (lines C through D), which becomes a pane in our smartphone layout. The m/index.html file has the following HTML code:

FIGURE 13.6: Multi-Pane Layout

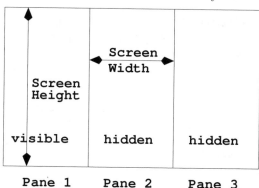

FIGURE 13.7: Individual Panes

```
<!DOCTYPE html>
<html xmlns="http://www.w3.org/1999/xhtml"
      lang="en" xml:lang="en">
<head><title>Paul S. Wang Mobile Homepage</title>
<meta charset="utf-8"/>
<meta name="viewport" content="width=device-width,        (A)
      initial-scale=1" />
<link rel="stylesheet" href="m.home.css"
      type="text/css" title="homepage style"/>
<link href="mystrip.css" rel="stylesheet"
      type="text/css" />
</head><body><div class="panestrip" id="thestrip"><?php    (B)
$last=4;                                                   (C)
require_once("page1.php"); require_once("page2.php");
require_once("page3.php"); require_once("page4.php");
require_once("page5.php"); ?></div>                        (D)
<script type="text/javascript" src="strippane.js">         (E)
</script></body></html>
```

Let's look at page2.php, a typical page. It consists of a top navbar (line G), the page content (line H), and a bottom navbar (line I).

```
<div id="pp-1" class="second pane">                        (F)
<?php $n=1; require('topnav.php'); ?>                      (G)
<div class="content">
```

FIGURE 13.8: Content Sharing

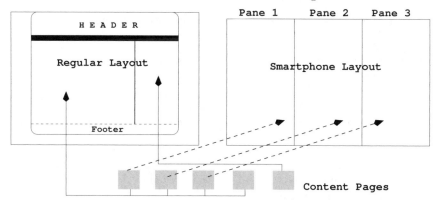

```
<?php require_once("../research.php"); echo "</div>";      (H)
  require('bottomnav.php');echo "</div>"; ?>               (I)
```

While one may choose to write a PHP loop based on the page number to include any number of panes, it is advisable to have no more than half a dozen panes.

Mobile browsers render pages in a virtual window called the *viewport* that can be significantly larger than the physical display screen. Mobile pages can use the `viewport` meta tag (line A) to configure viewport parameters. Non-mobile browsers ignore the `viewport` meta tag. See Section 13.6 for more on the mobile browser viewports and the `viewport` meta tag.

13.5.1 A Pane Layout

The smartphone version of Paul's homepage uses a multi-pane layout where users pan left-right to visit all five content sections. All panes (line E) are initially hidden except the first pane:

```
div.panestrip > div.pane
{ width: 100%; overflow: hidden; display: none; }
div.panestrip > div.first { display: block; }
```

Note: The `overflow: hidden` may be redundant because that usually is the default on mobile browsers.

Each `div.pane` sits in a horizontal strip styled this way:

```
div#thestrip.panestrip
{ width: 100%; vertical-align: top; }
```

Inside each pane, we have a header/navbar on top, followed by a content `div`. The content `div` provides some small left and right margins for the content materials.

```
div.panestrip > div.pane > div.content
{ padding-left: 4px; padding-right: 4px; }
```

The `topnav.php` file for the header/navbar (line F) is

```
<nav class="mobile">
<img <?php if ($n==0) echo "class='transparent'";?>
onclick="prevPane()" src="prev.png" alt="previous pane"/>
<span id="banner">Paul S. Wang     </span>
<img <?php if ($n==$last) echo "class='transparent'";?>
 onclick="nextPane()" src="next.png" alt="next pane"/></nav>
```

with centered white text and left/right nav icons against a dark blue background. Code for the bottom navbar is similar. For the first (last) pane, the prev (next) nav text and icon are made invisible (Figure 13.7) with the `transparent` class attribute.

JavaScript (line E) controls page panning, return to page top (line 1), and hiding the browser location box.

By keeping track of the current pane in display, we can easily hide it and show the next or previous pane (lines 2 through 3).

```
var cur_pane=0, pane_count=5;

function totop() { window.scrollTo(0, 1); }          // (1)

function nextPane()                                  // (2)
{ n=cur_pane; if ( n+1 >= pane_count) return;
   showPane(n+1); hidePane(n); setTimeout(totop, 1);
}
function showPane(n)
{ document.getElementById("pp-"+n).style
        .display="block";
   cur_pane=n;
}
function hidePane(n)
{ document.getElementById("pp-"+n).style
        .display="none";  }

function prevPane()                                  // (3)
{ n=cur_pane; if ( n-1 < 0) return;
   showPane(n-1); hidePane(n); setTimeout(totop, 1);   // (4)
}
```

The `setTimeout(totop, 1)` call (line 4) causes the mobile browser to hide the location box, saving valuable screen space. It is done onload and after each pane change.

The layout works equally well in portrait and in landscape mode and can be a good choice for any webpage with a handful of content sections.

See **Ex:** `DemoSite` at the DWP site to see the complete example of the author's homepage in both desktop and mobile versions.

13.6 The Viewport

The viewport concept is different for desktop browsers and for mobile browsers. The display area of a browser window is the viewport on desktops. This viewport can be resized and may have scrollbars. The viewport sets the boundaries for the layout width and height and for line wrapping.

On mobile browsers, the viewport becomes a virtual layout area with a browser-defined size. For example, Mobile Safari uses a viewport of 980×1090 pixels. A webpage is loaded and laid out onto this *viewport*, then displayed onto the screen in full-zoom-out mode (the default initial scale). The motivation for mobile browsers to do this is to display websites that are not designed for small-screen mobile devices, which may still be the majority of websites.

This explains why many pages are impossible to read on the mobile screen at first. Mobile users often immediately zoom in. However, reading materials by constantly scrolling and panning can be annoying. Fortunately, viewport parameters can be configured to avoid using the default size and scale. And configuring the viewport with the `viewport` meta tag is one of the first things a well-designed mobile page must do. Nonmobile browsers ignore the `viewport` meta tag.

For easy reading, it is recommended that a mobile webpage

- Use fluid rather than ice design.

- Configure the viewport width to device width and initial zoom to 100%. For example,

  ```
  <meta name="viewport" content="width=device-width,
                          initial-scale=1.0" />
  ```

- Assume a layout width of 100%. Any overflow, intentional hopefully, will be visible by panning and scrolling.

- Limit image files to a maximum width of about 640 pixels and use CSS to fit such images within screen width. The technique can provide crispier images on higher-resolution smartphones.

The `initial-scale` sets the zoom level when the page is first displayed. The `maximum-scale`, `minimum-scale`, and `user-scalable` (yes or no) properties control user zoom.

For example, the Birdsong Farm site uses the following viewport setting:

```
<meta name="viewport" content="width=device-width,
      initial-scale=1.0, user-scalable=yes,
      maximum-scale=1.5, minimum-scale=1.0"/>
```

In a CSS media query, the `width/height` value refers to the viewport. The value is available in JavaScript as DOM `documentElement.clientWidth` or `clientHeight`. The media query `device-width/device-height` value refers to the physical screen. The device value is available in JavaScript as `window.screen.width` or `window.screen.height`.

13.7   Automatic Redirection to Mobile Site

13.7.1   URL Redirection

With time and effort, it is not hard to make an attractive and functional mobile website. But we still need a way to automatically route incoming requests by mobile browsers from the main website, `HotSite.com` say, to the mobile site. The convention is to create the mobile site as a subdomain of the main site, giving it the URL `m.HotSite.com`.

With each incoming browser request, a website can first detect the client browser and device type, then redirect the request to an appropriate URL. This way, specialized pages can be delivered to different browsers and devices. It is not practical to construct pages for all the different device-browser combinations. It is usually sufficient to have a smartphone site for small screens and a main site for all larger devices.

The actual URL redirection is the easy part. From PHP the code,

```
header("Location: " . $full_url); exit();
```

does the job. From JavaScript, it is just as easy:

```
window.location= "url";     (full or relative url)
```

Another method is URL rewriting under Apache in `.htaccess` (Section 10.9) with

```
RewriteRule ^(.*)$ target_uri [L,R=302]
```

The hard part is the detection of the client browser and device, which is our next topic.

13.7.2   Browser/Device Detection

To detect the user's browser and device, either on the server side using PHP (say) or on the client side with JavaScript, the `userAgent` string is usually the key. For historical reasons, the format and content of the `userAgent` string is not standardized.

Here are some sample `userAgent` strings:

• For an Android browser on a Samsung Galaxy S 4G smartphone:

```
Mozilla/5.0 (Linux; U; Android 2.2.1; en-us;
SGH-T959V Build/FROYO) AppleWebKit/533.1
(KHTML, like Gecko) Version/4.0 Mobile Safari/533.1
```

- For a desktop Firefox on a Linux computer:

```
Mozilla/5.0 (X11; Linux i686 on x86_64; rv:13.0)
Gecko/20100101 Firefox/13.0
```

- For an Android browser on the Acer A500, a 10-inch touch-screen tablet:

```
Mozilla/5.0 (Linux; U; Android 3.2.1; en-us; A500
Build/HTK55D) AppleWebKit/534.13 (KHTML, like Gecko)
Version/4.0 Safari/534.13
```

- For Mobile Firefox (Fennec) on the same A500:

```
Mozilla/5.0 (Android; Tablet; rv:10.0.5)
Gecko/10.0.5 Firefox/10.0.5 Fennec/10.0.5
```

The site http://user-agent-string.info/list-of-ua has an extensive list of these strings. As you can see, it is a bit tricky using this information to determine the browser/device.

The recommended regular expression patterns to find inside the userAgent for most smartphones are listed here as a PHP array:

```
$keys = array(
    "android"       => "android.*mobile",
    "iphone"        => "(iphone|ipod)",
    "blackberry"    => "blackberry",
    "opera"         => "(opera\s*mini|opera\s*mobile)",
    "palm"          => "palmos",
    "windows"       => "iemobile");
```

It is perhaps better to first make sure that the client is not a 10-inch tablet. Then apply the above regular expressions to detect smartphones. The following PHP code, based on the userAgent string, can detect most 10-inch tablets.

```
$userAgent=$_SERVER['HTTP_USER_AGENT'];

function isTablet()
{ global $userAgent;
  if (strstr($userAgent, "iPad")        ||
      strstr($userAgent, "Kindle")      ||
      strstr($userAgent, "wOSBrowser")  ||
      (strstr($userAgent, "Android")&&
```

```
      strstr(userAgent, "Tablet"))     ||
      (strstr($userAgent, "Android")&&
       ! strstr(userAgent, "Mobile"))) return true;
  return false;
}
```

The `isTablet` and smartphone detection code are combined in working code at the DWP website (**Ex:** `Detect`).

To deliver webpages to small screens, it is perhaps more important to check the client features such as screen width and height. That is hard to do on the server side using just the `userAgent` string. And it ought to be easy to do on the client side using JavaScript. But, because there is no current standard to expose the device width and height, say from the `window` object, you are still left with examining the `userAgent` string. Perhaps this is why client-side redirection is rarely used.

In any case, it is advisable for websites performing redirection to provide a way for users to override and choose to go to any of the available alternatives. After all, our redirection conditions are not perfect and exceptions do exist.

Here is a link on Paul's mobile homepage (Section 13.5) that uses PHP for redirection, leading back to the regular site:

```
<a href="../index.html?nomobile">Go to full site</a>
```

The *Wireless Universal Resource FiLe* (WURFL) is an extensive device description repository (XML or database). If you download this data and use a language API to access it, you can perform very detailed device detection. The WURFL is constantly updated to include new devices and new features. If you decide to use WURFL, you may look into its PHP API (`wurfl.sourceforge.net/php_index.php`).

13.8   Touch Swipe Slide Show

On a touch-screen device, especially a smartphone, users are familiar with finger swiping to scroll pages and pictures. Mobile sites should take advantage of the touch-screen. One obvious application is in presenting a slide show. On a regular website, a user may click an icon to switch slides but on a touch-screen we can allow a swipe on the picture to move to the next/previous one.

The produce page on the Birdsong Farm site has a slide show (Figure 13.9). It works on smartphones with finger swipes.

In terms of implementation, the `slideshow` div displays a picture (line B) with a caption `span` (line A) above and next/prev buttons below (line D).

```
<div class="slideshow">
<span id="caption"> </span><br />                   (A)
<img class="slide" src="pictures/DSC00005-5.JPG"         (B)
     id="display" alt="Birdsong Farm Slide Show"
```

FIGURE 13.9: Produce Slide Show

```
        title="Birdsong Farm Slide Show"
        onclick='whichSide(event)' /><br />
<img class="nav left" title="backward"                      (C)
        src="pictures/backward_s.png" alt="previous slide"
        onclick="prevSlide()"/>
<img class="nav right" title="forward"                      (D)
        src="pictures/forward_s.png" alt="next slide"
        onclick="nextSlide()"/></div>
```

The click handlers `prevSlide()`, `nextSlide()`, and `whichSide(e)` change the `src` attribute of the image with id `display`. Calling the function

```
function whichSide(e)
{  var img=document.getElementById('display');
   var x=xPosition(img), half=img.clientWidth/2;
   if ( e.clientX-x > half ) nextSlide()
   else prevSlide();
}
```

switches to the previous or next slide, depending on whether the click location is on the left or right side of the picture. This works, touch-screen or not.

13.8.1 Touch Events

Now we will add finger swipe by handling these touch events on the displayed image:

`touchstart`

```
touchmove
touchend
```

On page load, `initSwipe()` is called to register handlers for the three touch events:

```
function initSwipe()
{  var s=document.getElementById('display');
   s.addEventListener("touchstart", startPosition, false);
   s.addEventListener("touchmove", moveHandler, false);
   s.addEventListener("touchend", endHandler, false);
}
```

The `touchstart` handler `startPosition` registers, in the global variable `tx`, the starting position of the finger touch, obtained from the `touch` event object:

```
var tx, dir;
function startPosition(e)
{  var touch = e.touches[0]; // touch event object
   tx=touch.pageX;
}
```

As the finger swipes, the `touchmove` handler gets called (multiple times). The `moveHandler` computes the move direction in the global variable `dir`, which is negative if the finger is moving to the right and positive to the left.

```
function moveHandler(e)
{  var touch = e.touches[0];
   dir=tx-touch.pageX;
   e.preventDefault();                 // (E)
}
```

The `preventDefault()` call (line E) prevents any preassigned default browser actions so the finger swipe will not cause unwanted actions such as page panning.

As the user finishes the finger swipe, the `touchend` event handler is called and it displays the next/previous slide, depending on the swipe direction.

```
function endHandler(e)
{  if ( dir > 35 )  nextSlide();
   if ( dir < -35 ) prevSlide();
   dir=tx=0;   // resets for next swipe
}
```

Finger swipes on the slide will not result in any page scrolling, vertically or horizontally. That is a consequence of `preventDefault()` (line E). If the user simply taps on the slide without swiping, the click event is still fired. This is because the `touchstart` event, together with the `touchend` event, constitutes a click.

For the lucky viewer, this slide show offers three ways to go through the pictures: clicking/tapping either icon, clicking/tapping left or right side of the slide, and finger swiping. Note that browser support for the new W3C touch event specification may not be uniform yet.

13.8.2 Momentum Scroll

The finger swipe technique can be applied in many other situations. For example, it can trigger the scrolling of thumbnail strips, start a ball bouncing, or shoot an object in a certain direction. More recent mobile browsers, including Android and iOS5 devices, actually support finger scrolling inside fixed-width/height block elements. On other devices, you can combine techniques here with JavaScript or CSS-supplied animation to simulate native-style momentum scroll, or use a package such as *Scrollability* or *iScroll*.

Here is a finger-scroll image strip (**Ex: IStrip**) that should work on newer touch-screen devices:

```
<div id="imgstrip" class="strip"><img ... />
<img ... /> ... <img ... /></div>
```

using the style

```
div.strip
{  background-color: darkgreen;
   overflow:auto; white-space:nowrap;
   width: 600px; margin: 0px auto;
   padding: 0; border: 0;
}
div.strip img
{  display: inline-block; height: 100px;
   vertical-align: top;
   margin:2px; border:0; padding:0;
}
</style>
```

13.9 Taking Advantage of the Smartphone

A mobile site can and should take good advantage of its host device, a smartphone, and offer features that are difficult or impossible to achieve on nonmobile devices.

Perhaps the most obvious are hyperlinks for phone calls, texting, fax, in addition to email. These URL schemes were described in Section 2.15.3:

```
mailto      tel      sms      fax      callto
```

We can also take advantage of the direct camera to form input API and the geolocation API. Both are discussed next.

13.9.1   Camera API

It is simple. When you use the file type `input` element for a file, devices equipped with a camera can let you choose to take a picture instead of finding an existing file. Either will work, of course. Our camera API demo (**Ex: CameraDemo**) is a simple HTML5 page (Figure 13.10)

FIGURE 13.10: Camera API Demo

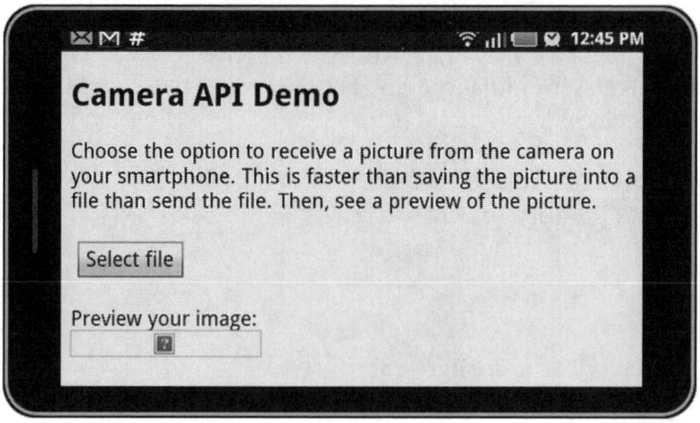

using the following code:

```
<!DOCTYPE html>
<p> <input type="file" onchange="preview(event)"
     id="thepic" accept="image/*"></p>          (1)

<p>Preview your image:<br />
<img src="something" alt="preview picture here"    (2)
    id="pic"/></p>
<p id="error"> </p></section>                  (3)
<script src="camera.js"></script></body></html>
```

where we accept an image file from the user (line 1) and immediately display the picture (line 2) or an error message if something goes wrong (line 3).

Figure 13.11 shows an Android dialog for choosing a file or using the camera directly.

The image input element (line 4) and the image preview element (line 5) are initialized on page load:

FIGURE 13.11: Choose Camera

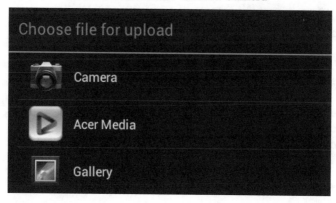

```
var pic_el, show_el;

function init()
{  pic_el = document.getElementById("thepic");   // (4)
   show_el = document.getElementById("pic");      // (5)
}
```

The function `preview` is called when the user picks an input file or takes a pictures with the camera.

```
function preview(e)
{  var files = e.target.files, file;
   if (files && files.length > 0)
   {  file = files[0];
      if (imgURL(file)||dataURL(file)) return;   // (6)
      else error();
   }
}
```

The target file is obtained from the input event, and an image URL or a data URL (line 6) is created and displayed in the previewing area by calling the following functions that apply the HTML5 URL and HTML5 File APIs. In case of failure, an error message is displayed.

```
function imgURL(file)   // Use HTML5 URL API
{ try
  { var URL = window.URL || window.webkitURL;
    var imgURL = URL.createObjectURL(file);
    show_el.src = imgURL;
    URL.revokeObjectURL(imgURL);
  } catch (ex) { return false; }
```

```
    return true;
}

function dataURL(file) // Use HTML5 File API
{ try
  { var fileReader = new FileReader();
    fileReader.onload = function (ev)
      { show_el.src = ev.target.result; };
    fileReader.readAsDataURL(file);
  } catch (ex) { return false; }
  return true;
}
```

Try **Ex:** CameraDemo on your smartphone at the DWP website.

13.9.2   Geolocation API

The HTML5 geolocation API provides a standard way to take advantage of geolocation information that many of today's mobile and nonmobile devices are capable of supplying automatically. This is particularly useful on mobile devices because most of them have built-in GPS.

To utilize geolocation from JavaScript, the key is the built-in `navigator.geolocation` object. You call its `getCurrentPosition` method and pass it a callback function to receive and process the *position* information returned asynchronously. The callback function can then process the position information in any way it sees fit, including passing the position to a map service to get back an area map or driving directions.

For example, the `getMap` function checks for geolocation support (line A), then passes two handlers in the `getCurrentPosition` call (line B).

```
function getMap()
{ if (navigator.geolocation)                      // (A)
  { navigator.geolocation.getCurrentPosition(     // (B)
      handlePosition, handleError);
  }
  else
  { alert("No browser support for geolocation."); }
}

function handleError(err) {  alert(err);  }
```

It is a security concern, when a user's location is passed to a program from some website. Therefore, the `getCurrentPosition` method will seek user permission before passing the position information to the `handlePosition` callback. If the user denies permission, the `handleError` callback is invoked.

The `handlePosition` function packs the position coordinates (lines C through D) in a `mapOptions` object (line E) for creating a Google map (line F). A current position marker overlay is also added to the map (line G).

```
function handlePosition(position)
  var lat = position.coords.latitude;                // (C)
  var lng = position.coords.longitude;
  var coords = new google.maps.LatLng(lat, lng);     // (D)
  var mapOptions =                                   // (E)
    { zoom:15, center:coords, mapTypeControl:true,
      navigationControlOptions:
      { style:google.maps.NavigationControlStyle.SMALL},
      mapTypeId:google.maps.MapTypeId.ROADMAP
    };
  var map = new google.maps.Map(                     // (F)
    document.getElementById("mapDisplay"), mapOptions);
  new google.maps.Marker(                            // (G)
      { position: coords, map: map,
        title: "You are here!" });
}
```

The Geolocation API standard can be found at the W3C site. The Google Map API is published by Google.

Let's put the preceding JavaScript in a file `geomap.js` and put it to use in a mobile page (**Ex:** GeoMap):

```
<!DOCTYPE HTML><html><head>
<meta name="viewport" content="width=device-width,
   user-scalable=yes, initial-scale=1.0,
   maximum-scale=3.0" />
<script type="text/javascript" src=
   "http://maps.google.com/maps/api/js?sensor=false">   (H)
</script>
<script type="text/javascript" src="geomap.js">
</script></head><body onload="getMap()">
<div id="mapDisplay">map here</div></body></html>
```

The Google Map API is included on line H. The map display area is styled different for mobile phone (Figure 13.12) and other (desktop/laptop/tablet) devices.

```
@media only screen and (min-width: 800px)
{ div#mapDisplay
  { width: 95%;  height: 800px;
    overflow: auto;
    border:8px solid darkgreen;
  }
}
```

FIGURE 13.12: Map with Geolocation

```
@media only screen and (max-width: 640px)
                  and (min-width: 320px)
{ div#mapDisplay
  {  width:100%; height: 640px;  }
}
```

With geolocation, a website can easily offer directions to the establishment from where the mobile user happens to be at any given time.

Let's take a look at mobile directions to the Davey Tree Expert Company in Kent, Ohio (**Ex: GeoDir**). The getDirection() called on page load obtains the geolocation and passes the current position to createMap:

```
function getDirection()
{ if ( navigator.geolocation )   // current position map
    navigator.geolocation.getCurrentPosition(
            createMap, errorHandler);
  else
    createMap(false); // fallback map
}

function errorHandler(err) { createMap(false);  }
```

In case of no geolocation support or no user permission, createMap(false) is called to display directions from downtown Kent:

```
var target="1500 North Mantua Street Kent, OH 44240",
    start="Kent, OH 44240";

function createMap(p)
{ var directionsService=new google.maps.DirectionsService(),
  directionsDisplay=new google.maps.DirectionsRenderer(),
```

```
from=p ? new google.maps.LatLng(
              p.coords.latitude, p.coords.longitude)
         : start;
var travel = { origin: from, destination: target,
   travelMode:google.maps.DirectionsTravelMode.DRIVING },
mapOptions = { zoom: 10, center:
   new google.maps.LatLng(41.154876,-81.357708),          // (1)
   mapTypeId: google.maps.MapTypeId.ROADMAP },
map = new google.maps.Map(document.getElementById(        // (2)
           "mapDisplay"), mapOptions);
directionsDisplay.setMap(map);                            // (3)
directionsDisplay.setPanel(document.getElementById(
              "dirDisplay"));
directionsService.route(travel,                           // (4)
    function(result, status)
    { if (status === google.maps.DirectionsStatus.OK)
       { directionsDisplay.setDirections(result); }
    });
}
```

The `createMap` function uses the Google Map API to set up a map centered at downtown Kent (line 1) and displays the driving route map (line 2) as well as the directions (lines 3 through 4).

Figure 13.13 shows the smartphone user views of this example. The `mapDisplay` and `dirDisplay` divs have different styles for different devices:

```
@media only screen and (min-width: 1280px)  /* desk/laptop */
{ div#dirDisplay
  {  width: 30%; overflow: auto;
     display: inline-block; float:left;
     border:1px solid black;
  }
  div#mapDisplay
  {  width: 750px; height: 700px; overflow: auto;
     display: inline-block; float:left;
     border:8px solid darkgreen;
  }
  section.mapdir {  display: block;  }
}
@media only screen and (max-device-width:1024px) /* 10" tablet */
     and (orientation:portrait)
     and (min-device-width: 700px)
{ div#dirDisplay
  {  width: 100%; overflow: auto;
     display: block; border:1px solid black;
  }
```

FIGURE 13.13: Mobile Driving Directions

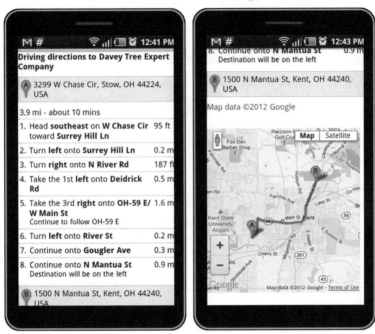

```
  div#mapDisplay
  {  width: 750px;    height: 700px;
     overflow: auto; display: block;
  }
}
@media only screen and (max-width: 640px)  /* smartphone */
        and (min-width: 320px)
{ div#dirDisplay
  {  width: 100%; overflow: auto; }

  div#mapDisplay
  {  height: 320px; width: 95%; overflow: auto;
     margin-top: 2em; margin-bottom: 2em;
  }
}
```

Try the ready-to-run example (**Ex: GeoDir**) on your regular and mobile browsers at the DWP website.

13.9.3 The Touch Icon

A *touch icon* takes the place of the favicon (Section 3.10.5) on mobile devices. It is a 57×57 or 114×114 PNG icon placed at the root of your site and/or linked by a `meta` tag. Can you spot the Birdsong Farm touch icon in Figure 13.14?

FIGURE 13.14: Touch Icons on Android Tablet

First introduced by Apple, the touch icon concept has been adopted by most mobile systems. It is recommended that you place a 114×114 touch icon under the site root folder named

```
pple-touch-icon-precomposed.png
```

and perhaps also name it `apple-touch-icon.png`. And add the meta tag

```
<link rel="apple-touch-icon-precomposed"
      href="apple-touch-icon-precomposed" />
```

That is it—simple and easy. The touch icon is used for bookmarks and for touch launching your page. For sites without a touch icon, mobile systems may use a thumbnail image of the page, the favicon, or a default icon instead.

13.10 Conversion to Mobile

Just like developing any website, designing and building it from scratch is the best approach. For those who already have a website, the task to develop a mobile version is made easier because the content, functionality, and navigation are mostly in place. Then the task becomes converting the existing website into one that works well on mobile devices, principally smartphones. Main conversion tasks include

- Use a fluid design and make sure all elements use percentage widths.

- Apply a single-column layout with a flexible width from 320 to 480 pixels.

- Size all images to fit comfortably within the column width.

- Get rid of all unnecessary margins, borders, and paddings.

- Simplify navigation and reduce page levels. Apply left-right touch scrolling liberally.

- Remove nonessential materials, avoid verbosity, and tiny or tightly packed links.

- Add media query controlled styles, set viewport to device width and initial zoom to 100%, and supply a touch icon.

- Take advantage of mobile features.

- Place mobile detection on the main site and redirect mobile requests to the mobile site.

- Be sure to have a link back from the mobile site to the main site.

- Test the mobile site on multiple mobile devices.

- Use HTML5-supported video/audio formats (Section 3.1 and Section 3.2) with fall-back to other popular formats such as mp3, mpeg4, and H.264.

- Maximize sharing of content files among main and mobile sites.

- Avoid iframes and pop-up windows by JavaScript, they do not work well or uniformly on mobile browsers.

To size images correctly for mobile sites, we can take all images, say under the folder `image` at the main site root, and scale them back to a width of about 640 pixels and put them under a separate `image` folder under the mobile site's root. This way, HTML image code stays the same for the main and mobile sites.

In this approach, you rescale once ahead of time. But for sites where the images are changing dynamically, you may prefer a dynamic approach where the image is rescaled, say with PHP (Section 8.11), as it is accessed. This means the `img src` attribute will be coded something like this:

```
<img src="mobileimg.php?img_name" ... />
```

A width setting can also be included in the HTTP query string to `mobileimg.php`.

It is also possible to use CSS to control image sizing on the client side:

```
@media only screen and (max-device-width: 480px)
{   img{ max-width:100%; height:auto; border: 0; }
}
```

Tools for automatically converting a site to mobile are available online. They may work well for some sites and not for others. Free tools include Google Mobile Optimizer, Mippin, Zinadoo, and mobiReady. More tools are being introduced all the time. Look for one that can help you do the conversion. At least, they can give you ideas and inspiration.

The Phpmobilizer tool from Google is nice. Phpmobilizer provides a handful of PHP5 files that you run on your own server to automatically serve your main site in a mobile-friendly format under a mobile domain m.*your_domain*. The tool relies on Apache URL rewriting (Section 10.16) and dynamic transformations of your regular webpages, including HTML code and image resizing. Because you have access to the tool's source, you can make adjustments to the way it works to suit your particular needs.

13.11   Webapps

Native apps run on particular platforms (hardware-OS combinations). They rely on OS-provided services. Webapps, on the other hand, are independent of any hardware-OS combination but instead need a Web browser to run. Native apps must be programmed for each different platform and be installed before use. Webapps should run on any platform with a Web browser and Internet connection. Native apps have access to all local files, networking, and other services. Webapps must use what is made available by the browser.

HTML5 makes webapps more agile through such client-side features as Rich-text Editing (Section 8.12), File API (Section 7.13), *Offline Caching*, and *Local Storage*.

13.11.1   Offline Caching

When you add a `manifest` attribute to your `html` element

```
<html manifest="manifest_file"
   xmlns="http://www.w3.org/1999/xhtml"
   lang="en" xml:lang="en">
```

you ask client browsers to cache the webpage and files listed in the given *manifest_file* for offline use. As a browser loads a page with a cache manifest, it will save the indicated files in a cache belonging to that page's domain. Provided that you list all the necessary files in the manifest, the given webpage will work in the browser afterward in *offline mode*[1]. In fact, the browser may

---

[1]Browser offline mode is independent of a computer's Internet connection status.

use the cache even in online mode to increase efficiency. It is advisable to add the manifest only after a webapp is tested and ready.

A manifest must be delivered under the content type

```
text/cache-manifest
```

Web servers such as Apache know about the `.manifest` file suffix and will serve such files correctly. For example, `TicTacToe` (Figure 13.15)

FIGURE 13.15: TicTacToe Webapp

is a webapp (**Ex: TicApp**) that works on regular and mobile devices. It uses the following cache manifest:

```
CACHE MANIFEST
####    Version 1.0   06-17-2012

CACHE:
tictactoe.html
o.png
tictactoe.css
tictactoe.js
version_update.js
x.png
```

Each and every resource needed by the webpage must be listed, with a URL relative to the manifest location or an absolute URL, individually in the cache manifest. Listing only a directory name will not work.

The game page has the following HTML code:

```
<!DOCTYPE html>
<html manifest="tictactoe.manifest"  lang="en"
 xmlns="http://www.w3.org/1999/xhtml" xml:lang="en">
<head><meta charset="utf-8"/>
<title>TicTacToe Game Webapp</title>
<meta name="viewport" content="width=device-width,
      initial-scale=1.0, maximum-scale=1.5">
<link rel="stylesheet" href="tictactoe.css"
      type="text/css"  title="Tic Tac Toe Game" />
<body><section id="game">
<h2>Welcome to TicTacToe</h2>
<?php include("ticgame.inc"); ?>
<p><button onclick="newgame()">New Game</button></p>
<p>You can play this game even in offline mode.</p>
</section>
<script type="text/javascript" src="tictactoe.js"></script>
<script type="text/javascript"
        src="version_update.js"></script></body></html>
```

Try this example (**Ex: TicApp**) at the DWP website.

When storing into the cache, a browser may ask for user permission (Figure 13.16) and remember the answer.

FIGURE 13.16: User Permission Dialog

Put your browser in offline mode, your smartphone in airplane mode, or disconnect from the Internet. Then try to play the Tic Tac Toe game, through a bookmark, for example; you will see that the game now works offline.

A browser with a cached version of a webapp may run the cached version, in both online and offline mode, regardless of changes made to the application on the server side.

Follow this procedure (using HTML5 Offline Cache API) to enable automatic update of your offline-capable webapp so you can easily release a new version:

1. Make sure all versions of your webapp contain JavaScript support for update (see `version_update.js` below).

2. Revise and improve the webapp and test it thoroughly.

3. Modify the manifest file; make sure it is different from the previous one. Even changing a comment line makes it different.

4. Do not change the URL of the webapp or its manifest file. Otherwise, it becomes a separate new webapp not a replacement (update) of the previous version.

For the required JavaScript support, our example uses

```
/////    version_update.js    /////

if ( window.applicationCache )
{ applicationCache.update();                        // (a)
  applicationCache.addEventListener('updateready',  // (b)
                        update_cache, false);
}

function update_cache()
{ if (applicationCache.status
      == applicationCache.UPDATEREADY)              // (c)
  { applicationCache.swapCache();                   // (d)
    if (confirm('An update for TicTacToe
            is available. Update now?'))
    { window.location.reload();  }                  // (e)
  }
}
```

With this JavaScript code, every time a user invokes the webapp, an updateready event handler is registered with the applicationCache object (line a) right after an asynchronous call to its update() method (line b), which checks and downloads updates to the offline cache.

If the manifest file has not changed, no update will be made. Otherwise, all files will be downloaded and then the updateready event is fired. The handler update_cache calls swapCache() to update the cache (lines c and d). A confirmation dialog (Figure 13.17) is then displayed, asking the user if the page should be reloaded to reflect the new version (line e).

See the W3C specification for *Offline Web Applications* for more information.

13.11.2 Local Storage

Native applications enjoy complete control over persistent data storage on the host system. While the HTML5 File API (Section 7.13) gives webapps read access to client-side files, saving data is still a challenge. HTML5 *Local Storage* (part of the HTML5 Web Storage API) evens the playing field by providing webpages local persistent store.

The HTML5 *Local Storage API* allows webpages to store and retrieve data

FIGURE 13.17: Offline Webapp Update Dialog

on the client side. This can be very useful for webapps. A game, for example, can save its status on the client side, to allow the game to be resumed at a later time. An education site can store student progress information on the client side to customize the learning experience for each student. All modern browsers support HTML5 local storage and provide at least 5MB of storage per domain.

The browser built-in object `window.localStorage` implements the local storage API which is very simple:

```
if ( localStorage )              // if supported
{ localStorage.setItem(key,value);  // to store value under key
  value=localStorage.getItem(key);  // to retrieve value of key
}
```

The `key` and `value` should be strings. If numbers are passed, they will be converted to strings. The value returned by `getItem` is a string. Alternatively the array notation can be used as long as both `key` and `value` are strings.

```
if ( localStorage )
{  localStorage["key"]="value";
   value = localStorage["key"];
}
```

Because we are dealing with persistent storage, all keys and values are automatically stored as character strings. Of course, you may use `Number()`, `parseFloat()`, and `parseInt()` to convert a string to a number; or `toString()` to convert a number to a string.

```
age = Number(localStorage["age"]);
str = age.toString();
```

Additionally, you may use

```
removeItem(key);        // to delete an item
localStorage.clear();   // to delete all items
```

and iterate over all stored items:

```
for (i=0; i<localStorage.length; i++)
{   key=localStorage.key(i);
    alert(localStorage[key]);
}
```

As an example, let's add a "Save Game" feature (Figure 13.18) to the TicTacToe webapp using local storage (**Ex: TicSave**).

To save a game, we store the current state **gstate**, a JSON object, into local store under a name given by the user (line I).

```
function saveGame(name)
{ name=document.getElementById(name).value;
  if ( ! confirm("Do you wish to save game as "
        + name +"?") )  return;
  if ( window.localStorage )
  {  if ( name ) localStorage.setItem(name,
               JSON.stringify(gState));        // (I)
  }
  else { alert("no localstorage"); }
}
```

To resume a game, we retrieve the specified game from local storage and set the game state accordingly (lines II and III). The the game display is updated (line IV).

```
function resumeGame(name)
{ name=document.getElementById(name).value;
  if (!confirm("Do you wish to resume saved game "
        + name +"?") )  return;
  if ( window.localStorage )
  {  if ( name=localStorage.getItem(name))     // (II)
     { eval("gState="+name+";");               // (III)
       displayGame();                          // (IV)
```

FIGURE 13.18: Save Game Feature

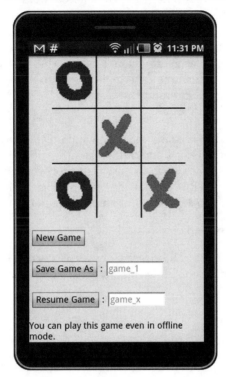

```
      }
   }
   else {   alert("no localstorage"); }
}
```

Because local storage deals with string keys and values, non-string quantities must be serialized before saving and deserialized after retrieving. The JSON.stringify() and JSON.parse(), supported in most browsers, perform these tasks.

Experiment with the example (**Ex: TicSave**) on the DWP website. Try Firefox and other browsers to see their local storage support.

The sessionStorage object works exactly the same way but will not persist beyond the end of a session when a browser is closed.

13.12   Mobile Site Testing and Debugging

Developing mobile websites that work well on many devices and across browsers is a challenge. While HTML5 standardizes many things, not all features are supported equally on different browsers. Thus, it is still a time-consuming task to test and verify that a mobile site works properly on different platforms—simply because there are so many of them.

It is perhaps wise to perform tests on major browsers running on Android and iOS devices first. Then test as many other environments as possible.

Device/browser emulators are offered online. But how well each emulates real devices is often difficult to determine. Some tools are worth looking at. Examples are Adobe Shadow and BrowserStack.

According to the Adobe site,

> Adobe Shadow is a new inspection and preview tool that allows front-end Web developers and designers to work faster and more efficiently by streamlining the preview process, making it easier to customize websites for mobile devices. The initial focus is on device pairing, synchronous browsing and remote inspection.

And according to the BrowserStack site,

> Instant access to all browsers across platforms. Forget clumsy virtual machine & mobile setup. Save time & money.

13.13   For More Information

For much more in-depth coverage of mobile site design and programming see various textbooks entirely dedicated to the topic.

HTML5 specifications related to touch-screen events, CSS3 media queries, local storage, cache manifest, and audio/video support are still evolving. Check the most current official standards for more details and up-to-date information.

13.14   Summary

Mobile Web browsing contexts are varied and very different from regular desktop/laptop environments. HTML5 offers standards to implement regular as well as mobile sites. With contemporary smart mobile devices and their support for HTML5, there is little reason to use specialized mobile languages.

The mobile browser viewport is a virtual layout area whose default settings are not suitable for mobile pages. The viewport meta tag, media query controlled CSS styles, automatic redirection, and fluid-width one-column design are important aspects of mobile sites. Detecting which browser on what device is accessing a page is the key to automatic redirection. Because the userAgent strings follow no standards, detection on the server side can be tricky. Client-side detection enjoys more information but can be much slower.

In addition to being concise and functional, a mobile site should take advantage of mobile device features, including touch-screen, momentum scrolling, camera, phone/microphone, SMS, and geolocation. Testing and debugging on multiple devices and browsers are necessary.

HTML5 Geolocation, File, Rich-text Editing, Local Storage, Cache Manifest, and other APIs help make webapps capable and work offline.

The mobile Web landscape is still evolving at a rapid pace. Mobile devices are fast becoming as powerful as yesterday's desktops. However, the day is still far away when a single website will work on mobile and nonmobile devices.

Exercises

13.1. Which are the most popular mobile platforms for Web browsing?

13.2. Name the major regular and mobile Web browsers.

13.3. Discuss the differences between desktop, laptop, tablet, and smartphone Web browsing environments.

13.4. Please list and explain the top-ten action items when converting a regular website to a mobile site.

13.5. What is a media query? How are they applied?

13.6. What is a viewport? A cache manifest? A touch icon?

13.7. Design the media query conditions for three style sheets intended for pages on a particular website: `desklap.css` for desk/laptops, `table.css` for 10-inch tablets, and `phone.css` for smartphones.

13.8. What is the relation between the `viewport` setting and media query width and height conditions? Please explain clearly.

13.9. Based on the Camera API demo (Section 13.9.1), build a picture upload page for mobile devices to handle multi-pictures upload using AJAX (Section 11.16).

13.10. Take the slide show example in Section 13.8 and animate (with CSS) the scrolling into view of the next/prev slide.

13.11. Find out how the `doubleclick` event works on touch-screen devices.

13.12. Discuss the wisdom of page-defined drag-and-drop operations on touch-screen devices.

13.13. How are orientation and orientation change events detected in JavaScript?

13.14. Discuss the pros and cons of mobile site redirection on the server side and on the client side.

13.15. Make an image strip that users can momentum scroll (Section 13.8.2) horizontally. Test it on smartphones and tablets.

13.16. Take the TicTacToe webapp (**Ex:** `TicSave` Section 13.11.2) and add a drop-down menu for the New Game, Save Game, Resume Game options. Remove the corresponding buttons if you will.

13.17. Continue from the previous exercise and improve the Resume Game feature so that it lists all saved games to choose. Then, improve the Save Game feature so it checks for name conflicts and consults the user for proper disposition.

13.18. Give a dozen ideas you find important, with justification, in designing and implementing websites so they become as easy to convert to mobile sites as possible.

Appendices Online

Appendices are online at the DWP website (`http://dwp.sofpower.com`).

Appendix: Secure Communication with SSH and SFTP

SSH lets you log in and access a remote computer securely from the Internet. SFTP lets you upload and download files securely to and from another computer. See the appendix at `dwp.sofpower.com/ssh.html`.

Appendix: Introduction to vim

There are many text editors, but **vim** (**vi** iMproved) is a visual interactive editor preferred by many. See the appendix at `dwp.sofpower.com/vimIntro.pdf`.

Appendix: Text Editing with vi

In-depth coverage of text editing concepts, techniques, and macros with the **vi** editor are provided. See the appendix at `dwp.sofpower.com/vi.pdf`.

Appendix: Vi Quick Reference

See the appendix at `dwp.sofpower.com/viQuickRef.pdf`.

Appendix: The emacs Editor

Rather than operating in distinct input and command modes like **vi**, **emacs** operates in only one mode: printable characters typed are inserted at the cursor position. See the appendix at `dwp.sofpower.com/emacs.pdf`.

Appendix: A Linux Primer

Introductory Linux materials useful for the modern Web developer. The primer is part of the author's *Mastering Linux* textbook which received superb reviews. See the appendix at `dwp.sofpower.com/linuxprimer.pdf`.

Website and Online Examples

Website

This book has a website (the DWP website) useful for instructors and students:

http://dwp.sofpower.com

The DWP website supplies code examples, the appendices, information updates, as well as other resources for the textbook.

Live Examples and Example Code Package

You can find ready-to-run examples for each chapter at the DWP website. Also, all examples in this book, and a few more, are contained in a code example package[1]. The entire package can be downloaded from the website in one compressed file: DWPH5_Wang.tbz2, DWPH5_Wang.tgz or DWPH5_Wang.zip. The access code for live examples and the example code package is **D2W0P12H5**.

The package contains the following files and directories:

```
exc01/   exc03/   exc05/   exc07/   exc09/   exc11/   guide.pdf
exc02/   exc04/   exc06/   exc08/   exc10/   exc12/   exc13/
```

corresponding to the chapters in the book. You can find descriptions for the examples in the textbook with cross-references to their file locations.

Unpacking

1. Place the downloaded file in an appropriate directory of your choice.

2. Go to that directory and, depending on the downloaded file, use one of these commands to unpack:

   tar jxpvf DWPH5_Wang.tbz2
   tar zxpvf DWPH5_Wang.tgz
   unzip DWPH5_Wang.zip

   This will create a folder **dwph5/** containing the example code package.

---

[1]Online examples and the example code package are distributed under license from SOF-POWER. The example code package is for the personal use of purchasers of the book. Any other use, copying, or resale, without written permission from SOFPOWER, is prohibited.

Bibliography

[1] Kurt Cagle. *HTML5 Graphics with SVG & CSS3*, O'Reilly Media, Sebastopol, CA, 2012.

[2] Ruben D'Oliveira. *Learn HTML5 Website Basic Features and Elements in 1 Day (Quick Guides for Web Designers in 1 hour!)*. Amazon (Kindle edition), 2012.

[3] Joe Fawcett, Danny Ayers, and Liam R. E. Quin. *Beginning XML*. 5th edition, Wrox Press, Hoboken, NJ, 2012.

[4] Gail Rahn Frederick and Rajesh Lal. *Beginning Smartphone Web Development: Building JavaScript, CSS, HTML and Ajax-Based Applications for iPhone, Android, Palm Pre, BlackBerry, Windows Mobile and Nokia S60*. Apress, New York, NY, 2010.

[5] Adam Freeman. *The Definitive Guide to HTML5*, Apress, New York, NY, 2011.

[6] Peter Gasston. *The Book of CSS3: A Developer's Guide to the Future of Web Design*, No Starch Press, San Francisco, CA, 2011.

[7] Jeremy Keith and Jeffrey Sambells. *DOM Scripting: Web Design with JavaScript and the Document Object Model*. 2nd edition, friendsofED, New York, NY, 2010.

[8] Sergey Mavrody. *Sergey's HTML5 & CSS3: Quick Reference*. 2nd edition, Belisso, Aurora, IL, 2012.

[9] Joel Murach. *Murach's MySQL*, Mike Murach & Associates, Fresno, CA, 2012.

[10] David Powers. *PHP Solutions: Dynamic Web Design Made Easy*, 2nd edition, friendsofED, New York, NY, 2010.

[11] Pavi Sandhu. *The MathML Handbook*, Charles River Media, Hingham, MA, 2002.

[12] Paul S. Wang and Sanda Katila. *An Introduction to Web Design and Programming*, Course Technology/Cengage Learning, Mason, OH, 2004.

[13] Paul S. Wang. *Mastering Linux*. CRC Press, Boca Raton, FL, 2010.

[14] Nicholas C. Zakas. *Professional JavaScript for Web Developers*, 3rd edition, Wrox Press, Hoboken, NJ, 2012.

Index